Lecture Notes in Artificial Intelligence 7375

Subseries of Lecture Notes in Computer Science

Preface

These proceedings contain the papers presented at *Living Machines: An International Conference on Biomimetic and Biohybrid Systems*, held in Barcelona, Spain, July 9–12, 2012. This new international conference is targeted at the intersection of research on novel life-like technologies inspired by the scientific investigation of biological systems — *biomimetics* — and research that seeks to interface biological and artificial systems to create *biohybrid* systems. We seek to highlight the most exciting international research in both of these fields united by the theme of "living machines."

Conference Theme

The development of future real-world technologies will depend strongly on our understanding and harnessing of the principles underlying living systems and the flow of communication signals between living and artificial systems.

Biomimetics is the development of novel technologies through the distillation of principles from the study of biological systems. The investigation of biomimetic systems can serve two complementary goals. First, a suitably designed and configured biomimetic artefact can be used to test theories about the natural system of interest. Second, biomimetic technologies can provide useful, elegant and efficient solutions to unsolved challenges in science and engineering. *Biohybrid* systems are formed by combining at least one biological component — an existing living system — and at least one artificial, newly engineered component. By passing information in one or both directions, such a system forms a new hybrid bio-artificial entity.

The development of either biomimetic or biohybrid systems requires a deep understanding of the operation of living systems, and the two fields are united under the theme of "living machines" — the idea that we can construct artefacts, such as robots, that not only mimic life but share the same fundamental principles; or build technologies that can be combined with a living body to restore or extend its functional capabilities.

Biomimetic and biohybrid technologies, from nano- to macro-scale, are expected to produce major societal and economical impacts in quality of life and health, information and communication technologies, robotics, prosthetics, brain–machine interfacing and nanotechnology. Such systems should also lead to significant advances in the biological and brain sciences that will help us to better understand ourselves and the natural world. The following are some examples:

- Biomimetic robots and their component technologies (sensors, actuators, processors) that can intelligently interact with their environments
- Active biomimetic materials and structures that self-organize and self-repair

- Biomimetic computers — neuromimetic emulations of the physiological basis for intelligent behavior
- Biohybrid brain–machine interfaces and neural implants
- Artificial organs and body parts including sensory organ-chip hybrids and intelligent prostheses
- Organism-level biohybrids such as robot–animal or robot–human systems

A key focus of Living Machines 2012 was on complete behaving systems in the form of *biomimetic robots* that can operate in or on different substrates, sea, land, or air, and inspired by the different design plans found in the animal kingdom — plants, invertebrates, and vertebrates. A further central theme was the physiological basis for intelligent behavior as explored through *neuromimetics* — the modelling of neural systems. Exciting emerging topics within this field include the embodiment of neuromimetic controllers in hardware, termed *neuromorphics*, and within the control architectures of robots, sometimes termed *neurorobotics*. Contributions from biologists, neuroscientists, and theoreticians that are of direct relevance to the development of future biomimetic or biohybrid devices were also included. We invited both full papers and extended abstracts. All contributions were assessed by at least two expert referees with relevant background (see list below). Following the conference, the journal *Bioinspiration and Biomimetics* will publish extended and revised versions of some of the best papers presented at the meeting.

A Brief History of Biomimetics

The ambition to mimic nature has been with us since ancient times. In the 4th century B.C., Archytas of Tarentum is said to have built a steam-driven model of a dove that could fly. Leonardo Da Vinci's designs for machines, which included a humanoid robot, were largely inspired by nature, and by his own detailed observations of natural systems and mechanisms. By the middle of the 17th century, Descartes was willing to assert that animals are complex automatons, and the extension of this radical idea to our own species came a century later with the book *L'Homme Machine* (1748), by Julien Offray de La Mettrie, which not only expanded Descarte's notion of the mechanistic nature of life to include the human species but also identified that machines — natural or otherwise — can be dynamic, autonomous, and purposive entities. In the 18th century the famous automatons of the French inventor Jacques de Vaucanson, the "ancestors" of modern theme-park animatronics, were emblematic of this emerging view of man and of nature. It is easy to imagine that the term "living machine" was applied to many of these early and awe-inspiring life-like artefacts, at the same time that people were beginning to take seriously the possibility that we, ourselves, might operate according to similar principles.

With the rise of cybernetics in the 1940s, it became clear that there was the possibility to create inventions that would realize La Mettrie's vision of machines that were both autonomous and purposive. At the same time, interest in nature as a source of inspiration was also gathering force. The term "biomimetics" was introduced by Otto Schmitt during the 1950s, and "bionics" by Jack Steel (popularized in Daniel Halacy's 1965 book *Bionics: the Science of "Living" Machines*) emerged as part of this growing movement in engineering that sought to build strong ties with the biological sciences and to make progress through "reverse engineering" natural systems. The biomimetic approach has since succeeded in overcoming many difficult challenges by exploiting natural design principles. Indeed, in the first decade of this century there has been an explosive growth in biomimetic research, with the number of published papers doubling every two to three years (see Lepora et al., this volume). The Living Machines conference sought to be a significant forum for this dialogue between nature and technology, and to be a place where people could discuss the biomimetic and biohybrid machines of tomorrow and what they might mean for understanding the biological machines of today.

The Living Machines Conference in Barcelona

The main conference, July 10–12, took the form of a three-day single-track oral and poster presentation program that included six plenary lectures from leading international researchers in biomimetic and biohybrid systems: Joseph Ayers (Northeastern University) on synthetic neuroethology; Dieter Braun (Ludwig Maximilians University) on synthetic life, Peter Fromherz (Max Plank Institute) on neuroelectronic hybrids; Toshio Fukuda (Nagoya University) on micro-nano biomimetic and biohybrid devices; David Lentink (Stanford University) on the biofluiddynamics of flight; and Barry Trimmer (Tufts University) on soft, invertebrate-inspired robots. The meeting was hosted in *La Pedrera*, a building designed by the modernist, nature-inspired Catalan architect Antoni Gaudi. La Pedrera is a world heritage site, one of the best known buildings in Barcelona, and a very fitting setting for the first Living Machines conference. The conference also included an exhibition of working biomimetic and biohybrid systems including several autonomous biomimetic robotic systems.

Organization and Sponsors

We take this opportunity to thank the many people that were involved in making LM 2012 possible. On the organizational side this included Carme Buisan, Mireia Mora, and Gill Ryder. Artwork was provided by Ian Gwilt. Organizers for the workshop program included Frank Grasso, Chiara Bartolozzi, Emri Neftci, and Stefano Vasanelli. We would also like to thank the authors and speakers who contributed their work, and the members of the International Program Committee for their detailed and considered reviews. We are grateful to the six keynote speakers who shared with us their vision of the future.

Finally, we wish to thank the sponsors of LM 2012: The *Convergence Science Network for Biomimetic and Biohybrid Systems* (CSN) (ICT-248986) which is funded by the European Union's Framework 7 (FP7) program in the area of *Future Emerging Technologies* (FET), the University of Pompeu Fabra in Barcelona, the University of Sheffield, and the Institució Catalana de Recerca i Estudis Avançats (ICREA). Additional support was provide by the FP7 FET Proactive Project BIOTACT (ICT-215910) and the ICT Challenge 2 project EFAA (ICT-270490).

July 2012

Tony J. Prescott
Nathan F. Lepora
Anna Mura
Paul F.M.J. Verschure

Organization

Organizing Committee

Co-chairs	Tony Prescott, Paul Verschure
Program Manager	Nathan Lepora
Local Organizer	Anna Mura
Treasurer	Carme Buisan
Art and Design	Ian Gwilt
Advisors	Stefano Vasanelli, Giacomo Indiveri

Program Committee

Andrew Adamatzky	Bristol Robotics Lab., UK
Sean Anderson	University of Sheffield, UK
Joseph Ayers	Northeastern University, USA
Yoseph Bar-Cohen	Jet Propulsion Lab., USA
Jennifer Basil	Brooklyn College, CUNY, USA
Frédéric Boyer	IRCCyN-Ecole des Mines de Nantes, France
Dieter Braun	Ludwig Maximilians University, Germany
Darwin Caldwell	Italian Institute of Technology, Italy
Federico Carpi	University of Pisa, Italy
Hillel Chiel	Case Western Reserve University, USA
Anders Lyhne Christensen	Instituto Universitario de Lisboa, Portugal
Noah Cowan	Johns Hopkins University, USA
Holk Cruse	University of Bielefeld, Germany
Mark Cutkosky	Stanford University, CA, USA
Danilo de Rossi	University of Pisa, Italy
Mathew Diamond	International School of Advanced Studies, Italy
Stephane Doncieux	Université Pierre et Marie Curie, France
Wolfgang Eberle	Imec, Leuven, Belgium
Charles Fox	University of Sheffield, UK
Michele Giugliano	University of Antwerp, Belgium
Frank Grasso	Brooklyn College, CUNY, USA
Roderich Gross	University of Sheffield, UK
Auke Ijspeert	EPFL, Switzerland
Akio Ishiguro	Tohoku University, Japan
Holger Krapp	Imperial College, London, UK
Jeff Krichmar	University of California, Irvine, USA
Maarja Kruusmaa	Tallinn University of Technology, Estonia
Andres Diaz Lantada	Universidad Politecnica de Madrid, Spain

Table of Contents

Full Papers

A Conserved Biomimetic Control Architecture for Walking, Swimming
and Flying Robots 1
  Joseph Ayers, Daniel Blustein, and Anthony Westphal

A Digital Neuromorphic Implementation of Cerebellar Associative
Learning 13
  Luis Bobo, Ivan Herreros, and Paul F.M.J. Verschure

Simulating an Elastic Bipedal Robot Based on Musculoskeletal
Modeling 26
  Roberto Bortoletto, Massimo Sartori, Fuben He, and Enrico Pagello

A Soft-Body Controller with Ubiquitous Sensor Feedback 38
  *Alexander S. Boxerbaum, Kathryn A. Daltorio, Hillel J. Chiel, and
  Roger D. Quinn*

Exploration of Objects by an Underwater Robot with Electric Sense.... 50
  Frédéric Boyer and Vincent Lebastard

Neuro-inspired Navigation Strategies Shifting for Robots:
Integration of a Multiple Landmark Taxon Strategy 62
  *Ken Caluwaerts, Antoine Favre-Félix, Mariacarla Staffa,
  Steve N'Guyen, Christophe Grand, Benoît Girard, and
  Mehdi Khamassi*

Bioinspired Tunable Lens Driven by Electroactive Polymer Artificial
Muscles.. 74
  Federico Carpi, Gabriele Frediani, and Danilo De Rossi

A Pilot Study on Saccadic Adaptation Experiments with Robots........ 83
  Eris Chinellato, Marco Antontelli, and Angel P. del Pobil

Jumping Robot with a Tunable Suspension Based on Artificial
Muscles.. 95
  Sanjay Dastoor, Sam Weiss, Hannah Stuart, and Mark Cutkosky

Static versus Adaptive Gain Control Strategy for Visuo-motor
Stabilization 107
  Naveed Ejaz, Reiko J. Tanaka, and Holger G. Krapp

Learning and Retrieval of Memory Elements in a Navigation Task...... 120
  Thierry Hoinville, Rüdiger Wehner, and Holk Cruse

Imitation of the Honeybee Dance Communication System by Means
of a Biomimetic Robot 132
  Tim Landgraf, Michael Oertel, Andreas Kirbach,
  Randolf Menzel, and Raúl Rojas

A Framework for Mobile Robot Navigation Using a Temporal
Population Code.. 144
  André Luvizotto, César Rennó-Costa, and Paul F.M.J. Verschure

Generalization of Integrator Models to Foraging: A Robot Study Using
the DAC9 Model... 156
  Encarni Marcos, Armin Duff, Martí Sánchez-Fibla, and
  Paul F.M.J. Verschure

The Emergence of Action Sequences from Spatial Attention: Insight
from Rodent-Like Robots 168
  Ben Mitchinson, Martin J. Pearson, Anthony G. Pipe, and
  Tony J. Prescott

How Past Experience, Imitation and Practice Can Be Combined
to Swiftly Learn to Use Novel "Tools": Insights from Skill Learning
Experiments with Baby Humanoids 180
  Vishwanathan Mohan and Pietro Morasso

Towards Contextual Action Recognition and Target Localization
with Active Allocation of Attention 192
  Dimitri Ognibene, Eris Chinellato, Miguel Sarabia, and
  Yiannis Demiris

Robot Localization Implemented with Enzymatic Numerical
P Systems 204
  Ana Brânduşa Pavel, Cristian Ioan Vasile, and Ioan Dumitrache

How Can Embodiment Simplify the Problem of View-Based
Navigation?... 216
  Andrew Philippides, Bart Baddeley, Philip Husbands, and
  Paul Graham

The Dynamical Modeling of Cognitive Robot-Human Centered
Interaction 228
  Mikhail I. Rabinovich and Pablo Varona

Internal Drive Regulation of Sensorimotor Reflexes in the Control
of a Catering Assistant Autonomous Robot 238
  César Rennó-Costa, André Luvizotto, Alberto Betella,
  Martí Sánchez-Fibla, and Paul F.M.J. Verschure

Incremental Learning in a 14 DOF Simulated iCub Robot:
Modeling Infant Reach/Grasp Development 250
  Piero Savastano and Stefano Nolfi

A True-Slime-Mold-Inspired Fluid-Filled Robot Exhibiting Versatile
Behavior 262
   Takuya Umedachi, Ryo Idei, and Akio Ishiguro

CyberRat Probes: High-Resolution Biohybrid Devices for Probing
the Brain 274
   Stefano Vassanelli, Florian Felderer, Mufti Mahmud,
   Marta Maschietto, and Stefano Girardi

Crayfish Inspired Representation of Space via Haptic Memory
in a Simulated Robotic Agent 286
   Stephen G. Volz, Jennifer Basil, and Frank W. Grasso

Parallel Implementation of Instinctual and Learning Neural Mechanisms
in a Simulated Mobile Robot 298
   Briana Young, Stefano Ghirlanda, and Frank W. Grasso

Distributed Control of Complex Arm Movements: Reaching
Around Obstacles and Scratching Itches 309
   David Zipser

Cerebellar Memory Transfer and Partial Savings during Motor
Learning: A Robotic Study....................................... 321
   Riccardo Zucca and Paul F.M.J. Verschure

Extended Abstracts

A Biomimetic Approach to an Autonomous Unmanned Air Vehicle 333
   Fotios Balampanis and Paul F.M.J. Verschure

Towards a Framework for Tactile Perception in Social Robotics 335
   Hector Barron-Gonzalez, Nathan F. Lepora,
   Uriel Martinez-Hernandez, Mat Evans, and Tony J. Prescott

A Locomotion Strategy for an Octopus-Bioinspired Robot 337
   Marcello Calisti, Michele Giorelli, and Cecilia Laschi

Design and Modeling of a New Biomimetic Robot Frog with the Ability
of Jumping Altitude Regulation................................... 339
   Sadjad Eshgi, Vahid Azimirad, and Hamid Hajimohammadi

Sensation of a "Noisy" Whisker Vibration in Rats 341
   Arash Fassihi, Vahid Esmaeili, Athena Akrami,
   Fabrizio Manzino, and Mathew E. Diamond

Integrating Molecular Computation and Material Production in an
Artificial Subcellular Matrix..................................... 343
   Harold Fellermann, Maik Hadorn, Eva Bönzli, and Steen Rasmussen

WARMOR: Whegs Adaptation and Reconfiguration of MOdular Robot
with Tunable Compliance 345
    Max Fremerey, Goran S. Djordjevic, and Hartmut Witte

Inverse and Direct Model of a Continuum Manipulator Inspired
by the Octopus Arm 347
    Michele Giorelli, Federico Renda, Andrea Arienti, Marcello Calisti,
    Matteo Cianchetti, Gabriele Ferri, and Cecilia Laschi

A Biomimetic, Swimming Soft Robot Inspired by the Octopus
Vulgaris 349
    Francesco Giorgio Serchi, Andrea Arienti, and Cecilia Laschi

Toward a Fusion Model of Feature and Spatial Tactile Memory
in the Crayfish Cherax Destructor 352
    Frank W. Grasso, Mat Evans, Jennifer Basil, and Tony J. Prescott

Development of Sensorized Arm Skin for an Octopus Inspired
Robot – Part I: Soft Skin Artifacts 355
    Jinping Hou, Richard H.C. Bonser, and George Jeronimidis

Development of Sensorized Arm Skin for an Octopus Inspired
Robot – Part II: Tactile Sensors 357
    Jinping Hou, Richard H.C. Bonser, and George Jeronimidis

Development of Sensorized Arm Skin for an Octopus Inspired Robot –
Part III: Biomimetic Suckers 359
    Jinping Hou, Richard H.C. Bonser, and George Jeronimidis

Decentralized Control Scheme That Enables Scaffold-Based Peristaltic
Locomotion 361
    Akio Ishiguro, Kazuyuki Yaegashi, Takeshi Kano, and Ryo Kobayashi

Autonomous Decentralized Control Mechanism in Resilient Ophiuroid
Locomotion 363
    Takeshi Kano, Shota Suzuki, and Akio Ishiguro

A Multi-agent Platform for Biomimetic Fish 365
    Tim Landgraf, Rami Akkad, Hai Nguyen, Romain O. Clément,
    Jens Krause, and Raúl Rojas

The State-of-the-Art in Biomimetics 367
    Nathan F. Lepora, Paul F.M.J. Verschure, and Tony J. Prescott

Action Development and Integration in an Humanoid iCub Robot 369
    Tobias Leugger and Stefano Nolfi

Insect-Like Odor Classification and Localization on an Autonomous
Robot   371
    Lucas L. López-Serrano, Vasiliki Vouloutsi, Alex Escudero Chimeno,
    Zenon Mathews, and Paul F.M.J. Verschure

Autonomous Viewpoint Control from Saliency   373
    Shijian Lu and Joo Hwee Lim

Bio-inspired Design of an Artificial Muscular-Hydrostat Unit for Soft
Robotic Systems   375
    Laura Margheri, Maurizio Follador, Matteo Cianchetti,
    Barbara Mazzolai, and Cecilia Laschi

Texture Classification through Tactile Sensing   377
    Uriel Martinez-Hernandez, Hector Barron-Gonzalez, Mat Evans,
    Nathan F. Lepora, Tony Dodd, and Tony J. Prescott

Bio-inspiration for a Miniature Robot Inside the Abdomen   380
    Alfonso Montellano López, Robert Richardson, Abbas Dehghani,
    Rupesh Roshan, David Jayne, and Anne Neville

Systematic Construction of Finite State Automata Using VLSI Spiking
Neurons   382
    Emre Neftci, Jonathan Binas, Elisabetta Chicca,
    Giacomo Indiveri, and Rodney Douglas

Self-burial Mechanism of *Erodium cicutarium* and Its Potential
Application for Subsurface Exploration   384
    Camilla Pandolfi, Diego Comparini, and Stefano Mancuso

Tragopogon dubius, Considerations on a Possible Biomimetic Transfer . . .   386
    Camilla Pandolfi, Vincent Casseau, Terence Pei Fu,
    Lionel Jacques, and Dario Izzo

Root-Soil Interaction Models for Designing Adaptive Exploring Robotic
Systems   388
    Liyana Popova, Alice Tonazzini, and Barbara Mazzolai

A Soft-Bodied Snake-Like Robot That Can Move on Unstructured
Terrain   390
    Takahide Sato, Takeshi Kano, Akihiro Hirai, and Akio Ishiguro

Direct Laser Writing of Neural Tissue Engineering Scaffolds
for Biohybrid Devices   392
    Colin R. Sherborne, Christopher J. Pateman, and
    Frederik Claeyssens

Biorobotic Actuator with a Muscle Tissue Driven
by a Photostimulation 394
     Masahiro Shimizu, Shintaro Yawata, Kota Miyasaka,
     Koichiro Miyamoto, Toshifumi Asano, Tatsuo Yoshinobu,
     Hiromu Yawo, Toshihiko Ogura, and Akio Ishiguro

Shape Optimizing of Tail for Biomimetic Robot Fish 396
     Majid Siami and Vahid Azimirad

Intuitive Navigation of Snake-Like Robot with Autonomous
Decentralized Control 398
     Yasushi Sunada, Takahide Sato, Takeshi Kano, Akio Ishiguro, and
     Ryo Kobayashi

Design of Adhesion Device Inspired by Octopus Sucker 400
     Francesca Tramacere, Lucia Beccai, and Barbara Mazzolai

Author Index 403

A Conserved Biomimetic Control Architecture for Walking, Swimming and Flying Robots*

Joseph Ayers, Daniel Blustein, and Anthony Westphal

Northeastern University Marine Science Center, Nahant MA 01908, USA
lobster@neu.edu
http://www.neurotechnology.neu.edu/

Abstract. Simple animals adapt with impunity to the most challenging of conditions without training or supervision. Their behavioral repertoire is organized into a layered set of exteroceptive reflexes that can operate in parallel and form sequences in response to affordances of the environment. We have developed a common architecture that captures these underlying mechanisms for implementation in engineered devices. The architecture instantiates the underlying networks with discrete time map-based neurons and synapses on a sequential processor. A common board set instantiates releasing mechanisms, command neurons, coordinating neurons, central pattern generators, and reflex functions that are programmed as networks rather than as algorithms. Layered exteroceptive reflexes mediate heading control, impediment compensation, obstacle negotiation, rheotaxis, docking, and odometry and can be adapted to a variety of robotic platforms. We present the implementation of this architecture for three locomotory modes: swimming, walking, and flying.

Keywords: robotics, biomimetics, electronic nervous system, exteroceptive reflex.

1 Introduction

The innate behavior of simple animals is controlled by central neuronal networks that self organize during development [1]. These central networks are comprised of command neurons, coordinating neurons and central pattern generators (CPGs) that are modulated by sensory feedback. The behavioral set is organized into layered exteroceptive reflexes that can operate in parallel and form sequences in response to affordances of the environment. We have developed a common architecture that captures these underlying mechanisms in engineered devices (Fig. 1). The initial goal in the development of all platforms was to capture the physics and biomechanics of the model organism. For example, the lobster has a three degree of freedom leg that enables it to walk in any direction [2]. The lamprey has an undulatory body axis that can provide high swimming maneuverability in the yaw plane [3]. A resonant flight motor can produce the high lift required for flight in honeybees [4], a mechanism that contrasts with

---

* Supported by NSF Grant ITR 0925751, ONR MURI in Synthetic Biology(2011).

T.J. Prescott et al. (Eds.): Living Machines 2012, LNAI 7375, pp. 1–12, 2012.

Fig. 1. *Biomimetic robotic platforms.* **A.** RoboLobster. **B.** RoboLamprey. **C.** RoboBee (photograph courtesy of Rob Wood).

the direct activation of antagonist muscle pairs in the lobster. While differences exist in the body structures and actuators, the underlying components and principles of the electronic nervous system (ENS) control architecture we use are conserved. Neuroethological observations drawn from the literature record and from our own work drive the formulation of hypothetical neural networks that derive from the command neuron, coordinating neuron and CPG framework [5]. Comparative study between animal and robot allows for the validation and improvement of the developed models. The generalized approach to control, along with a modularized robot assembly process, enables the production of a range of biomimetic vehicles that can operate in any environment. Here we present a common conserved architecture used in RoboBee, RoboLobster, and RoboLamprey that preserves the animals underlying biomechanics while allowing for the flexibility of control that the animal models display.

2   Electronic Nervous Systems

A key feature of our approach is to construct controllers from exteroceptive reflex networks based on known neurophysiology using simulated neurons and synapses. The discrete time map-based (DTM) model we use simulates neuron and synapse activity using a computationally efficient set of difference equations to allow for real time operation [6]. Detailed implementation of this model has been described elsewhere [3]. We have been able to realize all forms of neuronal integration including spiking, bursting, parametric modulation and perturbation using DTM neurons. Rapid network prototyping using DTM neurons and synapses is accomplished in a LabVIEW^{TM} simulation environment.

3   Central Pattern Generators

Organisms from across the animal kingdom use local neuron networks called *Central Pattern Generators* (CPGs) to drive locomotion. We have based our swimming CPGs on Buchanan's lamprey CPG model [7] and and their detailed operation has been described elsewhere. Similarly, we have implemented an omnidirectional walking CPG based on a four element CPG (elevation, depression, swing, stance) that uses presynaptic inhibition from commands for walking in different directions to alter patterns of coordination in bifunctional joint muscles [8,9]. The bee robot is myogenic and uses piezoelectric actuation [10,11].

4 Standard Board Set

Our walking and swimming robots are based on a standard board set. The basic modules are the main processor board that is home to the DTM neurons and synapses (Fig. 2a), a power board (Fig. 2b), a sensor board (Fig. 2c) that houses the compass, accelerometers and the DSP for a short baseline sonar array and a constant current driver board stack (Fig. 2d). Other circular sensor modules with their own power interfaces house specialized systems such as the filters for the sonar interface (Fig. 3). Our studies have shown [3] that numerical simulation of the swimming CPG network controlling the lamprey biomimetic robot can be implemented in real time using a TI DSP chip. The custom configuration[1] using a TI DSP TMS320C6727 running at 1.8GHz is capable of emulating networks of up to 2000 neuron models in real-time. We mediate excitation/contraction coupling in the underwater robots by using electric current to cause a heat-based state transition in a shape memory alloy, nitinol. The motor neuron action potentials from the brain DSP trigger current pulses generated by a driver board stack to heat the nitinol. Each driver board stack serves to isolate and amplify neuronal spikes as logic signals to allow for the control of 12 actuators.

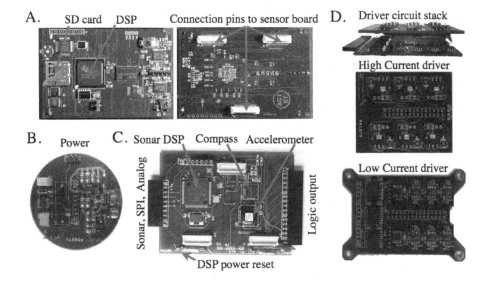

Fig. 2. *Overview of biomimetic robot board set.* **A.** Top and bottom views of the brain DSP. **B.** Power regulation board generates +/-5v and 3.3v from a 12v input. **C.** Sensor board containing sonar DSP processor, tilt-compensated compass for heading pitch and roll, and a three axis +/-1.5g accelerometer. **D.** Stacked constant current driver board set with Low Current board that isolates logic signals from the brain and performs current sensing and High Current board that contains the switching FETs delivering the high current pulse train to the actuators.

---

[1] Ariel, Inc. Santa Ysabel, CA 92070.

5   Releasing Mechanisms

Our goal of supervised reactive autonomy requires that we have a broad variety of sensors of unique modalities. Moreover, the sensors must discriminate particular affordances and environmental features while encoding the result in a neuronal spike train. Several of the platforms share modalities. For example, heading control is fundamental to all systems. Both the walking and flying systems will rely on optical flow for yaw stability and obstacle avoidance. Visually-mediated odometry is highly desirable and can be applied to all systems. A short baseline sonar array can mediate homing for docking or target acquisition. Our design goal is to therefore develop modular interfaces

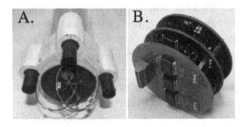

Fig. 3. *Sonar system of RoboLamprey.* **A.** Three AQ2000 hydrophones mounted in syntactic foam receive sonar inputs. **B.** The sonar processing stack is made of three stages. From left to right, the first stage is the buffer amplifier for the three hydrophone channels. The second stage consists of band-pass filters and programmable gain amplifiers for each channel. The third stage contains the power circuitry.

and DTM networks to emulate the innate releasing mechanisms found in the model organisms and encode sensor status in neuronal pulse trains. The existence of magnetotaxis has been demonstrated in a variety of species [12]. We provide our vehicles with a pitch and roll compensated compass (Fig. 4) implemented using the following equations:

$$Sig_1 = \begin{cases} |H_C - H_D| - (|H_C - H_D| - 180) \times 2, & |H_C - H_D| > 180, \\ |H_C - H_D|, & \text{otherwise} \end{cases} \quad (1)$$

$$Sig_2 = \begin{cases} |H_C - H_D| - (|H_C - H_D| - 180) \times 2, & (H_C - H_D) < -180, \\ (H_C - H_D), & \text{otherwise} \end{cases} \quad (2)$$

$$idc = \begin{cases} -1 \times Sig_1, & Sig_2 > -180, \\ Sig_2, & \text{otherwise} \end{cases} \quad (3)$$

$$idc_L = idc \times -1 \quad \text{and} \quad idc_R = idc \quad (4)$$

Based on the desired (H_D) and actual (H_C) compass heading, intermediate parameters Sig_1 (1) and Sig_2 (2) are calculated. Sig_1 and Sig_2 are then used to calculate the injected current (idc), constrained to a value between +/-180 and resulting in a bowl function with its base at H_D. Idc is then split into idc_L and idc_R (3,4) where a positive value results in excitation of the range fractionation network and negative values result in proportional inhibition (4). This excitation and inhibition governs the output of the compass neurons on either side (see Section 6). A short baseline hydrophone array and a sonar board stack provide the underwater robots with the capability for beacon tracking (Fig. 3). This stack, with its own power circuitry, is responsible for the buffering, filtering and amplification of the hydrophone signal (Fig. 3b). Microchip's dsPIC range of microcontroller was selected for sonar data processing as it offers a purpose built base for signal processing applications and meets the data storage, processing speed, and signal handling requirements to accurately sample the sonar signal. This allows for simultaneous real time sampling of the three hydrophone channels for calculation of elevation and azimuth angles to the transducer.

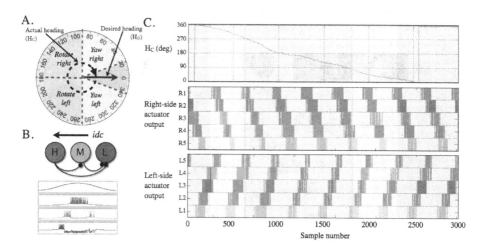

Fig. 4. *Heading-mediated orientation reflex.* **A.** Given a desired heading (H_D) of $0°$, three regions of heading deviation are defined: Blue for on target, green for yaw, and red for rotate. These regions are also broken into left and right sectors, in this case down the $0°$ to $180°$ midline. Depending on the error between the actual heading (H_C) and the H_D, the appropriate corrective maneuver is made. **B.** To transition from an error signal to a neuronal one, range fractionation is used with lateral inhibition in a network that consists of three neurons: L for low (blue), M for medium (green) and H for high (red). These neurons are tuned to respond to different thresholds of stimulus intensity. **C.** Demonstration of heading deviation modulation of the swimming pattern. **Top panel.** Onboard compass heading. **Middle and bottom panels.** Right and left side actuator outputs, respectively. There are five evenly spaced pairs of actuators with R1 / L1 at the head and R5 / L5 at the tail.

Flying insects use cumulative optical flow to determine distance flown [13]. We have developed a neural circuit to mediate odometry on RoboBee and RoboLobster (Fig. 5). A DTM leaky integrator neuron is used to accumulate bilateral optical flow sensory inputs. As the vehicle moves forward, optical flow accumulates and increases the voltage of the 1st-layer odometer neuron. Once the voltage increases past the firing threshold, the 1st-layer odometer neuron spikes and then resets. The frequency of the spikes of the 1st-layer odometer neuron indicates the rate of optical flow and the cumulative number of spikes indicates the total amount of optical flow that is used as a measure of distance traveled. This spike train of odometry information is passed to another DTM leaky integrator, the 2nd level odometer neuron, via an excitatory synapse. The synapse strength of the connection between the 1st- and 2nd-layer odometer neurons can be varied so that the 2nd-layer odometer neuron fires after different amounts of cumulative optical flow (Fig. 5c). As synapse strength is increased, the 2nd-layer odometer will spike more quickly indicating a lower amount of total optical flow, a signal of a shorter distance flown. The desired flight distance of a robot on a mission can be conveyed using a single number representing the synaptic strength for the optical flow-mediated odometry network.

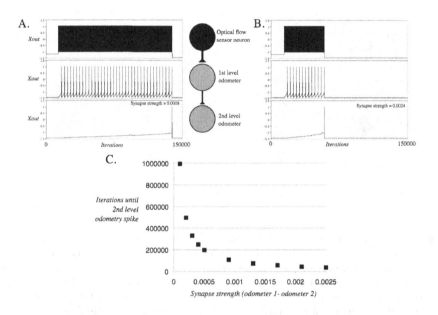

Fig. 5. *Neuronal optical flow-mediated odometry.* Voltage versus time traces for neurons of a two-layer odometer circuit with (**A**) weak and (**B**) strong synapses. With optical flow input (top traces), the first layer odometer fires regularly (middle traces). The synaptic strength between the first and second layer odometer neurons determines how quickly the second layer odometer (bottom traces) fires. The firing of this neuron can mediate behavioral transitions after the traversal of a specified distance. **C.** Plot of how varying synaptic strength leads to different firing times of the second layer odometer.

6   Layered Exteroceptive Reflexes

A common feature shared by all platforms is the employment of exteroceptive reflexes for heading control, beacon tracking, impediment compensation, and

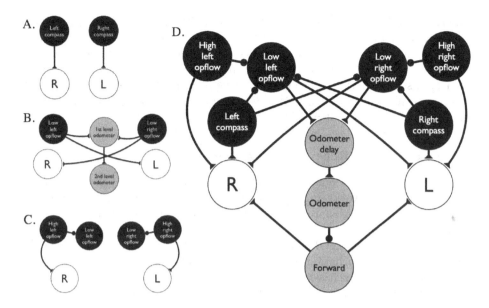

Fig. 6. *Exteroceptive reflex networks for RoboBee.* Unimodal reflex networks for (**A**) compass-mediated heading control, (**B**) optical flow-mediated odometry and compensation, and (**C**) optical flow-mediated obstacle avoidance. The complete exteroceptive network with fused inputs from multiple sensory modalities is depicted in (**D**). Black circles denote sensory neurons, grey circles are processing interneurons and commands, white circles are motor neurons. For Robobee, the R motor neuron elicits a turn to the right and the L motor neuron elicits a turn to the Left. Lines with triangle ends represent excitatory synapses; filled circle ends, inhibitory synapses.

obstacle avoidance[3,9,10]. These reflexes are generally mediated by command systems that integrate inputs from exteroceptors and activate the appropriate task groups. All exploratory behavior of the vehicles will be based on search vectors. The waggle dance of the bee is the biomimetic benchmark for the communication of search vectors [14]. The vector consists of a desired heading and odometry information and is computed by our compass (Fig. 4) and odometer (Fig. 5) neuronal systems.

6.1 Exteroceptive Reflexes for Flight Modulation

Several exteroceptive networks are layered to form the fused sensory network of RoboBee (Fig. 6). The compass network (Fig. 6a) uses the left or right compass error from a desired heading to mediate a corrective yaw response. For example, the left compass neuron fires when the actual heading is to the left of the target and excites the Right Command neuron that produces a yaw to the right. Low rates of optical flow mediate reflexive responses to visual perturbations through decussating excitatory synapses between the bilateral optical flow neurons and the Right/Left Commands (Fig. 6b). Optical flow is also used to perform odometry as described in Section 5. A high threshold optical flow circuit is used to mediate avoidance turns away from obstacles via excitatory connections to flight command neurons (Fig. 6c). Motors can be controlled by altering the discharge rate of premotor neurons to modulate the activating pulse-width duty cycle signal. There are a variety of strategies to produce yaw modulation in the flying bee that are being explored for implementation on Robobee including adjustments to wing stroke amplitude, wing angle of attack, and abdominal positioning [11].

6.2 Exteroceptive Reflexes for Walking

The fused exteroceptive reflex network of RoboLobster is also developed using layered unimodal networks (Fig. 7). Compass-mediated heading control is accomplished using range fractionated sensory input of the compass error (Fig. 7a). Yaw and Rotate interneurons of the compass network mediate reflexive turning towards a target heading based on the level of compass error. Translational optical flow elicits visual perturbation responses and obstacle avoidance maneuvers mediated by decussating excitatory connections (Fig. 7b). Bump sensory neurons receive input from mechanoreceptors on the claws to elicit a yaw response away from

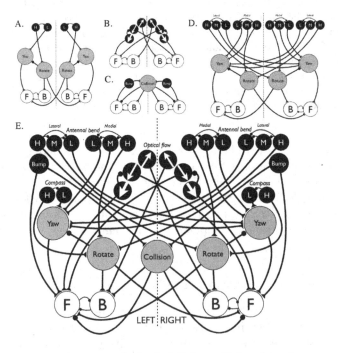

Fig. 7. *Exteroceptive reflex networks for RoboLobster.* Unimodal reflex networks for (**A**) compass-mediated heading control, (**B**) optical flow-mediated stabilization and compensation,(**C**) bump-mediated obstacle avoidance, and (**D**) antennal bending-mediated rheotaxis and surge compensation. The complete exteroceptive network is depicted in (**E**). Symbols are as described in Fig. 6.

obstacles via ipsilateral excitatory inputs to Forward walking commands (Fig. 7c). Response to hydrodynamic flow is mediated by antennal sensors that respond to high (H), medium (M), and low (L) bending medially and laterally and project to flow sensory interneurons (Fig. 7d). Under lateral surge conditions a high medial bend on one antenna and a high lateral bend on the other antenna excites the Rotate interneuron of the medially bent side to elicit rheotaxis. Off center axial surge produces less severe antennal bending that excites the Yaw interneurons causing the vehicle to align with the surge by producing bias between the bilateral Forward commands.

6.3 Exteroceptive Reflexes for Swimming

The compass-mediated heading network (Fig. 8a) in the lamprey operates on the principal of range fractionation with High or Medium compass error signals exciting the H and M sensory neurons, respectively. A large deviation from a target heading elicits a severe course correction via the Rotate interneuron. Moderate compass error elicits a more gradual corrective turn via the Yaw interneuron. A similar network structure produces sonar-mediated heading control with sonar inputs inhibiting compass inputs as sonar tracking predominates over compass heading control. In order to mediate collisions and impediments such as surge, the Collision and Impediment interneurons integrate data from the onboard accelerometer (Fig. 8b). High decelerations indicate a collision triggering the Collision neuron resulting in a bout of backward swimming. For decelerations during normal swimming the Impediment neuron is triggered causing the robot to increase its rate of swimming in order to mitigate surge. These networks are layered (Fig. 8c) with one another to achieve a hierarchal network capable of achieving goal-oriented locomotion.

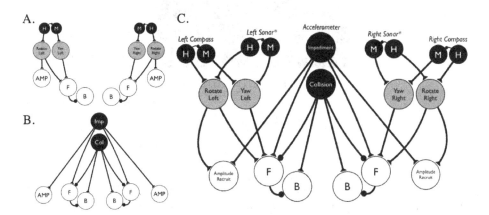

Fig. 8. *Exteroceptive reflex networks for RoboLamprey.* Unimodal reflex networks for (**A**) compass-mediated heading control, and (**B**) accelerometer-mediated obstacle negotiation. The complete exteroceptive network with fused inputs from multiple sensory modalities is depicted in (**C**). Symbols are as described in Fig. 6.

7   Behavioral Sequencing

The field behavior of these robots will be based on motivated behavioral sequences organized reactively relative to a supervisory motivation. Figure 9 shows a LabVIEW^{TM} software simulation of the networks underlying compass-mediated heading control and optical flow-mediated stabilization and compensation in a helicopter approximation of RoboBee (network diagrams: Fig. 8 ab). The behavioral sequence shown consists of three phases: I. Initiation and Clockwise yaw for target heading seek; II. Target heading overshoot and Counterclockwise yaw; III. Forward linear flight for a distance specified by the synaptic strength of the odometer circuit (Fig. 5). The resulting sensory milieu at the end of each phase elicits the next behavioral act. This chain-reflex sequencing to produce complex behaviors is common across a range of species [15,16].

8   Supervised Reactive Autonomy

The layered exteroceptive reflex architecture presented can form the basis of supervised reactive autonomy where the underlying motivation of the vehicle

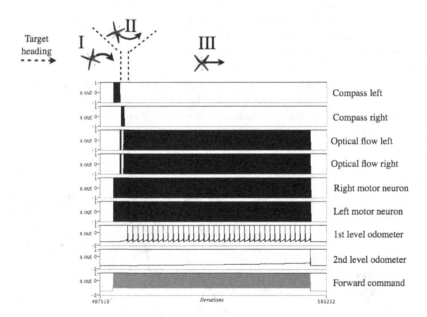

Fig. 9. *Simulation of fused compass and optical flow sensory network in LabVIEW^{T}M for a dual rotor helicopter.* Voltage versus time charts of the neurons of the simulated neural network described in Fig. 6 for a behavioral sequence of a helicopter taking off, obtaining a target heading (*I* and *II*) and using optical flow-mediated odometry to fly a specified distance (*III*). Once the Forward command (bottom trace) is activated in *I*, a yaw response is driven to obtain a specified heading. After a heading overshoot in *II*, the odometer neurons then integrate optical flow during forward flight in *III* to indicate when a desired distance has been flown.

is to pursue a search vector. Search vectors can be specified by two tokens, a direction and a distance, that could be communicated by sonar, LEDs, or wireless communication. The vehicle can also be provided with a propensity to negotiate or investigate obstacles that it encounters. We expect that ongoing efforts in synthetic biology will provide sensing capabilities and taxic behavior [17] for a broad variety of chemical sensors specific to agents of interest or harm [18,19]. Chemical sensing is key to the inter-individual exteroceptive reflexes that underlie stigmergy [20].

This category of vehicles will operate as marsupials dispensed by a delivery/ retrieval vehicle (perhaps a hive). Here, scaling becomes a challenge. Short-range communications between the vehicles and the delivery vehicle can be mediated by sonar underwater [21] but is problematic with ultra low power systems such as RoboBee. While sonar systems can be used to establish long baseline navigational arrays underwater to allow for precise positioning of the vehicles [22], relying on global communications is unrealistic given the myriad of environmental challenges and operational constraints; thus, a biomimetic localized communication system is needed. The ultimate challenge is to integrate stigmergy into the supervised autonomy of the vehicles. The potential use of odor trails and specific chemical sensors could revolutionize the control of both terrestrial and underwater robots.

9 Conclusion

The biomimetic robot architecture we have presented shows that it is feasible to control vehicles using neuronal integrative processes in engineered devices. This realizes the potential first developed by Braitenberg [23]. By mimicking the natural architecture of the animal nervous system, we can build adaptively behaving machines displaying aspects of reactive autonomy. This biorobotic approach gives the neuroscientist an amenable platform for hypothesis testing [24,25] while providing the roboticist with a way to capture some of the adaptive behavioral characteristics of animals in engineered devices.

References

1. Stein, P.S., Grillner, G.S., Selverston, A.I., Stuart, D.G.: Neurons, Networks and Motor Behavior. MIT Press, Cambridge (1997)
2. Ayers, J.: Underwater walking. Arthro. Struct. and Develop. 33(3), 347–360 (2004)
3. Westphal, A., Rulkov, N., Ayers, J., Brady, D., Hunt, M.: Controlling a lamprey-based robot with an electronic nervous system. Struct. and Sys. 8(1), 37–54 (2011)
4. Wood, R.: The First Takeoff of a Biologically Inspired At-Scale Robotic Insect. IEEE Transactions on Robotics (2008)
5. Kennedy, D., Davis, W.J.: Organization of Invertebrate Motor Systems. Handbook of Physiology. The organization of invertebrate motor systems. In: Geiger, S.R., Kandel, E.R., Brookhart, J.M., Mountcastle, V.B. (eds.) Handbook of Physiology, sec. I, vol. I, part 2, pp. 1023–1087. Amer. Physiol. Soc, Bethesda (1977)

6. Rulkov, N.F.: Modeling of spiking-bursting neural behavior using two-dimensional map. Phys. Rev. E 65, 041922 (2002)
7. Buchanan, J.T., Grillner, S.: Newly identified 'glutamate interneurons' and their role in locomotion in the lamprey spinal cord. Science 236(4799), 312–314 (1987)
8. Ayers, J., Davis, W.J.: Neuronal Control of Locomotion in the Lobster (Homarus americanus) 1. Motor Programs for forward and Backward walking. J. Comp. Physiol. A 115, 1–27 (1977)
9. Ayers, J., Witting, J.: Biomimetic Approaches to the Control of Underwater Walking Machines. Phil. Trans. R. Soc. Lond. A 365, 273–295 (2007)
10. Duhamel, P.E., Porter, J.A., Finio, B.M., Barrows, G.H., Brooks, D., Wei, G.Y., Wood, R.: Hardware in the loop for Optical Flow Sensing in a Robotic Bee. In: IEEE/RSJ Int. Conf. on Intelligent Robots and Systems, San Francisco, CA (2011)
11. Finio, B.M., Shang, J.K., Wood, R.J.: Body torque modulation for a microrobotic fly. In: IEEE International Conference on Robotics and Automation, pp. 3449–3456 (2009)
12. Lohmann, K., Pentcheff, N., Nevitt, G., Stetten, G., Zimmer-Faust, R., Jarrard, H., Boles, L.: Magnetic orientation of spiny lobsters in the ocean: experiments with undersea coil systems. J. Exp. Biol. 198(10), 2041–2048 (1995)
13. Srinivasan, M., Zhang, S., Lehrer, M., Collett, T.: Honeybee navigation en route to the goal: visual flight control and odometry. J. Exp. Biol. 199, 237–244 (1996)
14. Menzel, R., De Marco, R.J., Greggers, U.: Spatial memory, navigation and dance behaviour in Apis mellifera. J. Comp. Physiol. A 192(9), 889–903 (2006)
15. Dickinson, M.H., Lent, C.M.: Feeding behavior of the medicinal leech, Hirudo medicinalis L. J. of Comp. Physiol. A 154(4), 449–455 (1984)
16. Berridge, K.C., Fentress, J.C., Parr, H.: Natural syntax rules control action sequence of rats. Behavioural Brain Research 23(1), 59–68 (1987)
17. Grasso, F.W., Basil, J.A.: How lobsters, crayfishes, and crabs locate sources of odor: current perspectives and future. Curr. Opin. in Neurobiol. 12, 721–727 (2002)
18. Purnick, P.E., Weiss, R.: The second wave of synthetic biology: from modules to systems. Nat. Rev. Mol. Cell Biol. 10(6), 410–422 (2009)
19. Antunes, M.S., Morey, K.J., Smith, J.J., Albrecht, K.D., Bowen, T.A., Zdunek, J.K., Troupe, J.F., Cuneo, M.J., Webb, C.T., Hellinga, H.W., Medford, J.I.: Programmable Ligand Detection System in Plants through a Synthetic Signal Transduction Pathway. PLoS ONE 6(1), e16292 (2011)
20. Holldobler, B., Wilson, E.O.: The Superorganism: The Beauty, Elegance, and Strangeness of Insect Societies. W. W. Norton, Co., New York (2008)
21. Freitag, L., Grund, M., Singh, S., Partan, J., Koski, P., Ball, K.: The WHOI micromodem: an acoustic communications and navigation system for multiple platforms. In: Proceedings of MTS/IEEE, OCEANS 2005, vol. 2, pp. 1086–1092 (2005)
22. Freitag, L., Johnson, M., Grund, M., Singh, S., Preisig, J.: Integrated acoustic communication and navigation for multiple UUVs. In: MTS/IEEE Conference and Exhibition, OCEANS 2001, vol. 4, pp. 2065–2070 (2001)
23. Braitenberg, V.: Taxis, Kinesis and Decussation. Prog. Br. Res. 17, 210–222 (1978)
24. Webb, B.: Validating biorobotic models. J. Neural Engineering 3, 1–20 (2006)
25. Beer, R.D., Chiel, H.J., Quinn, R.D., Ritzmann, R.E.: Biorobotic approaches to the study of motor systems. Curr. Opin. Neurobiology 8, 777–782 (1998)

A Digital Neuromorphic Implementation
of Cerebellar Associative Learning

Luis Bobo[1,*], Ivan Herreros[1], and Paul F.M.J. Verschure[1,2]

[1] Synthetic, Perceptive, Emotive and Cognitive Systems Group (SPECS)
Technology Department, Univ. Pompeu Fabra (UPF), Barcelona, Spain
luis.bobo@upf.edu
[2] Institució Catalana de Recerca i Estudis Avançats (ICREA)

Abstract. The cerebellum is a neuronal structure comprising half the neurons of the central nervous system. It is essential in motor learning and classical conditioning. Here we present a digital electronic module, pluggable to an artificial autonomous system, designed following the neural structure of the cerebellum. It emulates the associative learning function as described in the context of classical conditioning. Building on our previous work we propose a neuromorphic implementation portable to a Field Programmable Gate Array (FPGA), capable of generating responses of variable amplitude. To validate our design we test it with the simulation of a robot performing a navigation task on a curvy track. Our digital cerebellum is able to make adaptively-timed rotations with variable amplitude suitable for the track. This suggests that the Purkinje cell dependent learning circuits of the cerebellum do not only time the triggering of actions but can also tune the specific response amplitude.

Keywords: Cerebellum, Digital, FPGA, Classical Conditioning, Neuroscience, Computation, Electronics, Robotics.

1 Introduction

In classical conditioning, a conditioned stimulus (CS) is paired with an unconditioned stimulus (US), until the former substitutes the latter in producing a conditioned response (CR) which anticipates an innate unconditioned response (UR). For instance if a tone (CS) is repeatedly elicited slightly before an air-puff (US) to the eye of a rat, then the animal will learn to anticipate the air-puff when it hears a tone by closing the eyelid (CR) at the time the air puff is expected to occur. This form of associative learning heavily depends on the cerebellum.

The site of convergence of CS and US information are the Purkinje cells (PU), with each PU receiving about 200000 parallel fibers [3]. The neural architecture of the cerebellum is quite regular, with two input pathways converging on the PUs (Fig. 1a). The first of the pathways is composed of the mossy fibers that originate in the pons (PN) and pass through the granule cell layer. From the granule cells, the pathway links with the PUs dendrites via the parallel fibers (pf). The pf input signal is also modulated by interneurons such as the Golgi cells (GC). The second input pathway originates in the inferior olive and terminates on the PUs, forming the, so called, climbing fibers (cf). PUs in turn innervate the cerebellar deep nuclei (DN). This

T.J. Prescott et al. (Eds.): Living Machines 2012, LNAI 7375, pp. 13–25, 2012.

circuit has negative feedback: when a PU is activated, it produces an inhibitory effect in the deep nucleus (DN). The DN, itself, when activated, inhibits the inferior olive (IO) input [5]. The cerebellum realizes associative learning by processing the CS information coming from the PN and the US information conveyed by the IO. When the CS arrives at a PU slightly before the US, there is a weakening in the efficacy of the synapses between parallel fibers and the PU. Eventually, the PU stops firing. This leads to a disinhibition of the DN and a rebound repolarization eliciting a CR [1] [9].

Behavioral experiments show that the cerebellum can simultaneously learn both the timing and the amplitude of the acquired response: In [8] it is shown that rabbits can be conditioned to produce CRs of a given target amplitude. In these experiments the magnitude of the eyelid closure is monitored online after the initial learning stage. If the closure still remains less intense than the target amplitude, the US is still triggered. In other words, in each trial the US is applied only if the CR was not strong enough to prevent it. With this protocol the authors show that the CR amplitude follows linearly the target amplitude.

But it is not clear how the cerebellum assists in the control of real-world action with respect to its timing and amplitude: we require efficient real-time implementations of the core learning principles implemented by the cerebellar micro-circuit that can be interfaced to real-world devices. In a previous study we developed a neuronal model based on the anatomy of the cerebellum [7]. This model has been generalized towards an analog implementation in the, so called, silicon cerebellum [6]. Subsequently this approach has been applied to a neuroprosthetic device for the cerebellum [11]. We demonstrated that a cerebellar microcircuit comprising PU, DN and IO (Fig. 1a) can acquire well timed conditioned responses. Using a robot obstacle avoidance task, we have shown that the combination of Long Term Depression (LTD) and Long Term Potentiation (LTP) in the cerebellar cortex is sufficient to precisely adjust the timing of the CRs. However, this model was focused on the accuracy of the timing, more than on the modulation of the output amplitude. Our aim is to extend this model by investigating whether the PU population response can account for modulation of amplitude and performing a real-world validation study: we design a physical device capable of performing associative learning to different target amplitudes in a real-world dynamic environment. We implement it as a cerebellar-like Digital Signal Processing (DSP) controller based on FPGA technology. It has several parallel microcircuits that control the CR in an additive manner, i.e. different amplitudes are achieved according to the superposition of CRs triggered by each microcircuit. Moreover, to allow for a wider range of learnable inter stimulus intervals (ISIs) and to increase the precision of the timing of the CRs, each of the microcircuits has different temporal dynamics: the CS is emulated as a decaying pulse memory trace in [7]; here, different microcircuits have these CS memory traces spanning slightly different ranges of ISIs. With DSP, mathematical operations can be computed more easily than with analog processing, while still keeping enough accuracy to emulate the biological cerebellum. Furthermore, FPGA is a standard that allows modularity and low cost reprogramming: this means that later updates of the model can be easily integrated with less time-to-market. FPGA also allows the transformation into structured Application-Specific Integrated Circuits (ASICs) for mass production. We evaluate this model of the cerebellum in a simulated robot navigating on a curvy track. This would show the robustness of this digital model while it remains simple enough from a computational point of view compared to others, where the eligibility trace has to be recalculated per synapse [10].

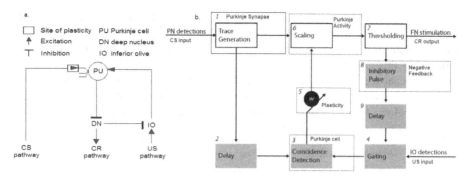

Fig. 1. a. Schematic representation of a cerebellar microcircuit comprising PU, DN and IO extracted from [7]. Coincidence of activity in the CS and US pathways, separated by the Inter-Stimulus Interval (ISI), induces plasticity in the point marked by the rectangle. b. Translation of this cerebellar model into an abstract model. The figure shows the effect of the negative feedback, the inputs integration (coincidence detection), the PU plasticity (the weight scaling the CS input) and the elicitation of the CR if the weighted CS is below a threshold.

2   Methods

Model Design. The computational model is based on the anatomical properties of the cerebellum (Fig. 1a). It is not based in a detailed neuronal model, but on an abstract design of the components and connections that contain the basic principles of the system proposed in [7] (Fig. 1b): The CS decaying pulses (memory traces) delivered to the PUs through the parallel fibers are translated as "PN" (pons) detections (block 1). The multiplicity of parallel fibers synapsing on the PU provides for eligibility traces considering several CS. The composition of a target output amplitude can be obtained by stimulating one or more microcircuits like the one in Fig. 1b, with slightly delayed CS inputs. The circuit measures the coincidence (block 3) between the signal read from this input (after an applied delay simulating the ISI -block 2-) and the signal read from the inferior olive (IO detections -block 4-). This coincidence produces a variable weight change which simulates the bi-directional plasticity effect (block 5) in the PU dendrites -LTD/LTP-. In the next step, the weight affects the input from the pons (CS) to the pf by applying a multiplying factor (block 6). Once scaled and in case that the modified signal is below a threshold (block 7), the PU activity can produce the output via stimulation of the DN and, eventually, the fastigial nucleus (FN). If this happens, apart from eliciting the output, the inhibitory pulse (the nucleo-olivary negative feedback) will be triggered (block 8) and transferred to the IO. The delay of the negative feedback is similar to the ISI (block 9). This allows for a synchronized arrival of both signals -the inhibitory pulse to the IO and the CS responsible of eliciting the CR- at the coincidence measure point (block 3).

Baseline Activity. Together with the USs, baseline activity is needed as a part of the IO activity. This allows keeping a stable learning state when there are no USs, by producing mutually cancelling LTD/LTP effects in coincidence with PN activity (CS traces). We recreate baseline activity in the IO with random signals auto-generated in the circuit, in order to generate a realistic scenario. The signals are introduced every 1 second with a jitter that follows a normal distribution with a standard deviation of 1. This high value simulates approximately white noise. The LTDs and LTPs randomly produced by this baseline activity should be mutually cancelled, in order to keep stability. This cancellation must be independent from the positive or negative correlation due to the CS and US stimuli -acquisition and extinction stages respectively-. This is calculated below, according to the CS pulse duration (DurationCS) and the IO activity (RateIO) and the sample time and frequency in the experimental simulation design:

$$
\begin{aligned}
ProbabilityIO &= RateIO \cdot SampleTime \\
LTPtotal &= DurationCS \cdot SampleFrequency \cdot LTPfactor \\
LTDtotal &= DurationCS \cdot SampleFrequency \cdot LTDfactor \cdot ProbabilityIO \\
LTPtotal &= LTDtotal \rightarrow LTPfactor = RateIO \cdot SampleTime \cdot LTDfactor
\end{aligned}
\tag{1}
$$

Adjusting Input Signals. The CS activity is assimilated to a memory trace. Since this is a linear decaying pulse, there is more accuracy about the moment of reaching the threshold (even after applying the weight to the CS signal) than if it was constant. The final accuracy will also depend on the sample rate of the comparison between the CS and the threshold. The threshold is reached following:

$$
Weight[InitialValue + (FinalValue - InitialValue)(\frac{t}{TraceDuration})] = Threshold
\tag{2}
$$

In our case, we set InitialValue to 1, FinalValue to 0.5, traceDuration to 0.25 and Threshold to 0.499. Each one of these microcircuits has its own noise. The LTD and LTP produced in this process are mutually compensated during the repetitive CS decaying pulses, as explained in (1). LTPfactor is 0.00015 and LTDfactor is 0.015, while RateIO is equal to 1 second and the sample time is 0.01 seconds. The US signals are also introduced with an Inter-Trial Interval (ITI) of 1 second. The weight is bounded between 0.001 and 2, emulating the saturation in the synapses.

Learning Acquisition and Extinction. During acquisition the weight decreases in every sample of the circuit until the PU inhibition over the DN is removed and a CR is elicited. This induces a subsequent stable learning phase stage with the negative feedback enabled, in which USs cannot arrive to the PU coincidence detector, except if they are not cancelled out by IO inhibition. Extinction can be produced just keeping the CSs activity, but without correlated USs. At each sample occurrence, the pf-PU weight is reprocessed as follows:

$$
newWeight = oldWeight + \begin{pmatrix} LTDfactor, & CS>0 \ and \ IO \ activity>0 \\ LTPfactor, & CS>0 \ and \ IO \ activity=0 \\ 0, & otherwise \end{pmatrix}
\tag{3}
$$

Generation of the Response and Control of the Output Amplitude. The composition of the output for time adaptation is generated with 5 microcircuits excited with a CS composed by different phases. In the circuit, this is emulated with different delays in the CS pulses entering the different microcircuits. Following the same process used in [8], US is used as a single binary input. The output is composed by iteratively adding the signal of the different microcircuits. The amplitude of the composed output is compared with the target CR amplitude. This target amplitude is externally defined by the environment: if more output amplitude is needed to avoid the source of the US elicitation, then US pulses are still applied. Consequently, acquisition continues and more microcircuits are used. The single signals of the microcircuits have a fixed duration of 20 ms. and the final output approximates to a triangle pulse. Therefore, the comparison with the target amplitude is done with a delay equal to half of the circuit's internal delay (20 ms). In this way, the target amplitude is compared with the peak of the output pulse.

Software. We use the MATLAB Simulink (Mathworks Inc.) software to create the computational design of the digital circuit. This software has the advantage of being capable of exporting the system into several vendor´s FPGAs, including the Hardware Description Language (HDL) for the card description and the Tool Command Language (TCL) script file for the card programming. If the results of the circuit simulations are valid, then exporting the design into the physical FPGA becomes possible, via the Joint Test Action Group (JTAG) standard interface and by using the FPGA-vendor software. Communication between MATLAB and the additional software for the experiments is done via UDP transport protocol. MATLAB Simulink serves as a workshop to create the model and manage the scaling, thresholds, rates and the delays, in order to reproduce the cerebellum associative learning behavior. Figure 2 shows a design for this microcircuit by using standard elements: arithmetic operations, delay, comparisons, multiplexors and counters.

Experimental Design. The circuit is tested with two simulations:

- A simulation showing the general functioning of the associative learning in the model. The simulation is 60 seconds long with a sample time of 0.01 seconds, but a clock of 1 millisecond, in order to control the synchronizations.
- Along the line of this experiment, we designed a simulation of a navigation task where a robot has to perform a turn within a narrow track, mimicking classical conditioning. It communicates with the running circuit via UDP. The CS is represented by a colored patch in the ground readable by the robot. Whenever the robot hits the wall the collision signal acts as a US triggering a predefined reactive turn (UR) that allows the robot to proceed. After repeated runs along the track, the robot will start learning the association between the patch (CS) and the wall (US) and eliciting CRs. At the behavioral level this will be reflected by a change in trajectory just before hitting the wall. More precisely, the distance between the patch and the wall determines the ISI between CS and US, and the angle of the curve determines the CR amplitude. As in the target eyelid-closure protocol, in this robot navigation task the binary US will be avoided only when the amplitude of the robot turn with the onset at a correct time is sufficient to avoid hitting the wall (Fig.8).

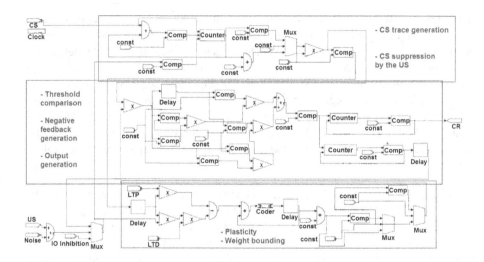

Fig. 2. Cerebellum microcircuit with simple elements. The upper group controls the CS input generating the pulse. The middle group is the implementation of the PU activity of integration and output generation.The bottom group computes the coincidence detection and the plasticity.

3   Results

Demonstration of Learning in the Model with Multiple Microcircuits. Figure 3 illustrates the behavior of five microcircuits during a simulated experiment where we set an ISI of 250 ms and a target amplitude of 3 microcircuits eliciting CR. We see that all microcircuits implement a same linearly decaying memory trace with differently delayed onsets (Fig. 3a). To avoid late occurrences of CRs to be produced by the decaying traces passing below threshold after the occurrence of the US, the traces are reset by the US signal (Fig. 3a and 3b). However, this is irrespective of whether the activation is elicited by a US or if it stems from the intrinsic IO spontaneous activity (Fig. 3, IO activity in b. and interrupted CS traces in a.). In addition, the coincidence of IO activity with an active CS trace yields LTD, which is reflected by a decrease of the associative weight (Fig. 3c). Such a decrease continues until the execution of well-timed CRs activating the negative-feedback loop that, by suppressing IO activity stops the acquisition of LTD. After trial 40 we observe the associative weight of all microcircuits fluctuating around the same stable value.

Figure 4 shows a comparison of delayed CS pulses before and after learning. In Fig.4a(1 trial), the threshold has not yet been reached. In Fig.4b (40 trials), the threshold has been reached for 3 microcircuits (target amplitude) at the ISI time (250ms).

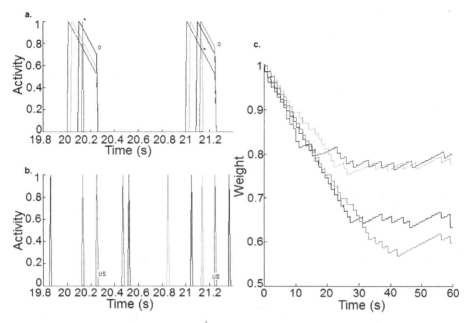

Fig. 3. Detail of the signals resulting from the simulation. Each color represents a different microcircuit. **a.** CS decaying pulses in Purkinje synapses (trials at 20s and 21s) **b.** IO activity, including USs with ISI=250ms and baseline. This activity stops the CS pulses of its respective microcircuit: by US(marked "o") or baseline("*") **c.** Weight adaptation during learning (eq. 3).

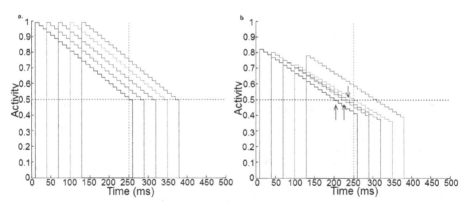

Fig. 4. a. The 5 mutually delayed pulses of CS entering into the microcircuits before learning. After learning, some of the pulses have gone below the threshold at the target time (marked)

Adaptability to Different ISIs and Target Amplitudes. Next, we tested the system to see if it can adapt to a range of ISIs and target amplitudes. We used two different ISIs (150 and 250 ms) and two target amplitudes (1 and 3). In each condition, the average CR adapts both its timing and amplitude after training. Figure 5 shows the model behavior in cases with shorter and higher ISI and target amplitude. It can be seen how the response adapts in time and amplitude. A lower accuracy in the amplitude is shown in higher ISIs due to the bigger effect of noise activity in the pulses composing the output.

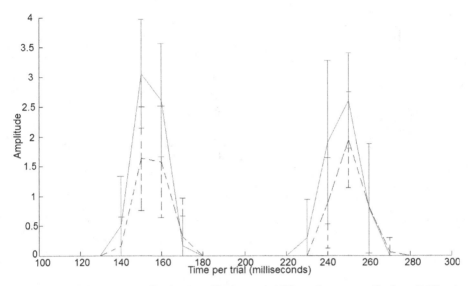

Fig. 5. Peak of the mean amplitude of the CR for several ISIs and target amplitudes:solid line is target amplitude of 3, dashed line, target amplitude of 1, ISI is 150ms (left) and 250ms (right). The graph shows the error bars with the standard deviation in the 60 trials of the simulation.

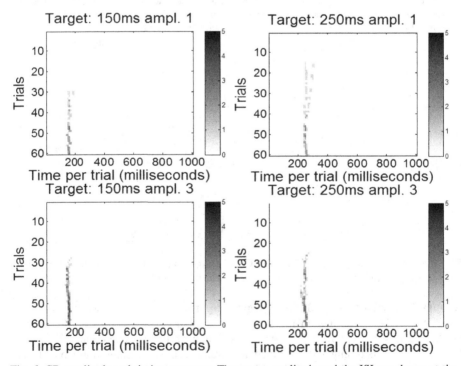

Fig. 6. CR amplitude and timing response. The target amplitude and the ISI are shown at the top part of each figure. The grayscale shows the amplitude, the vertical axis shows the trials, and the horizontal axis shows the timing for a trial.

Final Results at a Glance. The response in amplitude and timing with all the trials can be seen in Fig. 6: ISI values of 150 ms and 250 ms together with target amplitudes of 1 and 3. Due to the decaying CS input pulses, lower ISI values (Fig. 6a and 6b) need more trials in order to reach the threshold of the microcircuit. So, the first CRs appear slightly later than in the cases with higher ISI values (Fig. 6c and 6d).

Table 1 shows the success ratio (percentage of trials when the target amplitude was reached). Regarding asymptotic performance, we observe that it is higher for shorter ISIs than for longer. This is explained by the stochasticity introduced by the inferior olive activity, since the longer a CS trace has to be maintained, the higher the possibilities for it to be interrupted by a spontaneous IO activity. We also observe that the higher the target amplitude, the lower the success ratio. This result can also be interpreted as an increased difficulty of activating several microcircuits simultaneously.

Table 1. Success ratio

Amplitude/ISI	150ms	250ms
1	95%	85%
3	90%	50%

To better assess the performance in terms of the target amplitude, we compare the relation between CR amplitude and target amplitude. We observe that the linear relation holds regardless of the ISI for a range between 150 and 250 ms (Fig. 7), an outcome that closely matches the results in [8].

Application of the Digital Circuit Model to the Obstacle Avoidance Task. In the case of the experiment with the robot simulation, the results are similar qualitatively, but small improvements are needed in the circuit. For example, if the turn produced is still not accurate enough, the robot will still hit the wall, but at a later point. And this later US needs to be captured by the circuit, so the CS pulses width is slightly augmented. Furthermore, the target amplitude here needs to really adapt to the circuit, since a stronger CR would yield a turn too wide.

Figure 8 shows a schema of the track simulation. In Fig. 8a it is shown the stabilization of the number of USs to a low level after 200 trials: it can be seen how baseline conditions are translated into some oscillation in the output, but the stable tendency is avoiding USs.

4   Discussion and Conclusion

We have presented a real-time implementation of the cerebellum addressing the question of how cerebellar microcircuits can learn both the timing and the amplitude of CRs. We have shown that both can be accounted for when considering on one hand the depression of the excitatory drive onto the PU leading to the acquisition of a pause of their activity and disinhibition of the DN. On the other the number of converging microcircuits that cooperate in triggering the CR defines the amplitude of the CR. We have shown that this learning system could effectively control a robot that has to acquire the correct turn angles to navigate a curvy track.

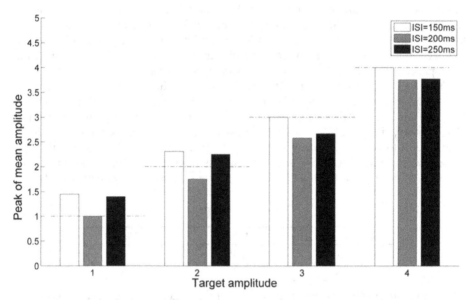

Fig. 7. Adaptation of the output amplitude to the target: The figure shows the peak of the mean amplitude for different target amplitudes and different ISIs. In this ISI range (150-250 ms), the amplitude of the response adapts to the target, regardless of the ISI.

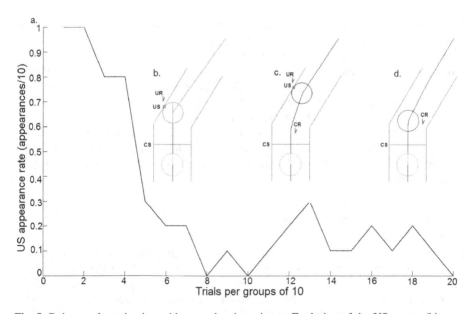

Fig. 8. Robot track navigation with several trajectories. a. Evolution of the US events (hits on the wall) per groups of 10 trials. b. The robot hits the wall before learning. After some learning (**c.**), the robot starts to turn, but not enough, so it still hits the wall. After stable learning (**d.**) the robot performs the good trajectory.

Adjusting the Inputs of the Model to Test the Functionality. Our design incorporates several microcircuits, so as to generate output of variable amplitude. Moreover, having a population of delayed memory traces allows the model to span more time. The granularity of the CR responses is determined by the number of microcircuits employed. For the current cerebellar model, this contrasts with the granularity in timing adaptation that is not directly linked to the number of microcircuits, i.e.: a single microcircuit can adapt the timing of its response within the range spanned by its CS trace. In our model we have 5 microcircuits and we select values close to the real nature: CS pulses 250ms long, and each pulse separated 30ms allows for a maximum output amplitude to be reached from within the 150ms to 250ms range.

Giving Accuracy to the Response in Time and Amplitude. Steeper slopes in the CS input pulses can give more precision to the timing adaptation of the response: there is more definition of the moment when a scaled CS signal goes below the microcircuit threshold. Furthermore, small steps in LTD and LTP select more accurately which neurons activate and, thereby, what inputs are inhibited. This also adds granularity to the output. The threshold of the microcircuit is also adapted in the model, so as to have short-term learning results in simulations, allowing a quick visualization of the functioning of the circuit. Logically, shorter clock periods simulating and sampling the system would give even higher precision.

Controlling the Pf-PU Weights. The suppression of the CSs when a US or when noise arrives, allows a different behavior between the different microcircuits: the CS pulses start to decrease at different times, but they can be suppressed at the same time (if it is by a US pulse) or at different times (if it is by noise). So, in the long term, the separation of the different weight values is feasible.

Generation and Validity of the Output Pulse. If the inhibitory pulses are not wide enough -for example, shorter than the CS pulses-, or if the inhibition delay is not the same as the CS signal comparison delay, some unwanted responses at the end of the inhibition could be filtered. In fact, depending on the inhibition delay and the duration of these inhibitory pulses, some pulses can contribute to some inhibitions or others. With our CR pulses of 20ms the adequate response in amplitude adapts to the required time. The model works well enough, because with CR pulses even shorter than the CS pulses separation, the desired output response amplitude can be reached at the expected time.

The Delay Effect. In Fig. 6c, a delay in the adaptation in time for the first responses can be perceived: when the weighted CS signal is close to the threshold, some noise in the IO input could elicit the output. This is more likely for large ISIs, when minor LTD effect is needed and there are more possibilities of IO activity coinciding with the CS pulse. This delay effect can be removed by filtering the output, so as it is elicited only if the value overpasses a given threshold. During extinction, the delay effect would be symmetrical to the one during acquisition: the pulses are lower and wider until disappearing. They become wider because low amplitudes can be generated by different microcircuits, as long as they are already close to the threshold. On the other hand, higher absolute values of LTD and LTP make faster responses, but less stable or accurate, with less precise and more disperse results.

CS suppression by the US Avoids Late CRs. As explained in Fig. 3, in order to avoid late CRs, the appearance of a US pulse suppresses the CS trace. The delays of the nucleo-olivary inhibitory pulses are important because they suppress the possible IO activity entering into the US pathway of the corresponding microcircuit. And, as explained, this activity could inhibit the CS entering into the microcircuit as well.

Conclusion. In general, the adaptation in the output amplitude can be slow and works like an oscillator until it can fit better to the goal CR amplitude. But in the long term it can be seen that the mean of the output adapts significantly in time and amplitude: the feedback signal from the environment, when applied as US activation, can control the output amplitude. Evidently, the more the microcircuits, the higher the granularity level and precision provided by the model. Nonetheless, the expected cerebellar association learning behavior is demonstrated in this study and the model executes all the designed operations. These include variation of weight in the adequate microcircuit according to the target amplitude, suppressing of CS by US in order to avoid late CRs, inhibition control by the negative feedback and control of the US according to the goal CR amplitude in the real world. Hence, according to our experimental results, by sending US and CS to the circuit and receiving the CRs, we achieve adaptable associative learning as if it were a natural cerebellum: the simulated robot turns enough in the track before hitting the walls. Our model is also biologically based on plasticity and emulates reinforcement learning. This makes it congruent with [12]. We can therefore claim that we have designed a valid neuromorphic cerebellar model that is implementable in FPGA.

Transfer of the Model into FPGA. In [2] the authors were capable of controlling a robot by using a cerebellar model based on a neural network simulated with analog circuitry. This neuronal model was a modified version of the Spike-Response Model [4]. Here, we use a MATLAB Simulink Altera library to build the circuit. The debugging of this model in Altera Quartus software gave no errors. This validates the possibility of transferring the designed cerebellar model into a FPGA digital implementation. Since VHDL code is generated during the process, the model can be tested and programmed into any other FPGA vendor. Furthermore, the circuit is scalable to any number of microcircuits, with the signals running in parallel through the components.

Acknowledgments. This work is supported by the European Union 7th Framework Programme project in Information and Communication Technologies "Extending Sensorimotor Contingencies to Cognition" (eSMCs FP7-ICT-270212). The research has also been done with the financial assistance of "La Caixa Foundation" and the Technology Department of the Univ. Pompeu Fabra. We also want to acknowledge gratitude to the Altera Corporation for the license of the software DSPBuilder, which has been used to transform the computational model into the electronic layout design.

References

1. Albus, J.S.: Theory of Cerebellar Function. Mathematical Biosciences 10(1/2), 25–61 (1971)
2. Carrillo, R.R., Ros, E., Boucheny, C., Coenen, O.J.-M.D.: A real-time spiking cerebellum model for learning robot control. Bio Systems 94(1-2), 18–27 (2008)
3. Eccles, J.C., Ito, M., Szentágothai, J.: The Cerebellum as a Neural Machine (1967)
4. Gerstner, W., Kistler, W.M.: Spiking Neuron Models. Cambridge University Press (2002)
5. Hesslow, G., Yeo, C.: The functional anatomy of skeletal conditioning. In: A Neuroscientist's Guide to Classical Conditioning. Springer, New York (2002)
6. Hofstötter, C., Gil, M., Eng, K., Indiveri, G., Mintz, M., Kramer, J., Verschure, P.F.M.J.: The cerebellum chip: an analog VLSI implementation of a cerebellar model of classical conditioning. Advances 17 (2005)
7. Hofstotter, C., Mintz, M., Verschure, P.F.M.J.: The cerebellum in action: a simulation and robotics study. European Journal of Neuroscience 16(7), 1361–1376 (2002)
8. Kreider, J.C., Mauk, M.D.: Eyelid conditioning to a target amplitude: adding how much to whether and when. The Journal of Neuroscience: the Official Journal of the Society for Neuroscience 30(42), 14145–14152 (2010)
9. Marr, D.: A Theory of Cerebellar Cortex. J. Physiol. 202, 437–470 (1969)
10. McKinstry, J.L., Edelman, G.M., Krichmar, J.L.: A cerebellar model for predictive motor control tested in a brain-based device. Proceedings of the National Academy of Sciences of the United States of America 103(9), 3387–3392 (2006)
11. Prueckl, R., Taub, A.H., Herreros, I., Hogri, R., Magal, A., Bamford, S.A., Giovannucci, A., Almog, R.O., Shacham-Diamand, Y., Verschure, P.F.M.J., Mintz, M., Scharinger, J., Silmon, A., Guger, C.: Behavioral rehabilitation of the eye closure reflex in senescent rats using a real-time biosignal acquisition system. In: 2011 Annual International Conference of the IEEE on Engineering in Medicine and Biology Society, EMBC, August 30-September 3, pp. 4211–4214 (2011)
12. Wörgötter, F., Porr, B.: Temporal sequence learning, prediction and control - A review of different models and their relation to biological mechanisms. Neural Comp. 17, 245–319 (2005)

Simulating an Elastic Bipedal Robot Based on Musculoskeletal Modeling

Roberto Bortoletto[1], Massimo Sartori[2], Fuben He[3], and Enrico Pagello[1]

[1] Intelligent Autonomous Systems Laboratory
Department of Information Engineering (DEI)
University of Padua, Italy
{bortolet,epv}@dei.unipd.it
[2] Department of Neurorehabilitation Engineering
Bernstein Focus Neurotechnology Goettingen
Georg-August University
Von-Siebold-Str. 4, 37075 Goettingen, Germany
massimo.sartori@bccn.uni-goettingen.de
[3] School of Mechanical Engineering
Dalian University of Technology, Dalian, China
hefuben@mail.dlut.edu.cn

Abstract. Many of the processes involved into the synthesis of human motion have much in common with problems found in robotics research. This paper describes the modeling and the simulation of a novel bipedal robot based on Series Elastic Actuators (SEAs) [1]. The robot model takes inspiration from the human musculoskeletal organization. The geometrical organization of the robot artificial muscles is based on the organization of human muscles. In this paper we study how the robot active and passive elastic actuation structures develop force during selected motor tasks. We then compare the robot dynamics to that of the human during the same motor tasks. The motivation behind this study is to translate the mechanisms underlying the human musculoskeletal dynamics to the robot design stage for the purpose of developing machines with better motor abilities and energy saving performances.

Keywords: Flexible Robotic, Musculoskeletal Model, OpenSim Simulation.

1 Introduction

There are many reasons for exploring robotic solutions for bipedal legged locomotion [2]. These include the development of robot devices that can move on uneven and rough terrain, the understanding of human and animal locomotion mechanics [3], or the need to build better artificial legs for amputees. In the past years studies about elastic bipedal robots (i.e. robot actuated by elastic actuators) produced many bipedal robot models based on compliant legs that utilize mono-articular and bi-articular arrangement of tension springs. Experiments using simulation as well as real robotic platform showed that these models

T.J. Prescott et al. (Eds.): Living Machines 2012, LNAI 7375, pp. 26–37, 2012.

can provide eminent features that could not be explained by the previously developed simpler models such as the ballistic walking, generally known as passive dynamic walking. In [4] it was shown that the compliant elements in the proposed robot leg structure made it possible to generate both walking and running gaits. Furthermore, the proposed model was able to achieve human-like leg movements compared to those of ballistic walking. This demonstrated the role of bi-articular muscle arrangements in supporting energy transfers between joints for the ultimate self-stabilization of walking and running. In [5] a motion simulation model in the two-dimensional sagittal plane was based on the extended series actuation principle demonstrating a reduction in the energy requirements. In [6] a biologically inspired robot was presented showing to be capable of both energy-efficient and human-like walking and jogging gaits.

This paper presents a three-dimensional model of a novel bipedal robot based on the organization of the human musculoskeletal system. The structure of the robot and its actuation system is here described. Both robot and human were simulated using the open-source musculoskeletal simulation software Open-Sim[1] [7]. The dynamics of the robot torque actuators and artificial muscles was then compared to the muscle and joint dynamics of the human body. The aim of this work is an initial study intended as a proof of concepts towards the final goal of providing a novel methodology to replicate the mechanisms underlying the human musculoskeletal dynamics into artificial anthropomorphic systems. This was motivated by the recent achievements obtained in the state of the art bipedal robots in which biologically inspired designed led to better walking robots. In Section 2, a brief description of musculoskeletal human model used and the structure of our robot is provided with a detailed description of methods and tools adopted during the modeling and simulation phases. In Section 3, the evaluation approach to validate our work is described. Section 4 provides the results obtained with the comparison between human and humanoid. Section 5 concludes the paper with discussion and plan for future work.

2   Methods

We started our work by modeling the human movement at the musculoskeletal level. We used a computer model of the musculoskeletal system that represents the kinematics of joints [8], the geometry of bones, the three-dimensional organization and force-generating properties of lower limb muscles [9–11]. Each leg in the model had seven Degrees Of Freedoms (DOF) including: hip internal-external rotation (HRO), hip adduction-abduction (HAA), hip flexion-extension (HFE), knee flexion-extension (KFE), ankle subtalar flexion-extension (ASA), and ankle plantar-dorsi flexion (AFE). Line segments approximated the muscle-tendon path from the origin to insertion of the 86 muscles of the lower limb included into the model. We then modeled and simulated our proposed elastic bipedal robot [12, 13] which is characterized by a mass of about 2Kg and an height of about 40cm, in order to study a small humanoid robot, according to the dimensions suggested

---

[1] Freely available from: `https://simtk.org/home/opensim`

for participating into the kidsize class of the RoboCup Humanoid League. It is a biologically inspired four-segment robot with active and passive elastic actuation structures. These ones can be subdivided into: uni-directional joint and bi-directional joint. In the first case, the joint is describes as a single rotational servo motor of one DOF only which produces the rotational motion in one direction only. A bi-directional joint is modeled as a servo motor that can rotate in two opposite directions. Both types of articulation are connected in series to active springs that are stretched or compressed as the servo motor rotates and produce passive resistive force to the rotational movement. Springs in the model can cross one (i.e. mono-articular) or two joints (i.e. bi-articular).

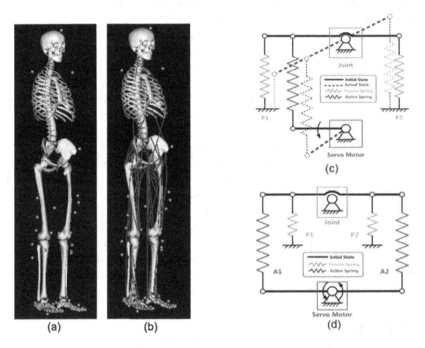

Fig. 1. Human musculoskeletal model used. (a) Torque-driven model of a human subject; (b) Muscle-driven model of a human subject; (c) The schematic of elastic mono-articular joint; (d) The schematic of elastic bi-articular joint.

There is a strong relationship between the main human muscle groups and the robot elastic cables and springs as depicted in Fig. 2. The iliopsoas (ILIO) and gluteus (GL) drive the hip. The rectus femoris (RF) and the biceps femoris (BF) are to keep the equilibrium position of the thigh. The vastus (VAS) is active at the knee only while the gastrocnemius (GAS) also operates at the ankle. The tibialis anterior (TA) and the soleus (SOL) are antagonistic muscles acting at the ankle only. Among these muscles, VAS and SOL are mono-articular; the others act as bi-articular. This mapping between human and humanoid is principally based on the movement in which each muscle is involved.

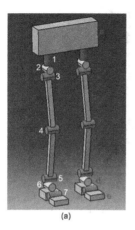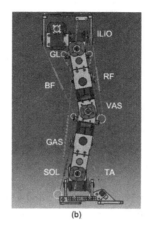

Fig. 2. Elastic Bipedal Robot Modeling. (a) Kinematic model of robot; (b) The main muscle-tendon groups compartments considered in our modeling; green lines stand for passive springs, while the red ones are active springs. (c) The corresponding model developed with OpenSim.

Every joint into the model connects a parent body to a child body. A joint defines the kinematic relationship between two frames each affixed to a rigid-body (the parent and the child) parameterized by joint coordinates. Every body has a moving reference frame in which its Center-Of-Mass (COM) and inertia are defined. Servo motors are defined by parameters including: minimum and maximum allowed control values, an associated coordinate frame, and an optimal force value. With regards to the passive elastic structures, we chose a simple point-to-point spring model with a resting length and stiffness. Each spring starts creating resistive force when the length exceeds the resting lengths: 3.2cm (ILIO), 3.5cm (GL), 13.8cm (RF), 14.6cm (BF), 6.4cm (VAS), 20cm (GAS), 8.6cm (SOL), and 7cm (TA). The stiffness coefficients were experimentally set between the range of 15-27 Newton x meters (Nm), based on the mechanical development of the robot [12, 13].

2.1 Scaling

In this study we scaled the human model to the anthropometry of a subject based on marker locations. Optimal fiber length and tendon slack length were scaled with the total muscle-tendon length so that they maintain the same ratio. This process started with the unscaled OpenSim musculoskeletal model and placed a set of virtual markers on the model to match the locations of the experimental markers. Marker trajectories were collected during static trials from one healthy, male subject volunteered for this investigation (age: 28years, height: 183cm, mass: 67kg). Data were collected at the Gait Laboratory of the School of Sport Science Exercise and Health of the University of Western Australia. The subject was taken through the testing protocols and informed consent was obtained prior

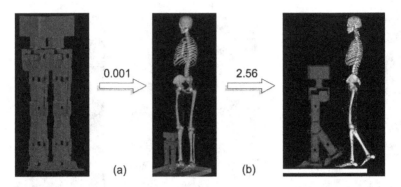

Fig. 3. Scaling procedure. (a)From a generic model of robot, to the real-dimension biped; (b)From the real-dimension biped to a thereotical model of our robot scaled to human dimension.

to data collection. To record the human body kinematics was used a 12 camera motion capture system (Vicon, Oxford, UK) with sampling frequency at 250Hz. Body kinematics was low-pass filtered with cut-off frequencies ranging from 2 to 8Hz depending on the trial. In this way, starting from the original unscaled model (Weight 75Kg, Height 175cm) we obtained the new one referred to the real subject. The scaling process is an OpenSim built-in tool.

Similarly, as depicted in Fig.3, we started with a generic model of the robot. The scaling was then performed with the aim to obtain a model of the robot in real size (Weight: 2.272 Kg; Height: 40 cm), preserving the distribution of mass across bodies. This tool was also used in the opposite direction: from the real robot, we got a version of the same at "human dimension". The scale factor was calculated by dividing the distance of the human pelvis from ground with the distance of the waist of the robot from ground. The factor obtained is equal to: 2.56.

2.2 Inverse Kinematics

Marker trajectories recorded during dynamic movements were used by an Inverse Kinematics (IK) algorithm to calculate 3D joint angles. Similarly to scaling process, also this step was done using the OpenSim Inverse Kinematics tool in which an IK algorithm solves for the joint angles that minimized the difference between the experimentally measured marker positions and the virtual markers on the model. A more detailed description of this tool is provided in [7].

2.3 Computed Muscle Control

In this step we produced a dynamic simulation of muscle-tendon dynamics during walking and was applied both to the robot and human models. We prescribed muscle-activation patterns and joint kinematics, and calculated the muscle forces and fiber lengths that satisfy these constraints. This was performed using the Computed Muscle Control (CMC) [7] tool in OpenSim.

With regard to human model, we used the Thelen Muscle Model that is a slight modification of the Hill-type muscle model. A detailed description of the force-length, force-velocity and tendon force-strain relationships implementation can be found at [14]. On the other hand, in the robot we used elastic cables and extension springs instead of muscle contraction to generate the elastic forces. Based on the generalized coordinate system and according to Newton-Euler Equations, we can formulate the dynamics of the system as:

$$M(q)\ddot{q} + C(q,\dot{q})\dot{q} + F(\dot{q}) + G(q) = \tau \tag{1}$$

where q represents the angular positions of joints, $\dot{q}$ is the vector of angular velocities and $\ddot{q}$ is the vector of angular accelerations, $M(q)$ is the mass matrix, $C(q,\dot{q})$ stands for the Coriolis and centripetal forces, $F(\dot{q})$ is the friction force, $G(q)$ is the gravity force, and τ is the torque of motor.

According to the spring characteristics and elasticity theory, the elasticity effects, $E(q,\dot{q})$, of springs can be represented as follow:

$$E(q,\dot{q}) = D(\dot{q}) + K(q) \tag{2}$$

where $D(\dot{q})$ is the coefficient matrix of damping, $K(q)$ is the coefficient matrix of stiffness. The status of the elastic actuator can be described as follow, by combining (1) and (2):

$$M(q)\ddot{q} + C(q,\dot{q})\dot{q} + E(q,\dot{q})q + F(\dot{q}) + G(q) = \hat{o} \tag{3}$$

This dynamic equation is practical and integrated considering the influences of frictions and inertia items in robot locomotion. By varying values of stiffness and damping the dynamic situations can be modified and consequently the performances of the elastic actuators which have the effects of storing and releasing energy. Based on the principles of feed-forward controller in the original and extended SEAs we build up the expected control structure in which inputting the reference trajectory to the robot system, the joint positions, velocities and accelerations can be obtained by differential equations. To compensate the desired motions, a PID controller is introduced to deal with the values integrated with measurement data of sensors. Furthermore, a feed-forward controller is also used to reduce the errors produced by the characteristics of springs. Through the process of control, the robot can perform adjustable and compliant locomotion. The study of control will be continued in the future.

3   Evaluation

Experiments were performed using both the real-size and human-size robot. This was done for the purpose of better comparing the magnitude of torques, powers and passive elastic forces that were simulated. The robot movement was first simulated using servo motors only. Then, we added the passive elastic springs. We developed some simulations in order to estimate our proposed robotic system capacities. We used the experimentally recorded human kinematics during one walking trial to directly drive our robot joints[2]. Three different type of analysis were conducted. In the first one we analyzed the kinematics of bodies that make up the models in order to obtain positions (center of mass position and orientation), velocities (linear and angular) and accelerations (linear and angular) of the COM of each segment. In the second analysis we compared the actuation force and power in joint and muscles between robot and human. The actuation power in Watt (W) was intended as the rate at which an actuator produces work. Positive work means that the actuator is delivering energy to the model; negative work means that the actuator is absorbing energy from the model. In the third analysis, we performed a validation on the passive forces developed by the springs with respect to the passive forces generated by the human muscles during the same movements. For all springs we obtained Pearson Correlation Coefficient (PCC) values which vary into the range 0.64 - 0.85. In Fig.6 we report the results referring to three of the major muscles: Tibialis Anterior, Gastrocnemius, Vastus Medialis. Forces and torques were scaled by the weight of the subject and the robot respectively in order to account for the different arrangements of the reference systems of the lower limb parts, together with the different position and orientation of the COM, and the resulting different distribution of mass and anthropometric characteristics of the human subject and robot.

Both, during the robot and human simulations a *weld constraint* was introduced between the waist and ground in order to keep the robot and human, respectively, raised from the ground and exclude the effects of Ground Reaction Forces (GRFs) on the models for this early stage of work. This contributed to produce the joint dynamics depicted in Figs 4 and 5. In Fig. 4a the actuation torque developed by the robot at hip joint is higher in magnitude with respect to that of the human during the first 12% of the gait cycle; on the other hand, in Fig. 4b-4c the robot develops a less torque than the human at knee and ankle joints in the same range. In the subsequent part of the cycle, from 12% to about 50% the actuation torque are very similar in all three joint: hip, knee and ankle. The main differences are observed during the swing phase of the cycle, from 62% to the next heel strike of the right foot. This is in agreement with what illustrated by the PCC that shows a poor correlation between the actuation torques developed by the robot with those produced by the human. Discrepancies in the joint dynamics are mainly due to the different distribution of mass that characterizes the robot with respect to human. Fig.5 shows results on joint actuation power confirming the major differences during the swing phase of the gait cycle. Fig. 6 shows results on muscles and springs force dynamics during a walking gait through one cycle. The operating principle that governs a human muscle, characterized by active and passive forces, is not replicable in the springs which in the robot provide only passive forces. On the other hand, the active component in the actuation of the robot is represented by the engines. This justifies the comparison between passive forces developed by the human muscles and forces developed by the springs. We took as reference the values obtained from the human, depicted as a dotted line in the graphs, and compare them with those

---

[2] Video Available at: http://youtu.be/-w4sEkckWYO

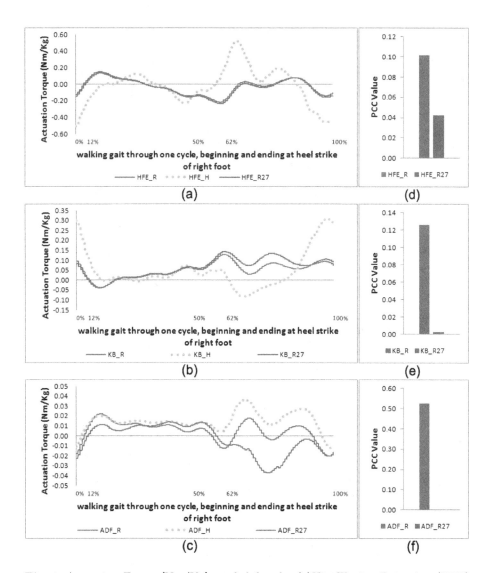

Fig. 4. Actuation Forces [Nm/Kg] needed for the (a)Hip Flexion-Extension (HFE), (b)Knee bending (KB), and (c)Ankle Dorsi-Flexion (ADF) movements during walking gait. Pearson Correlation Coefficient (PCC) values for each case are reported on the right column of the image, respectively called (d)-(e)-(f). In each graph two configuration of robot (without springs: HFE_R, KB_R, ADF_R; with springs characterized by a stiffness coefficient of 27Nm: HFE_R27, KB_R27, ADF_R27) are plotted with the reference values of human model (HFE_H, KB_H, ADF_H).

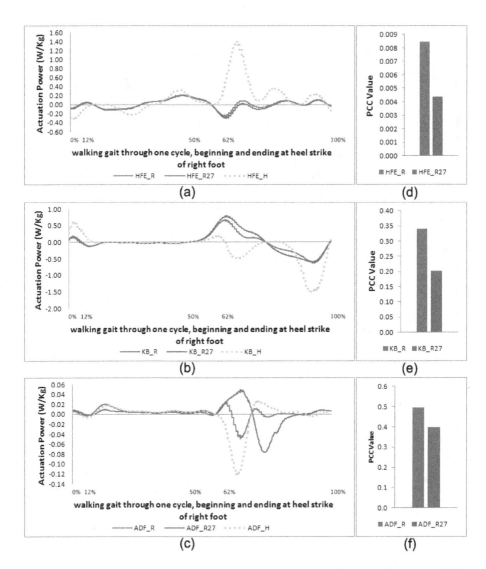

Fig. 5. Actuation Powers [W/Kg] generated for the (a)Hip Flexion-Extension (HFE), (b)Knee bending (KB), and (c)Ankle Dorsi-Flexion (ADF) movements during walking gait. Pearson Correlation Coefficient (PCC) values for each case are reported on the right column of the image, respectively called (d)-(e)-(f). In each graph two configuration of robot (without springs: HFE_R, KB_R, ADF_R; with springs characterized by a stiffness coefficient of 27Nm: HFE_R27, KB_R27, ADF_R27) are plotted with the reference values of human model (HFE_H, KB_H, ADF_H).

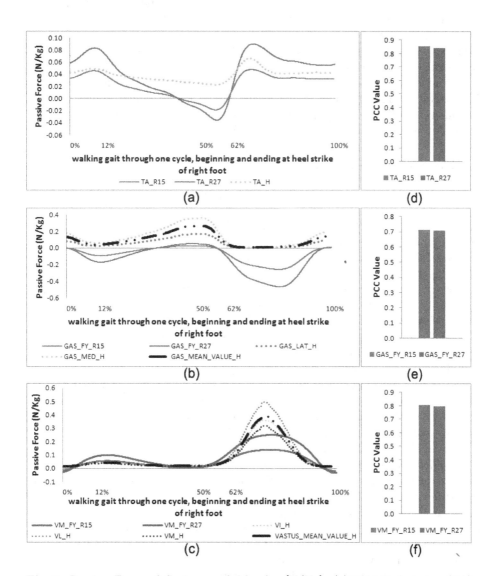

Fig. 6. Passive Force of Springs and Muscles [N/Kg]. (a)Tiabiali Anterior (TA); (b)Gastrocnemius (GAS); (c)Vastus Medialis (VM). Pearson Correlation Coefficient (PCC) values for each case are reported on the right column of the image, respectively called (d)-(e)-(f). For each muscle/spring two configuration of robot (with springs characterized by a stiffness coefficient of 15Nm: TA_R15, GAS_FY_R15, VM_FY_R15; with springs characterized by a stiffness coefficient of 27Nm: TA_R27, GAS_FY_R27, VM_FY_R27) are plotted with the reference values of human model (TA_H, GAS_LAT_H, GAS_MED_H, VI_H, VL_H, VM_H).

obtained with robot, simulated by using two different configuration also in this case: in the first one we moved the robot in which we introduced springs with a stiffness coefficient of 15Nm, meanwhile in the second one we took into account springs with a stiffness coefficient of 27Nm. Fig. 6a shows the force produced by the tibialis anterior muscle, Fig. 6b shows the force produced by the gastrocnemius, and Fig.6c is about the vastus muscle, both in the robot and in the human. The activation timing and the shape of the curves are extremely similar despite the different dynamics observed at the joint level and the different operating principles between human muscle and spring.

4   Conclusion and Future Work

This work produced an initial study on the design and simulation of a bio-inspired bipedal robot actuated by servo-motors and passive elastic springs. The introduction of springs was motivated by the potential of making the motion of the robot more compliant and energy efficient. Furthermore, the bio-inspired arrangement of the artificial muscles allowed reproducing in the robot the mechanisms underlying the passive elasticity of human muscles (Fig. 6). We want now to further improve our study by introducing a contact model between foot and ground in order to fully appreciate the compliant effects due to springs and simulate the effect of the robot body weight on the dynamics of both joints and artificial muscles. With the introduction of the contact we expect significant increases in this direction, precisely due to the action of GRF. We want also to improve our model by changing the stiffness coefficient of each spring based on the ratio between Tendon Slack Length (TSL) and Optimal Fiber Force (OFF) which characterize the compliant behavior of human muscles. This will most likely allow us to obtain a much closer relationship between human and humanoid but also to refine the control strategy and reduce the differences illustrated in Fig.4-5.

The development of our proposed methodology will also provide insights into the design of robotics prosthetics and powered orthoses [15, 16]. This work is only the starting point of a wide range of other possible future works: from the control structure completion and whole-body control application, to imitation learning and reinforcement learning for human locomotion, from motion test on flat ground to motion test on rough ground, to the introduction of a trunk in order to take into account that man is a vertebrate and its locomotion is clocked and driven by his trunk, and obviously the transition from simulation to practice with a real elastic bipedal robot biologically-inspired that can move like a human being.

References

1. Pratt, G., Williamson, M.: Series Elastic Actuators. In: IEEE/RSJ Int. Conf. on Intelligent Robots and Systems, vol. 1, pp. 399–406 (1995)
2. Omer, A.M.M., Ghorbani, R., Lim, H., Takanishi, A.: Semi-Passive Dynamic Walking for Humanoid Robot Using Controllable Spring Stiffness on the Ankle Joint. In: Proceedings of the 4th International Conference on Autonomous Robots and Agents, Wellington, New Zeland (February 2009)

3. Geyer, H., Herr, H.: A Muscle-Reflex Model that Encodes Principles of Legged Mechanics Produces Human Walking Dynamics and Muscle Activities. IEEE Transaction on Neural Systems and Rehabilitation Engineering 18(3), 263–273 (2010)
4. Iida, F., Rummel, J., Seyfarth, A.: Bipedal walking and running with spring-like biarticular muscles. Journal of Biomechanics 41 (2008)
5. Radkhah, K., Lens, T., Seyfarth, A., von Stryk, O.: On the influence of elastic actuation and monoarticular structures in biologically inspired bipedal robots. In: Proceedings of the 2010 IEEE International Conference on Biomedical Robotics and Biomechatronics (2010)
6. Radkhah, K., Maus, M., Scholz, D., Seyfarth, A., von Stryk, O.: Toward Human-Like Bipedal Locomotion with Three-Segmented Elastic Legs. In: 41st Int. Symp. on Robotics / 6th German Conf. on Robotics, pp. 696–703 (June 2010)
7. Delp, S.L., Anderson, F.C., Arnold, A.S., Loan, P., Habib, A., John, C.T., Guendelman, E., Thelen, D.G.: OpenSim: Open-Source Software to Create and Analyze Dynamic Simulations of Movement. IEEE Transactions on Biomedical Engineering 54(11) (November 2007)
8. Delp, S.L., Loan, J.P., Hoy, M.G., Zajac, F.E., Topp, E.L., Rosen, J.M.: An interactive graphics-based model of the lower extremity to study orthopaedic surgical procedures. IEEE Transactions on Biomedical Engineering (1990)
9. Anderson, F.C., Pandy, M.G.: A dynamic optimization solution for vertical jumping in three dimensions. Computer Methods in Biomechanics and Biomedical Engineering 2, 201–231 (1999)
10. Anderson, F.C., Pandy, M.G.: Dynamic optimization of human walking. Journal of Biomechanical Engineering 123, 381–390 (2001)
11. Yamaguchi, G.T., Zajac, F.E.: A planar model of the knee joint to characterize the knee extensor mechanism. Journal of Biomechanics 22(1), 1–10 (1989)
12. Bortoletto, R.: Simulating a Flexible Robotic System based on Musculoskeletal Model. M.Sc Thesis - Department of Information Engineering, University of Padua (December 2011)
13. He, F., Liang, Y., Zhang, H., Pagello, E.: Modeling and Dynamics of Extended Elastic Actuator Applied to Robot. Submitted to IAS 2012 (February 2012)
14. John, C.T.: Complete Description of the Thelen 2003 Muscle Model, http://simtk-confluence.stanford.edu:8080
15. Sartori, M., Reggiani, M., Lloyd, D.G., Pagello, E.: A neuromusculoskeletal model of the human lower extremity: Towards EMG-driven actuation of multiple joints in powered orthoses. In: Proceedings of IEEE International Conference on Rehabilitation Robotics (ICORR 2011), Swizzerland (June 2011)
16. Sartori, M., Lloyd, D.G., Reggiani, M., Pagello, E.: Fast Runtime Operation of Anatomical and Stiff Tendon Neuromuscular Models in EMG-driven Modeling. In: Proceedings of IEEE International Conference on Robotics and Automation (ICRA 2010), USA (May 2010)

A Soft-Body Controller with Ubiquitous Sensor Feedback[*]

Alexander S. Boxerbaum[1], Kathryn A. Daltorio[1], Hillel J. Chiel[2], and Roger D. Quinn[1]

[1] Department of Mechanical and Aerospace Engineering, Case Western Reserve University, Cleveland, OH 44106-7222 USA
{asb22,kam37,rdq}@case.edu
[2] Departments of Biology, Neurosciences, and Biomedical Engineering, Case Western Reserve University, Cleveland, OH 44106-7080 USA
hjc@case.edu

Abstract. In this paper, we investigate control architectures that combine implicit models of behavior with ubiquitous sensory input, for soft hyper-redundant robots. Using a Wilson-Cowan neuronal model in a continuum arrangement that mirrors the arrangement of muscles in an earthworm, we can create a wide range of steady waves with descending signals. Here, we demonstrate how sensory feedback from individual segment strains can be used to modulate the behavior in desirable ways.

1 Introduction

Soft-bodied invertebrates, such as leeches, worms, and slugs move by peristaltic crawling, anchor-and-extend, swimming, and dexterous manipulation. This large and flexible behavioral repertoire could be extremely useful if transferred to a robotic platform. The control is challenging because many actuators may have the same effect on the relative position of the end-effort (hyper redundancy) [1] and because the actuator and load forces induce large nonlinear deformations [2].

Robotic designs often reduce or group the degrees of freedom [3]–[5], or approximate soft bodies with rigid joints [6][7]. These simplifications for the sake of control come at a cost to multifunctional flexibility and performance. Hyper-redundant robotic snakes [8][9] and salamanders [10] have been successful at navigating rough terrain, but cannot adapt to tunnel diameter like a soft-bodied animal.

Animals do not have to avoid large deformations or singular Jacobian matrices. Soft-bodied or not, their bodies are controlled using distributed arrays of neurons to coordinate the many degrees of mechanical freedom[11][12].

1.1 The Mechanics of Peristalsis

Earthworms have continuous sheets of both longitudinal and circumferential muscle fibers that create waves of peristaltic motion [13]. The circumferential muscles cause lengthwise elongation, and move the segments forward, whereas the longitudinal

[*] This work was supported by the NSF grant IIS-1065489 and by Roger D. Quinn and Hillel J. Chiel.

muscles shorten the segments. During forward locomotion, these two muscle groups are antagonistically coupled by segments of hydrostatic fluid, creating a hydrostatic skeleton, and are typically activated in alternation at a given location along the body. The waves of expansion and contraction flow in the opposite direction of the motion. When a section leaves the ground, a new ground contact point forms directly behind it. The contracting section will accelerate outward, but that motion is constrained on the rear side by the new ground contact point, so the segment must move forward. Our recent analysis has shown that continuously deformable mechanisms prevent slip by ensuring good transition timing among segments [14][15].

In our recent robot design, we use a braided mesh similar to that used in pneumatically-powered artificial muscles [16] to create this coupling between longitudinal and radial motion (Figure 1). Our large-scale prototype uses a cam mechanism at the end of the robot to power all twelve actuators with a single offset waveform. The resultant deformation wave has many of the properties that our analysis shows are desirable: it is periodic, it has a constant shape and speed, any given segment has a short ground phase, and the waveform moves rapidly, traveling the length of the robot in 1.3 seconds. Our prototype achieved a speed of 6 body lengths per minute [14], a very fast speed for peristaltic locomotion. In comparison, earthworms travel at speeds of 1.2 to 3.6 body lengths per minute [13], and our own previous segmented robot traveled at 0.8 body lengths per minute [17].

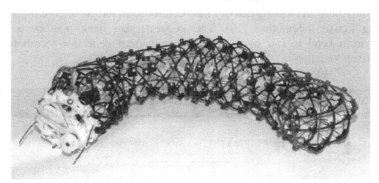

Fig. 1. A robot that creates peristaltic motion with a continuously deformable exterior surface. The outer wall consists of a single continuous braided mesh that interpolates the positions of circumferential actuators spaced at intervals along the long axis. In lieu of longitudinal actuators, return springs are used along the length of the body to cause lengthwise contraction in the absence of activation. When the actuators are activated in series, a smooth continuous waveform travels down the length of the body. The result is a fluid motion more akin to peristaltic motion than that generated by previous robots.

With the twelve circumferential actuators effectively coupled into a single degree-of-freedom for initial testing, the current robot cannot actively adapt to environmental changes. For this, we need more actuators, which greatly complicates the control problem.

1.2   Biologically Inspired Control

Many groups have used neuroethological studies and neurally-inspired control to tackle the problem of generating robust motor coordination for hyper-redundant systems. By coupling central pattern generator (CPG) controllers to different degrees of freedom, it is possible to create coordinated rhythmic behavior in hyper-redundant robots (e.g., swimming in robotic lamprey [18] and knifefish [19], walking and swimming in a salamander robot [20], and undulatory locomotion in a snake robot [21]). The CPGs are based on limit cycle oscillators that drive each segment with a stable rhythmic pattern. Adjacent segments are then coupled via a phase offset to achieve waves along the length of the body. The approach of Wadden, et al. [22] is similar to our own in that it forgoes discrete segments and coupled CPGs to develop a new model of undulatory swimming in the lamprey. Recent work by Boyle et al have also shown how swimming behavior in *C. Elegans* can be reproduced with a full neuro-mechanical simulation [23]. Several of the assumptions in this model parallel our own, and we comment on this in the Discussion.

The neuroethology of earthworms has been studied in some detail. In an early experiment it was shown that if a worm was cut in two sufficiently far from the head, and then a string was tied between the two parts, that the wave of motion would propagate uninterrupted from head to tail through the string [24]. Similarly, a worm suspended from its head by a string will naturally generate waves of motion. However, if the worm is dipped into a solution such that it is neutrally buoyant, this motion will stop [25]. These findings strongly suggest that strain receptors play a direct role in transmitting the neuro-muscular control signal down the length of the worm. Furthermore, decreasing activity has been found in the nerve cord from head to tail during fictive locomotion [26]. This suggests that there may be a gradual transition from head to tail between top-down pattern generator locomotion and sensory-driven locomotion.

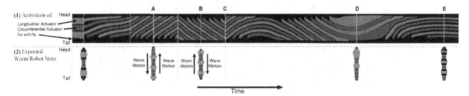

Fig. 2. The first row shows the activation level along the length of the body of the two opposing actuator groups as a function of time, where the vertical axis is the position of the muscle group, and the horizontal axis is time. At t=0, there are 1.5 waveforms along the body (row two). The waveform initially travels towards the head, causing backwards motion, then switches between A and B, causing forward motion. **A** and **B** and correspond to the greatest positive and negative temporal frequency, and have the greatest difference between anterior and posterior inhibition weights. **D** and **E** correspond to the least and greatest spatial frequency and are the local maximum and minimum of the *sum* of the inhibitory weights. First 50 time-steps not shown.

We recently built a waveform generator [27] out of a parallel structure of excitatory and inhibitory state representations (*i.e.* Wilson-Cowan neuron population model [28][29]). This controller used only two descending signals that modulated the strength of the anterior and posterior inhibitory connections. We demonstrate that the network can quickly, robustly and accurately adjust both the shape of the deformation wave and the speed of the wave, two key parameters to controlling peristaltic locomotion (Figure 2). Further, this arrangement can easily tune its period and even come to a complete stop for indefinite amounts of time.

1.3 Aims

This paper aims to demonstrate that this method of modeling neuronal behavior can be integrated with ubiquitous sensory feedback, and will be well suited to controlling complex soft-bodied structures. Towards this end, we outline several principles of peristaltic locomotion, and then describe the architecture of our new modified Wilson-Cowan network. A basic model of worm body posture is developed that can be controlled with the network, and proprioceptive feedback from simulated strain receptors is introduced into the control loop, and results are presented.

2 Methods

Designing neural control networks is very challenging. We have found three things are essential to the process: a familiarity with the desired behavior, some knowledge of the neuroethology of animals that exhibit the behavior, and lastly a familiarity with the dynamic properties of neural control circuits. Our experience studying peristaltic motion has led to principles for effective motion outlined in the introduction. Combined with the limited knowledge of earthworm neural circuitry, we will proceed to develop a neural model for the adaptive control of peristaltic motion.

2.1 Using a Wilson-Cowan Model to Create Adaptive Waves

The dynamics of our system are based on the spatially-extended Wilson-Cowan model of the primary visual cortex [29]. The Wilson-Cowan model is a way of modeling the discrete firing responses of a whole population of neurons as continuous state variables (typically in pairs, one to represent inhibition and one to represent excitation) [28]. In the classical model, these excitatory and inhibitory populations are arranged in two stacked 2-D arrays, and populations near each other innervate one another more strongly. This has been shown to produce statically stable patterns, as opposed to oscillatory behaviors. In our example, the two excitatory arrays are E_{circ} and E_{long}, which drive the circular and longitudinal actuators, and have a spatial configuration that mirrors the muscle architecture of a worm (Figure 3). We modify the classic cortical arrangement by replacing each single inhibitory population array with two separate arrays of inhibitory populations, I_a and I_p, for each excitatory array, whose connections are shifted *anteriorly* and *posteriorly*, respectively. The simulation presented here consists of a column of 150 actuator pairs along the body length, represented by 6 x 150 state variables. We will outline the structure of this network, while focusing on the recent changes that allow for integration of sensory feedback. Where not stated otherwise, the network constants and structure remain identical to those presented in [27].

The general form of both the excitatory and inhibitory state variable equations can be described as:

$$\tau \dot{U} = -U + \sigma\left(\sum_{i=1}^{n} w_i\left(U_i \otimes \hat{U}_i\right)\right),$$

where $\dot{U}$ is the time rate of change of the vector U, which represents the average level of activity in a given excitatory or inhibitory array, which stretches along the body of the worm. Within an array, all constants are the same, and connections are shift-symmetrical. Under no other external input, an active region of U will decay at a rate of $-U/\tau$, where τ is a time constant of the population. The function σ is a generalized logistic, which has a y-intercept of $\sigma(0) = 0$ and is bounded by the range $-0.1 < \sigma(x) < 1$; strong negative inputs yield weak bounded negative outputs, which contributes to the stability of the system. The $\otimes$ operator convolves a population vector U_i with an associated kernel, $\hat{U}_b$, a Gaussian distribution that defines the spatial connectivity of each population. In general, the excitatory kernels ($\hat{E}$) have a much tighter spread than the inhibitory kernels ($\hat{I}$), corresponding to the more localized excitation in the network (Figure 3). The constant n is the total number of arrays in the simulation. All kernels are normalized, and the total weight of the connection is determined by the scalar constant, w_b, which provides the easiest way to tune the behavior of the system. The sign of w indicates whether the connection is excitatory or inhibitory.

Fig. 3. An excitatory (blue) and inhibitory (red) Gaussian kernel combine when excited to produce a "Mexican hat" influence (magenta) (a). Similar influences can be created with two offset inhibitory kernels (b). If the inhibitory populations have asymmetrical influence, a variety of patterns can emerge (c). In a and b, motion is not sustained without external influence. In c, waves of motion naturally occur.

Based on what is known about the neuroethology of earthworms, we theorize that as the wave travels smoothly down the body, the control dynamics shift from being primarily driven by the neural dynamics to being driven by sensory feedback. There are many ways that sensory input can be integrated into the system. The effects can be excitatory, inhibitory, or even logic-style operators, and they can innervate any layer of the network. To select among the many alternatives, we compared simulated strain receptor data to the specific contributions of various network components (Figure 4) and found a strong correlation between strain receptor activity of a given kind of muscle and its excitatory signal. This is not surprising since when the body is not under load, muscle excitation causes a deformation to the soft body.

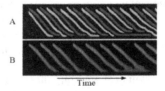

Fig. 4. (A) is the excitatory array activation of E_{hoop} (blue) and E_{circ} (Green) over time. (B) is the signal from the hoop muscle strain receptor (see Methods, section B). When not under external loading, the hoop muscle strain receptor correlates strongly to the E_{hoop} array.

We had success by blending the effects of sensory input and excitatory dynamics in different ratios as a function of position along the body. Compared to our previous model, all of the innervations of an excitatory array have been adjusted by adding sensory feedback from the strain sensor corresponding to that array. The amount of feedback depends on the position of the neuronal population along the body. This blending of inputs is achieved by applying separate weighted arrays that are the length of the simulated body and in the range of 0 to 1 ([G] and [1-G]) to both the sensory feedback and the excitatory connections (Figure 5 and Figure 6). [G] scales the local strength of the excitatory connection via the ° operator, which is the Hadamard product, or entrywise product. [G] and [1-G] are defined such that they sum to unity at any point along the length of the worm. [G] can have many different characteristic profiles, and several are explored in the Results. The new state equation for a single excitatory layer now takes the form:

$$\tau \dot{E} = -E + \sigma\left(w_1 [G] \circ \left(E \otimes \hat{E}\right) + w_2 [1-G] \circ \left(S \otimes \hat{S}\right) + \sum_{i=3}^{n} w_i \left(I_i \otimes \hat{I}_i\right)\right),$$

where S is the strain receptor input, which has broad connectivity similar to an inhibitory connection in scale. While we currently only have a linear muscle model, this broad connectivity provides a similar low-pass filter that a more elaborate muscle model would.

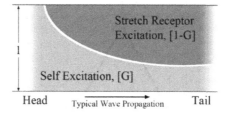

Fig. 5. One possible blending of excitatory connections ([G]) and strain receptor inputs ([1-G])

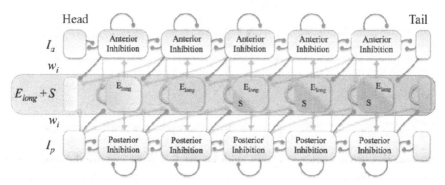

Fig. 6. For the purpose of showing several connection trends, the nearly continuous dynamics of a single muscle activation group are represented as five nodes along the body. It is important to note that each synaptic connection shown here is applied to multiple populations through a Gaussian distribution to its neighbors. The excitatory populations (E_{long}, green middle layer) directly stimulate the lengthwise actuators (not shown), while also locally stimulating the anterior and posterior inhibition networks (blue). These two networks in turn inhibit the excitatory network more broadly, but are shifted in opposite directions along the network. Towards the tail, the excitatory connections are combined with strain sensor input (S, in red) from the muscle that the network controls. The weights of the inhibitory connections, w_i, can control the direction, speed and wavelength of the naturally occurring waves in a top-down manner. The weak inhibition across the E_{long} and E_{circ} networks that keeps the networks phase locked is not shown.

2.2   Model of Worm Posture and Strain Receptors

In order to model the effect of strain receptors on our control system, we have made a basic model of worm posture as a function of a given set of muscle activations and vertical environmental constraints, i.e., under the assumption that the environment is essentially frictionless. Our previous work [14][27] indicates that tangential ground reactions are small if the wave is symmetric and steady state, so this model does not consider tangential forces and coupling effects between segments.

Each segment's position is solved by balancing the forces on a braided mesh element (Figure 7). We assume that the excitatory layers E_{long} and E_{circ} correlate linearly with muscle force generation, F_x and F_y. These forces are resisted by a non-linear spring with a rest length, L. We assume the weight of the structure can be ignored. If there are no other opposing forces, the posture of each mesh element can be solved independently since all the forces are internal.

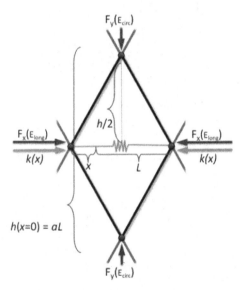

Fig. 7. A single braided mesh element used to derive robot pose as a function of muscle activity

Because this is a constrained system with known forces we can use the principle of virtual work. We use the Pythagorean theorem on the rigid lengths to define the coupling, where h is the height of the element and x is the extension of the spring. L is the resting length of the element and aL is the resting height of the element.

$$h = \sqrt{(1+a^2)L^2 - (x+L)^2}$$

Note that if we want to be able to extend to $x = L$, a must be greater than sqrt(3) in order to maintain nonzero height. Taking the variation and rearranging results in

$$-\left(\frac{x+L}{h}\right)\delta x = \delta h$$

If $k(x)$ is the spring force (shown schematically in Figure 7), we can model it as $k(x) = A \tan(x/L \cdot pi/2)$, where A is a constant. This nonlinear spring ensures that the position of the resultant element is bounded regardless of the input forces. Thus the equation to be numerically solved at each time step is:

$$-F_x \delta x - k(x) \delta x - F_y \delta h = 0$$

$$-F_x - A \tan\left(\frac{x}{L}\frac{\pi}{2}\right) + \left(\frac{x+L}{h}\right) F_y = 0$$

We assume the strain receptor measures x. E_{long} is only stimulated by positive values of x, and E_{circ} only by negative values. We used constants $L = .55$ mm, $a =$sqrt(3.1) The resulting strain on the braided mesh element based on the force balance between the nonlinear spring and controller forces is graphically represented in Figure 8.

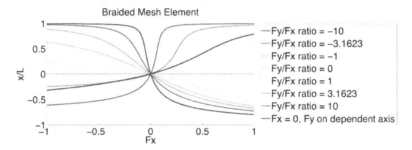

Fig. 8. The extension in the length of an element x normalized by the rest length L is demonstrated for several constant ratios of Fx to Fy. Fx is plotted on the lower axis, and Fy is directly proportional. This provides a characterization of the effective properties of the element. Due to the anisotropic properties of the braided mesh and the non-linear spring, the kinematics of very large forces are bounded. In the case of the constraining tube in Figure 9, the length was not permitted to shrink below the rest length (x/L>0).

3   Results

We tested several different techniques for integrating strain sensor information into the control of a wormlike robot. Based on our knowledge of worm neuroethology, we have focused first on strain sensors. While we tried many different arrangements, the only family of arrangements we found that were stable and desirable were when the strain sensor information from a given muscle control network was integrated with the excitatory connections, E_{long} and E_{circ}, surrounding that network. Within this family, we tested many techniques of blending excitatory connections and strain sensor input (Figure 8). The dynamics of the network without sensory input had desirable properties, but was inherently open-loop (Figure 9, upper left). However, sensory input alone was not enough to generate stable wave patterns (Figure 9, upper right), even when the wave started under the natural dynamics of the system (Figure 9, lower left). As an alternative, we reduced the excitatory connections from head to tail, such that in the absence of any sensory input, the system dropped to zero activity about half way down the body (Fig 8, lower right). However, in the second half of the body, there are still excitatory connections, and combined with the presence of strain sensor feedback, the wave stays above threshold and continues down the body (Figure 9, center).

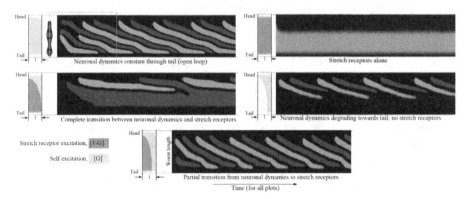

Fig. 9. Five different [G] filters and their resultant behaviors. To the left is a representation of the ratio of strain sensor (red) to neuronal dynamics (teal) from the head to the tail. While the neuronal dynamics alone produce waves, they are inherently open loop (label). Conversely, no wave was generated based only on strain sensor input without an excitatory layer (label). Instead, both the neuronal dynamics and sensory input are needed throughout the body. This can be seen as an advantageous behavior, where the head is doing its best to expand the tube it is creating, but the rest of the body conforms to the given tube diameter to save energy.

We have also demonstrated this network's ability to respond appropriately to a given sensory input. An external constraint was applied to the kinematic simulation such that the final expanded diameter of the robot would not exceed a fixed value. This in turn affected the strain sensors. Near the tail where the wave propagation is driven by the strain sensors, the lengthwise actuator wave drops off early (Figure 10). Because of the new kinematics of the tube constraint, this does not cause a loss in speed, and therefore it is more efficient than its open loop counterpart (Figure 11).

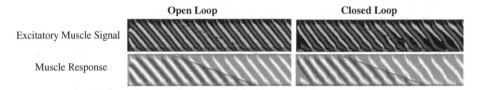

Fig. 10. The system responds to a simulated environment by changing its activity. The top row shows the open loop and closed loop excitatory muscle signals, where E_{hoop} is shown in blue, and E_{circ} in green. The bottom row shows the respective body positions over time (horizontal axis), where dark gray is fully expanded axially, and white is fully contracted. The red line in each trace indicates the gradual transition from a large tube to a smaller tube, and to the right of the line, the robot can no longer fully expand, as indicated by the absence of dark gray lines. Only in the closed loop case does the simulation respond to the constrained environment and reduce activation of the circular hoop actuators.

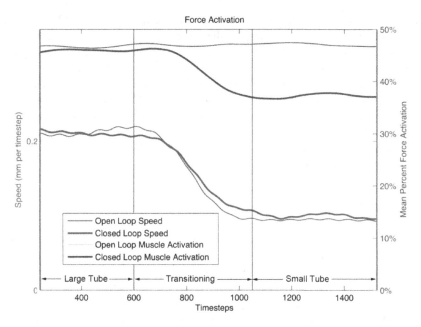

Fig. 11. The speed for both open and closed loop control decreases when the constraint is applied (in between vertical lines). Only in the closed loop case does the muscle activity decrease in response to the constraint. The speed is determined by assuming the worm does not slip at the widest point, and the mean percent force activation is the sum of all activations across E_{circ} and E_{long} averaged across all segments.

4   Discussion

The system we have developed has some properties that are commonly found in biology: opposing muscles are applied to each segment, each repeated segment process is networked to its neighbors, there is no explicit environmental model other than the connected network dynamics. The way in which the strain sensor propagates the wave is almost the opposite of a proportional-derivative (P-D) controller. When an axially expanded segment senses that the one in front of it is contracting, instead of fighting this change, the segment begins to contract. This is how the wave of motion propagates in simulation, and is consistent with observations of the role of strain receptors in worms [25]. This implies that if a segment is attempting to expand into a rigid tube, it will only exert so much force before stopping. If the worm is burrowing, but making little or no progress, this would be a desirable behavior, and potentially more energy efficient. However, the worm must also generate enough outward force to provide friction for tasks such as burrowing. We suspect that for this reason, other sensory input, such as pressure and touch sensors may have different dynamic effects on the system, more akin to a P-D control loop, to ensure that enough burrowing force can be generated. With our next generation worm robot prototype and simulation, we will begin to explore these more complex relationships by adding internal forces and improved muscle modeling.

The work Boyle et al. on a neuro-mechanical simulation of nematode behavior has several parallels to our model, despite the differences in resultant motion. They weakly inhibit opposing muscle groups to prevent co-contraction (and probably help wave propagation). The stretch receptors directly excite the local group of neurons when stretched. However, the stretch receptor influence is highly asymmetrical, in that it affects only neurons posterior to the stretch [23]. The reach of a given stretch receptor then determines the spatial wavelength. While a good deal is known about the neuro-morphology of nematodes there is no specific evidence regarding these design choices, so it is very interesting that both their group and ours have come to similar design choices in different animals.

5   Conclusions

In this paper, we have explored unconventional control architectures that combine implicit models of behavior with ubiquitous sensory input. By combining models of peristalsis, the neuroethology of earthworms and the dynamic properties of neural control circuits, we have demonstrated a novel control architecture that can respond appropriately to simulated sensory feedback. We believe that this kind of system will be well suited to controlling the continuously deforming outer body of our novel robot.

Acknowledgment. We would like to thank Andrew Horchler, Kendrick Shaw and Nicole Kern for their countless insights and support.

References

[1] Ostrowski, J., Burdick, J.: Gait Kinematics for a Serpentine Robot. In: Proc. IEEE Int. Conf. Robotics and Automation (ICRA), Minneapolis, MN, vol. 2, pp. 1294–1299 (1996)

[2] Hannan, M.W., Walker, I.D.: Analysis and Initial Experiments for a Novel Elephant's Trunk Robot. In: Proc. IEEE Int. Conf. Intelligent Robots and Systems (IROS), pp. 330–337 (2000)

[3] Menciassi, A., Gorini, A., Pernorio, G., Dario, P.: A SMA Actuated Artificial Earthworm. In: Proc. Int. Conf. Robotics and Automation, ICRA (2004)

[4] Tsakiris, D.P., Sfakiotakis, M., Menciassi, A., La Spina, G., Dario, P.: Polchaete-like Undulatory Robotic Locomotion. In: Proc. IEEE Int. Conf. Robotics and Automation (ICRA), Barcelona, Spain, pp. 3018–3023 (2005)

[5] Trimmer, B., Takesian, A., Sweet, B.: Caterpillar locomotion: a new model for soft-bodied climbing and burrowing robots. In: Proc. 7th Int. Symp. Technology and the Mine Problem, Monterey, CA (2006)

[6] Wang, K., Yan, G.: Micro robot prototype for colonoscopy and in vitro experiments. J. Med. Eng. & Tech. 31(1), 24–28 (2007)

[7] Omori, H., Nakamura, T., Yada, T.: An underground explorer robot based on peristaltic crawling of earthworm. Industrial Robot 36(4), 358–364 (2009)

[8] Hirose, S.: Biologically Inspired Robots: Snake-Like Locomotors and Manipulators. Oxford University Press, Oxford (1993)

 [9] Hatton, R.L., Choset, H.: Generating gaits for snake robots: annealed chain fitting and keyframe wave extraction. Auton. Robot. 28, 271–281 (2010)
[10] Ijspeert, A.J., Crespi, A., Ryczko, D., Cabelguen, J.M.: From swimming to walking with a salamander robot driven by a spinal cord model. Science 315(5817), 1416–1420 (2007)
[11] Brusca, R.C., Brusca, G.J.: Invertebrates. Sinauer Associates, Sunderland
[12] Ekeberg, Ö., Griller, S.: Simulations of neuromuscular control in lamprey swimming. Philos. Trans. R. Soc. Lond. B. Biol. Sci. 354, 895–902 (1999)
[13] Quillin, K.J.: Kinematic scaling of locomotion by hydrostatic animals: ontogeny of peristaltic crawling by the earthworm lumbricusterrestris. J. Exp. Biol. 202, 661–674 (1999)
[14] Boxerbaum, A.S., Chiel, H.J., Quinn, R.D.: Continuous Wave Peristaltic Locomotion. International Journal of Robotics Research (January 2012)
[15] Boxerbaum, A.S., Chiel, H.J., Quinn, R.D.: A New Theory and Methods for Creating Peristaltic Motion in a Robotic Platform. In: Proc. Int. Conf. Robotics and Automation (ICRA), pp. 1221–1227 (2010)
[16] Quinn, R.D., Nelson, G.M., Ritzmann, R.E., Bachmann, R.J., Kingsley, D.A., Offi, J.T., Allen, T.J.: Parallel Strategies for Implementing Biological Principles Into Mobile Robots. Int. J. Robotics Research 22(3), 169–186 (2003)
[17] Mangan, E.V., Kingsley, D.A., Quinn, R.D., Chiel, H.J.: Development of a peristaltic endoscope. In: Proc. IEEE Int. Conf. Robotics and Automation (ICRA), pp. 347–352 (2002)
[18] Ayers, J., Cricket, W., Chris, O.: Lamprey Robots. In: Proc. Int. Symp. Aqua Biomechanisms (2000)
[19] Zhang, D., Hu, D., Shen, L., Xie, H.: Design of an artificial bionic neural network to control fish-robot's locomotion. Neurocomputing 71, 648–654 (2008)
[20] Ijspeert, A.J.: Central pattern generators for locomotion control in animals and robots: A review. Neural Networks 21, 642–653 (2008)
[21] Matsuo, T., Yokoyama, T., Ueno, D., Ishii, K.: Biomimetic Motion Control System Based on a CPG for an Amphibious Multi-Link Mobile Robot. J. Bionic Eng. Suppl., 91–97 (2008)
[22] Wadden, T., Hellgren, J., Lansner, A., Grillner, S.: Intersegmental coordination in the lamprey: simulations using a network model without segmental boundaries. Biol. Cybernetics 76, 1–9 (1997)
[23] Boyle, J., Berri, S., Cohen, N.: Gait Modulation in C. Elegans: An Integrated Neuromechanical Model. Frontiers in Computational Neuroscience (March 2012)
[24] Moore, A.R.: Muscle tension and reflexes in earthworm. Journal of General Physiology 5, 327 (1923)
[25] Gray, J., Lissmann, W.: Studies in Animal Locomotion VII: Locomotory reflexes in the Earthworm. Journal of Experimental Biology 15, 506–517
[26] Collier, H.: Central nervous activity in the earthworm. Journal of Experimental Biology (1939)
[27] Boxerbaum, A.S., Horchler, A.D., Shaw, K., Chiel, H.J., Quinn, R.D.: A Controller for Continuous Wave Peristaltic Locomotion. In: International Conference on Intelligent Robots and Systems, IROS (2011)
[28] Wilson, H.R., Cowan, J.D.: Excitatory and Inhibitory Interactions in Localized Populations of Model Neurons. Biophys. J. 12, 1–24 (1972)
[29] Ermentrout, G.B., Cowan, J.D.: A Mathematical Theory of Visual Hallucination Patterns. Biol. Cybernetics 34, 137–150 (1979)

Exploration of Objects by an Underwater Robot with Electric Sense

Frédéric Boyer and Vincent Lebastard

IRCCyN-Ecole des Mines de Nantes, 4, rue Alfred Kastler B.P. 20722, 44307, Nantes Cedex 3, France

Abstract. In this article, we propose a solution to the underwater exploration of objects using a new sensor inspired from the electric fish. The solution is free of any model and is just based on the combination of elementary behaviors, each of these behaviors being achieved through direct feedback of the electric measurements. The solution is robust, cheap and easy to implement. After, stating and interpreting it, the article ends with a few experimental results consisting in exploring small and large unknown objects.

1 Introduction

In spite of its high potential interest for applications as the exploration of deep seas or the rescue missions in catastrophic conditions, underwater navigation in confined unstructured environments wetted by turbid waters is till today a challenge for robotics. Obviously due to the fluid opacity, vision cannot be used while the jamming caused by the multiple interfering reflections as well as the diffraction by small floating particles considerably increase the problems posed to echolocation by sonar. Pursuing a bio-inspired approach in robotics, we can question nature to learn which solutions could be implemented to solve this difficult problem. In fact, evolution has discovered an original sense well adapted to this situation: the electric sense. Developed by several hundreds of fish who have evolved in parallel on both African and South-American continents, electric sense have been discovered by Lissman in 1958 [1]. In the African fish Gnathonemus Petersii for instance (pictured in figure 1-(a)), the fish first polarizes its body with respect to an electric organ discharge (EOD) located at the basis of its tail. This polarization which is applied during a short time-pulse, generates around the fish a dipolar shaped electric field which is distorted by the objects present in its surroundings (see figure 1-(b)). Then, thanks to many electro-receptors distributed along its body, the fish "measures" the distortion of the electric fields and infers an image of its surroundings. Thus, electric sense has a quite narrow but relevant niche since none other sense as vision or sonar can work in these conditions. Thus, understanding and implementing this bio-inspired sense on our technologies would offer the opportunity to enhance the navigation abilities of our today under-water robots. Based on these potential interests, Mc. Iver and co authors have recently exploited a sensor based on the

T.J. Prescott et al. (Eds.): Living Machines 2012, LNAI 7375, pp. 50–61, 2012.

measurement of the electric potential through electrodes in order to address the problem of electrolocation of small objects through off-line particle filtering [2]. In this case, the sensor reduces its body to a so a small surface (two points electrodes between which the difference of potentials is measured) that it does not perturb the electric field produced by another pair of punctual (emitting) electrodes between each the voltage is imposed. In Angels[1][3], another technological solution is proposed to the electric sensing. In this case, the sensor is embedded on a realistic 3D body on which each electrode can be polarized with respect to the others through a given vector of voltage U. The electric field distortions are then measured through the vector I of the currents flowing across the electrodes and the measurement mode is then qualified of $U - I$. The first letter standing for the emission (here, a vector of voltage U), the second, for the reception (here a vector of currents I), and distinguished from the $U - U$ mode of [4]. In the article here presented we address the problem of the underwater exploration of objects electrically un-transparent with respect to the ambient water. Based on the morphology of the sensor (its symmetries in particular), the solution does not require any model of the environment. At the end, it consists of sensor based feedback loops, or in the language of neurobiology of sensory-motor loops. Each of these loops ensure to the sensor a simple behavior whose the combination allows it to find and explore the objects. By exploration, we here means an orbiting motion around the object. Finally, the article is structured as follows. In section 2 we introduce our experimental set up including a family of sensor said slender and the electro-location test-bed on which the algorithms are developed. In section 3, the electric model of the slender sensors is given, while a control strategy suited to object exploration is dealt with in section 4. Finally, experiments based on this control law are detailed in section 5. The article ends with a conclusion in section 6.

2   The Electrolocation Test-Bed

2.1   Sensor

Based on the $U - I$ measurement principle previously explained, we built a first generation of sensors named "slender probes" due to their high aspect ratio (length/thickness) morphology. On these probes, the macro-electrodes $\mathcal{E}_\alpha$ are rings or hemispheres which are azimuthally divided into an even number of identical measurement electrodes. As an illustrative example, Fig. 2 shows one of this probes where each of the macro-electrodes (except $\mathcal{E}_0$ located in the tail) is divided into a pair of two identical left-right measurement electrodes (it is consequently named the 7-electrode probe and such that $\mathcal{E}_1 = e_1 \cup e_2, \mathcal{E}_2 = e_3 \cup e_4, \mathcal{E}_3 = e_5 \cup e_6$. In all the following, $\mathcal{E}_3$ will be named the head electrode, $\mathcal{E}_2$, the neck electrode while $\mathcal{E}_0$ is the tail electrode.

---

[1] The ANGELS project is funded by the European Commission, Information Society and Media, Future and Emerging Technologies (FET) contract number: 231845.

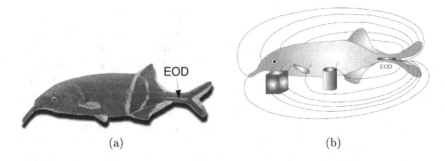

(a)   (b)

Fig. 1. (Left) The electric fish *Gnathonemus Petersii*. (Right) The electric field is distorted by the presence of an object (for instance, an insulating cylinder pushes the field lines away whereas a conductive cube funnels them).

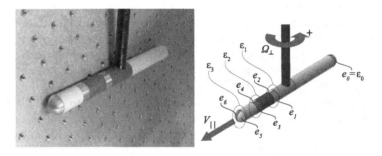

Fig. 2. Picture (left) and schematic view (right) of a 7-electrode sensor organized in 4 polarizable rings, 3 of them being divided in two half rings allowing two lateral (left and right) current measurements

2.2   Tank and Cartesian Robot

In order to test our electrolocation sensors and algorithms in controlled and repeatable conditions, an automated test bench consisting in a tank of one cubic meter side with insulating walls and a three-axis cartesian robot has been built (see Fig. 3). The robot fixed on top of the aquarium allows probes positioning in translation along X and Y with a precision of $1/10$mm and the orientation in the (X, Y) plane is adjusted in $0.023°$abs using an absolute yaw-rotation stage. All probes tested are positioned in the aquarium at adjustable height using a rigid glass epoxy fibre tube ($\oslash$14mm). This vertical insulating tube allows the passage of electrical cables dedicated to the signals coming from the electrodes to readout electronics (analogue chain + ADC board) without compromising the measurements. The maximum speed available is 300mm/s ($\simeq 1$km/h) for both translations and $80°$/s (13.5tr/min) for rotation.

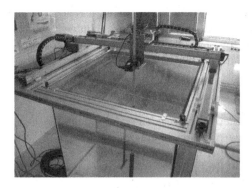

Fig. 3. Electrolocation test bench

2.3   Objects and Scenes

To investigate navigation algorithms in quite complex scenes using electric sense, a set of test objects has been fabricated with conductive and insulating materials. By insulating (or conductive), we mean an ideal material with a conductivity γ such that $\gamma/\gamma_0 = 0$ (or $\gamma/\gamma_0 = \infty$), γ_0 being the conductivity of the ambient fluid (for instance, an ordinary tap water). In practice, plastics (metals) are a good approximation of insulating (conductive) materials. To these simple shaped bodies, we can also add some removable insulating walls. Finally, in the following any scene will be constituted of a combination of these objects and removable walls arbitrarily configured between the four fixed (insulating) walls of the tank.

3   Model

We here restrict our investigations to the case of resistive phenomena. In these conditions, the electric state of a scene can be entirely parameterized by an electric potential ϕ satisfying Laplace equations with boundary conditions imposed on the sensor and wall boundaries as well as crossing boundaries through the objects. In these conditions, reconsidering the principle of the sensor, we first polarize the $\mathcal{E}_\alpha$ ($\alpha \neq 0$) with respect to the ground electrode $\mathcal{E}_0$. This is done by imposing the vector of voltages $\mathbf{U} = (U_0, U_1...U_m)^T = (0, U, U...U)^T$ with $U_0 = 0$ fixing the ground and U defining the controlled voltage. Then, we measure the vector $\mathbf{I}$ of the currents I_k crossing the e_k for $k = 0, 1, 2...n$. According to the laws of electrostatics [5], they detail as:

$$I_k = \gamma_0 \int_{e_k} \boldsymbol{\nabla}\phi.\mathbf{n} \ ds \tag{1}$$

where we used the Ohm law relating the vector field of currents in the medium $\mathbf{j}$ with the electric field $-\boldsymbol{\nabla}\phi$ and where by convention $\mathbf{n}$ being the inward normal to the sensor, one current is considered positive when it flows out from the sensor.

To this first decomposition of the total currents vector we can add another one which is based on the morphology of the sensor. Indeed, thanks to the symmetries of the sensor, the vector of total currents can be decomposed as a sum of two components named *"lateral currents"* and *"axial currents"* as follows:

$$\mathbf{I} = \mathbf{I}_{ax} \oplus \mathbf{I}_{lat}. \tag{2}$$

In (2), $\mathbf{I}_{ax}$ is axi-symmetric whereas the component $\mathbf{I}_{lat}$ is axi-skew-symmetric. In other words, for any of the e_i belonging to a same $\mathcal{E}_\alpha$, the components $I_{ax,i}$ are identical whereas for a couple of opposed (e_i, e_{i+1}) on a same ring $\mathcal{E}_\alpha$, we have $I_{lat,i} = -I_{lat,i+1}$. From these considerations of symmetry, it will be easy to extract $\mathbf{I}_{lat}$ and $\mathbf{I}_{ax}$ from $\mathbf{I}$ using the relations:

$$\mathbf{I}_{ax} = \mathbf{D}_+\mathbf{P}_+\mathbf{I}, \tag{3}$$

Where $\mathbf{P}_+$ projects the currents crossing the e_k on those crossing the $\mathcal{E}_\alpha$, by simply adding all the I_k of a same ring $\mathcal{E}_\alpha$, that being done ring by ring. Regarding $\mathbf{D}_+$, it is the $(n+1) \times (m+1)$ matrix defined by $D_{+(i\alpha)} = A_i/A_\alpha$ if $e_i \subset \mathcal{E}_\alpha$ and $D_{+(i\alpha)} = 0$ otherwise, with A_i and A_α the area of e_i and $\mathcal{E}_\alpha$ respectively. This matrix allows one to equi-distribute the I_α onto the e_k with $e_k \in \mathcal{E}_\alpha$.

In the same manner, we can write:

$$\mathbf{I}_{lat} = \mathbf{D}_-\mathbf{P}_-\mathbf{I}. \tag{4}$$

where, the projection matrix $\mathbf{P}_-$ is such that once applied to $\mathbf{I} = (I_0, I_1, I_2, ...I_n)^T$, it gives the reduced vector : $(0, I_2 - I_1, I_4 - I_3, ...I_n - I_{n-1})^T$, in the case (always verified) where the numbering of the electrodes e_i $(i > 1)$, is such that the (e_i, e_{i+1}) are opposed pairs in the same $\mathcal{E}_\alpha$. In addition, $\mathbf{D}_-$ allows one to recover the vector of currents $\mathbf{I}_{lat}$ by distributing the $|I_{i+1} - I_i|$ onto the $(n+1)$ e_i slots.

From an expansion in perturbations with respect to the aspect ratio (radius of the sensor over its length) we obtained in [6], the model of the lateral currents as following:

$$\mathbf{I}_{lat} \simeq \mathbf{P}_\perp\nabla_\perp\mathbf{\Phi}_1, \tag{5}$$

with:

$$\nabla_\perp\mathbf{\Phi}_1 = \left(\int_{e_0} \nabla\phi_1.\mathbf{n}_\perp ds, ..., \int_{e_n} \nabla\phi_1.\mathbf{n}_\perp ds\right)^T \tag{6}$$

the vector of lateral excitation fluxes in which ϕ_1 is the potential reflected by the object once polarized by the sensor, $\mathbf{n}_\perp$ is the lateral component of the normal to the sensor, and $\mathbf{P}_\perp$ is a tensor named tensor of lateral polarizability of the sensor which is diagonal positive at the leading order. As regards the axial currents, we have shown in [6], that we also have at the leading order:

$$\mathbf{I}_{ax} \simeq \mathbf{I}^{(0)} - \mathbf{D}_+\overline{\mathbf{C}}^{(0)}\mathbf{\Phi}_1. \tag{7}$$

with: $\mathbf{\Phi}_1 = (\phi_1(\mathbf{x}_{c0}), ... \phi_1(\mathbf{x}_{cm}))^T$, the vector of iso-potentials applied by the reflected field by the object on the sensor's electrodes; $\mathbf{I}^{(0)}$ the vector of basal currents, i.e. with no object in the scene; $\overline{\mathbf{C}}^{(0)}$ the conductivity matrix of the sensor with no object in the scene (i.e. such that $\mathbf{I}^{(0)} = \mathbf{D}_+\overline{\mathbf{C}}^{(0)}\mathbf{U}$). Finally, $\mathbf{I}^{(0)}$, $\mathbf{P}_\perp$, and $\overline{\mathbf{C}}^{(0)}$ are data intrinsically related to the sensor and the ambient medium with no object. As such, they can be computed once before all through a numerical code as the Boundary-Elements-Method or directly through a preliminary off-line calibration of the sensor. On the other hand, $\nabla_\perp \mathbf{\Phi}_1$ and $\mathbf{\Phi}_1$ are vectors depending on the geometry of the object and its situation (position-orientation) with respect to the sensor frame.

4 Control Law

In this section, we present a control strategy allowing the sensor to explore the objects in its surroundings. This strategy is stated in this section, while we postpone to the next section its physical interpretation.

4.1 General Architecture

The general architecture of the underwater robot is decomposed into three elementary blocks as depicted on figure 4. The first block is directly related to the electric measurements of the electric sensor. The second is related to the navigation, and is consequently named "navigator". The last one controls the locomotion of the underwater vehicle on which the sensor is embarked. As a result, it is named "the locomotor". In the following we will concentrate our attention onto the second block which allows the robot to compute the desired (axial) linear and angular (lateral) velocities $V_\parallel$ and $\Omega_\perp$ as a function of the electric measurements $\mathbf{I}$. The navigator, can be decomposed into the following three sub-blocks:

- 1°) A sub-block "axialisation" which extracts the axial component of the currents $\mathbf{I}_{ax}$ from the vector of total currents $\mathbf{I}$.
- 2°) A sub-block "lateralisation" which extracts the lateral component of currents $\mathbf{I}_{lat}$ from the vector of total currents $\mathbf{I}$.
- 3°) A sub-block "memorization" which memorizes certain values taken by the measurements along the motion of the sensor.

4.2 Case of the 7-Electrode Sensor

Before going on with the implementation of this general structure, let us illustrate these three operations with the 7-electrode sensor of figure 2. In this case, the axialisation operator consists of:

$$\mathbf{I}_{ax} = \mathbf{D}_+\mathbf{P}_+\mathbf{I} = (I_5 + I_6, I_5 + I_6, I_4 + I_3, I_4 + I_3, I_2 + I_1, I_2 + I_1, I_0)^T \quad (8)$$

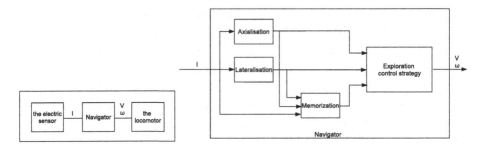

Fig. 4. (Left) The general architecture of the underwater robot. (Right) The navigator.

Alternatively, we also introduce the vector of reduced axial currents:

$$\bar{\mathbf{I}}_{ax} = (I_5 + I_6, I_4 + I_3, I_2 + I_1, I_0)^T \tag{9}$$

where the over-barre indicates that the concerned currents are those flowing across the m rings $\mathcal{E}_\alpha$. The lateralisation operator allows to obtain the vector of lateral currents :

$$\mathbf{I}_{lat} = \mathbf{D}_-\mathbf{P}_-\mathbf{I} = (I_6 - I_5, I_5 - I_6, I_4 - I_3, I_3 - I_4, I_2 - I_1, I_1 - I_2, 0)^T \tag{10}$$

As a first application of the "memorization block" the basal component of the currents $\mathbf{I}^{(0)}$ is memorized in a preliminary calibration phase. By virtue of the symmetry properties of the sensor, this component is purely axial. Thanks to this preliminary calibration, we can compute at any time of the experiment the perturbative axial currents $\delta\mathbf{I}_{ax}$:

$$\delta\mathbf{I}_{ax} = \mathbf{I}_{ax} - \mathbf{I}^{(0)}, \tag{11}$$

which only result from the presence of an object in the scene. We introduce $\delta\bar{\mathbf{I}}_{ax}$ the reduced vector of perturbative currents flowing across the rings. In the same way, the vector of lateral perturbative currents is directly given by $\delta\mathbf{I}_{lat} = \mathbf{I}_{lat}$ since, due to the lateral symmetry of the sensor, the vector of basal currents has no lateral component, i.e.: $\mathbf{D}_-\mathbf{P}_-\mathbf{I}^{(0)} = \mathbf{0}$. With these definitions in hand, we are now going to introduce a simple control law ensuring the exploration as previously defined.

4.3   Exploration Control Strategy

Using the three elementary operations previously introduced, one can derive a control strategy consisting in exploring the objects present in the scene. By "exploring" we here mean "to seek the objects" and "to follow their boundaries without touching them". Such a law is the natural resultant of the combination of three basic behaviors:

- 1°) Seek any object electrically contrasted with respect to the ambient medium.
- 2°) Flee from the electric influence of the object.
- 3°) Follow the boundaries of the object.

Now, let us reveal how these three behaviors can be easily obtained through simple sensory-motor loops. The two first ones are in fact produced by applying a feedback control law of the head lateral currents of the general form:

$$V_\| = cte > 0, \quad \Omega_\perp = K \, \delta I_{lat,1}, \tag{12}$$

with K a gain ensuring the sensor to be attracted by any un-transparent object when $K = k/\delta\overline{I}_{ax,1}$ and to be repulsed by any object when $K = k/\vert \delta\overline{I}_{ax,1} \vert$ with $k > 0$ in both cases. As regards the third behavior, it is simply obtained by applying the control law:

$$V_\| = cte > 0, \quad \Omega_\perp = k' \, (\delta\overline{I}_{ax,1} - \delta\overline{I}_{ax,1}^{(mem)}), \tag{13}$$

where the axial velocity $V_\|$ is ruled by the same constant value as in (12), while $k' > 0$ is a constant gain. In (13), $\delta\overline{I}_{ax,1}^{(mem)}$ represents a value of the axial head current stored by the memorizing block of figure 4 at a time of the motion where it has a remarkable evolution (typically when it attains an extremum). Thus, beyond this time, $\delta\overline{I}_{ax,1}^{(mem)}$ represents a desired value that the sensor will track along its motion.

4.4 Application to the 7-Electrode Sensor

We are now going to see how we can combine the three behaviors of section 4.3 in order to achieve the exploration task. Indeed, it suffices to sequentially order the three behaviors along with the three following phases symbolized through the graph of transitions of figure 5-(a) whose the application is schematized on figure 5-(b):

- First phase (from A to B in figure 5): It consists in seeking an electrically non-transparent object by applying the attractive behavior (12) with $K = k/\delta\overline{I}_{ax,1}$ and $k > 0$.
- Second phase (from B to B' in figure 5): This second phase corresponds to the initialization of the orbiting motion of the sensor around the object. It is obtained by applying the repulsive behavior, i.e. (12) with $K = k/\vert \delta\overline{I}_{ax,1} \vert$ and $k > 0$.
- Third phase (from B' to C in figure 5): This is the orbiting phase obtained by applying the law (13), in which $\delta\overline{I}_{ax,1}^{(mem)}$ is the value of $\delta\overline{I}_{ax,1}$ measured at the last time of the previous phase.

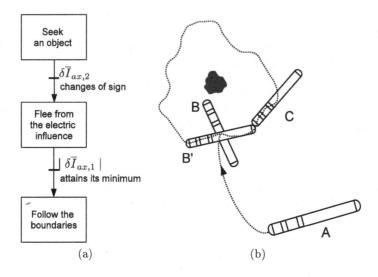

Fig. 5. (a) Sequential graph of exploration task. (b) typical scenario of the object exploration.

As indicated on the sequential graph of figure 5-(a), the commutation between these phases is ruled by the following events which only depend time variation of the measurements (and not of their magnitude). Indeed, the commutation from phase 1 to phase 2 is activated by the change of sign of $\delta \bar{I}_{ax,2}$, i.e. when the current flowing across the second ring (neck ring) change of sign. The second transition (from phase 2 to phase 3) is activated when $| \delta \bar{I}_{ax,1} |$ attains its minimum value.

5 Interpretation

Let us now interpret the previous law. First, since due to equations 5 and 6, the lateral current $\delta I_{lat,1} = I_{lat,1}$ is proportional to the lateral flux of the electric field reflected by the object, any law of the general form (12) ensures the sensor to align its head on the electric field lines. This first condition is ensured while the sensor moves with the constant axial velocity. As a result, since all the electric lines (integral of the reflected electric field) emanate from the objects close to the sensor, the sensor is attracted or repulsed by them depending of the sign of the gain K in (12). Finally, the presence of $|\delta \bar{I}_{ax,1}|$ in (12) allows to normalize the control as the sensor approaches a targeted object. When it is applied in the first phase, the sensor is attracted by the objects. When getting closer and closer to an object, the sensor faces it and more and more electric field lines are captured by the object if it is conductive. Thus, the electric field lines are funneled on the front (head) electrode. On the contrary, if the object is insulating the electric field lines are pushed backward along the sensor axis. As a result, when the sensor approaches an object, the current lines go from the head to the neck electrode if the object is and insulator and from the neck to the head if it is a conductor (see figure 6). Thus, in both cases it happens a time when $\delta \overline{I}_{ax,2}$ changes of sign. This is at this time that the control law commutes from the attractive to the repulsive behavior. When this occurs, the sign of K in (12) is changed and the sensor starts to avoid the object, this is done till the $\mid \delta \overline{I}_{ax,1} \mid$ attains its minimum which means that the tail ($\mathcal{E}_0$) and the head electrodes are equidistant from the object. Indeed, when one of these two electrodes is closer to the object than the other, the electric lines emitted by the tail or the lines received by the head are more perturbed (than in the equidistant case) and the corresponding measurement $\mid \delta \overline{I}_{ax,1} \mid$ is larger (see figure 6). Finally, as this condition is satisfied, $\delta \overline{I}_{ax,1}$ is memorized as a reference that the sensor try to maintain. That corresponds to follow an iso current around the sensor, this last condition ensuring the following of the object boundaries.

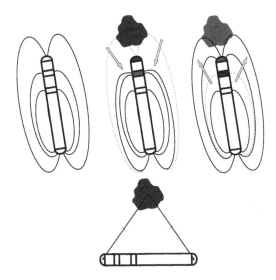

Fig. 6. (Top) The electric lines are pushed forward and backward depending if the object facing the sensor is conducting or insulating. (Bottom) When the head and tail electrodes are equidistant of the object, the head currents attain a minimum.

6   Experimental Results

In this section, we report some of the experimental results illustrating the previous exploration control strategy. The electric feedback loops (12) and (13) are tuned once for all with $k = 50 V_{\parallel}$ and $k' = 2, 5 V_{\parallel}$, and then applied to all the tests. The experimental conditions are those described in section 2. For each experiment the sensor first starts and seek the object, second, it flees from the electric influence of the object and third, follows the boundaries of the object.

Many trials have been successfully achieved with this single simple law. We here
report three of them, the first one illustrating the exploration of a small object
while the second and the third are applied to the exploration of large objects.
In the first case, the explored object is a conducting ellipsoid. Figure 7 (Top)
displays the trajectory of the sensor when the sensor orbits around the ellipsoid.
Note that while the sensor trajectory is near to be elliptic, this elliptic orbit displays an angular shift with respect to the elliptic boundary of the object. This
shift is probably due to the initial conditions of the orbit while the aspect ratio

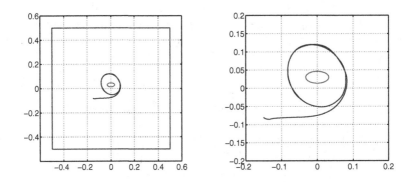

Fig. 7. Exploration of a conducting ellipsoid

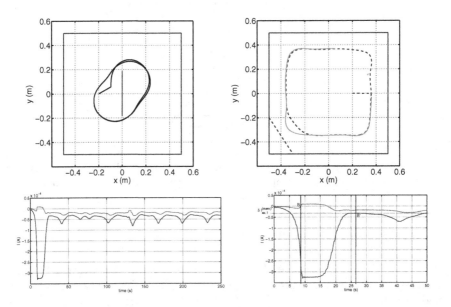

Fig. 8. (Top) Exploration of large objects. (Bottom) Time evolution of the currents
$\delta \overline{I}_{ax,1}$ (blue) and $\delta \overline{I}_{ax,2}$ (red) along the sensor path when the sensor follows the tank
walls.

of the elliptic trajectory (great over small axis) should be an image of that of the object. Finally, the case of the exploration of large objects is illustrated on the two last examples figure 8, the first object is an insulating wall placed in the middle of the tank and the second is simply the bound walls of the tank. In the last case, while the sensor is orbiting around the tank, a piece of insulating wall is placed on one of the tank corners (bottom-left). When moving past this corner, the sensor does follow the new piece of wall. Figure 8 (Bottom) shows the time evolution of the measured currents along the sensor trajectory in the case where it follows the tank walls. The two time transitions of the sequential graph (5-a) are pointed by B and B' as in figure (5-b).

7   Conclusion

In this article we have addressed the problem of object exploration using a sensor bio-inspired of electric fish. The solution is free from any model of the sensor electric interactions with the surrounding. It allows seeking any object electrically contrasted with respect to the ambient water and then to turn around. This behavior will be exploited in future to infer a geometric model of the object. Today, it has shown its robustness in many situations with many different objects. In spite of these first success, many other questions today remain. In particular, since the use of the model is replaced by the morphology of the sensor, the following question arises. In which proportion the sensor morphology determines its behavior with respect to the explored objects. For instance, the distance between the head and neck rings determines the distance at which the sensor detects the presence of an object.

References

1. Lissmann, H.W., Machin, K.E.: The mechanism of object location in gymnarchus niloticus and similar fish. The Journal of Experimental Biology 35, 451–486 (1958)
2. Solberg, J., Lynch, K., MacIver, M.: Active electrolocation for underwater target localization. The International Journal of Robotics Research 27, 529–548 (2008)
3. The angels project (2009), http://www.theangelsproject.eu
4. Solberg, J.R., Lynch, K.M., MacIver, M.A.: Robotic electrolocation: Active underwater target localization. In: International Conference on Robotics and Automation, pp. 4879–4886 (2007)
5. Jakson, J.D.: Classical Electrodynamics, 3rd edn. John Wiley and Sons (1999)
6. Boyer, F., Gossiaux, P., Jawad, B., Lebastard, V., Porez, M.: Model for a sensor bio-inspired by electric fish. IEEE Transactions on Robotics 28(2), 492–505 (2012)

Neuro-inspired Navigation Strategies Shifting for Robots: Integration of a Multiple Landmark Taxon Strategy

Ken Caluwaerts[1,2], Antoine Favre-Félix[1], Mariacarla Staffa[1,3],
Steve N'Guyen[1], Christophe Grand[1], Benoît Girard[1], and Mehdi Khamassi[1]

[1] Institut des Systèmes Intelligents et de Robotique (ISIR) UMR7222, Université
Pierre et Marie Curie, CNRS, 4 place Jussieu, 75005 Paris, France
[2] Reservoir Lab, Electronics and Information Systems (ELIS) department, Ghent
University, Sint-Pietersnieuwstraat 41, 9000 Ghent, Belgium
[3] Dipartimento di Informatica e Sistemistica, Università degli Studi di Napoli
Federico II, Via Claudio 21, 80125, Naples, Italy
mehdi.khamassi@isir.upmc.fr

Abstract. Rodents have been widely studied for their adaptive navigation capabilities. They are able to exhibit multiple navigation strategies;
some based on simple sensory-motor associations, while others rely on
the construction of cognitive maps. We previously proposed a computational model of parallel learning processes during navigation which could
reproduce in simulation a wide set of rat behavioral data and which could
adaptively control a robot in a changing environment. In this previous
robotic implementation the visual approach (or taxon) strategy was however paying attention to the intra-maze landmark only and learned to
approach it. Here we replaced this mechanism by a more realistic one
where the robot autonomously learns to select relevant landmarks. We
show experimentally that the new taxon strategy is efficient, and that it
combines robustly with the planning strategy, so as to choose the most
efficient strategy given the available sensory information.

1 Introduction

Neurorobotic researches provide a multidisciplinary approach that can both (i)
contribute to robotics by taking inspiration from the computational principles
underlying animals' behavioral flexibility and (ii) contribute to neurobiology by
using robots as platforms to test the robustness of current biological hypotheses [1, 2, 3]. Several neurorobotic projects have in particular focused on the
study of spatial cognition inspired by rodents' neurophysiological substrates for
navigation. Indeed rats are able to show highly adaptive behaviors whose repro-
duction could help improve current robots' decisional autonomy. Thus several
previous robots have been endowed with biomimetic models to enable them to
build a cognitive map of the environment and to efficiently plan trajectories in
it [4, 5, 6, 7, 8, 9].

T.J. Prescott et al. (Eds.): Living Machines 2012, LNAI 7375, pp. 62–73, 2012.

However, an ability that has not yet been thoroughly investigated in Neuro-robotics is the coordination of multiple navigation strategies. Indeed, rats and more generally mammals are able to learn to select the most appropriate strat-egy for a navigation problem, to avoid costly computations associated with their cognitive map when a simple sensorimotor strategy is enough, and to shift from one strategy to another in response to environmental changes [10]. Among the numerous possible strategies, experimental neuroscience studies of strategy in-teractions favored two main families:

– Response strategies, resulting from the learning of direct sensorimotor as-sociations (like moving towards a cue indicating the goal, which is called a *taxon strategy*).
– Place strategies, where the animal builds an internal representation (or cog-nitive map) of the various locations of the environment, using the configura-tion of multiple allocentric cues. It can then use this information to choose the direction of the next movement by planning a path in a graph connect-ing the places with the actions allowing the transitions from one place to another (*topological planning strategy*).

It has been shown that the multiple navigation strategies of rodents are operated by parallel independent memory systems [11, 12], which can result in *coopera-tive* or *competitive* behaviors, depending on the experimental protocol. Place strategies would rely on the Hippocampus, with its ability to encode places in the so-called *place cells* [13] and to contribute to trajectory planning compu-tations within a cognitive map [14]. Lesions of the hippocampal system impair place strategies while sparing response strategies [15, 16]. In contrast, lesions of the striatum impair the expression of response strategies while sparing place strategies [16, 11].

In previous work, we proposed a modular computational model that can ex-plain a wide range of behavioral data recorded in rodent laboratory maze tasks [17]. We implemented the model in a robot and showed that it enabled the robot to benefit from the particular advantages of each strategy (the planning strat-egy being efficient far from the goal, the taxon strategy being more precise in adjusting the robot's trajectory near the goal based on vision) by autonomously learning which strategy was the most efficient in each part of the environment [18]. Moreover, the robot could efficiently adapt to changes in the goal location by detecting contextual changes that require new place-strategy associations.

However, in our previous work the taxon strategy was simplified by having visually access only to the single intramaze cue that indicates the goal location, as in rat experiments, and had to learn how to orientate itself towards this landmark. In contrast, the planning strategy had full access to all landmarks in order to learn a cognitive map of the environment. This was biologically acceptable since biologists assume that taxon strategies rely only on intramaze cues while planning strategy rely on extramaze cues, and since a single intramaze cue is used in most protocols [15, 16]. However in a situation with multiple intramaze cues, the real issue is to learn which cue leads to reward. Moreover, the capability to orient toward a chosen cue is probably hardwired in the Superior

Colliculus [19]. From a robotic point of view, this is also not satisfying because one would want the robot to autonomously learn to select relevant landmarks within its visual field.

In this work, we extended the taxon strategy of our model by having it learn by reinforcement which visual landmarks it should orient toward. We first made an experiment where the taxon strategy competed with a random exploration strategy to show that the taxon could succesfully learn and progressively win the competition against the exploration strategy. Then we combined this taxon strategy with the whole model, thus competing with the planning and exploration strategies. We found that the robot successfully learned to rely less and less on the exploration strategy through learning. Moreover, at the end of learning the robot learned to prefer the taxon strategy in subparts of the environment where the robot could robustly perceive salient landmarks near the goal, while it learned to prefer the planning strategy in parts of the environment where individual landmarks were less reliably associated with goal reaching.

2   Computational Model

The model (Fig. 1) is composed of three navigation strategy experts (the taxon, the planning and the exploration), which propose directions for the next movement Φ_k, and a gating network, which learns to choose the most efficient strategy depending on the current position. As stated in the introduction, it is identical to the one described in detail in [18], except for the taxon expert.

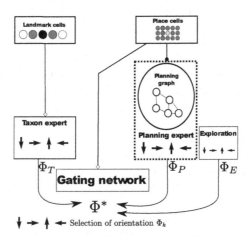

Fig. 1.   Overview   of   the   computational   model.   Different   strategies (taxon/planning/exploration) are connected to the gating network. Each strategy has a dedicated expert which proposes actions (Φ_T for the taxon, Φ_P for the planning, Φ_E for the exploration). The gating network decides which of the experts is the winner in the current situation and then the action Φ^* from this strategy is performed.

The exploration expert just provides a randomly chosen direction of movement, redrawn every three timesteps, so as to ensure that the selection of this strategy on successive timesteps results in coherent movements. It does not learn at all.

The planning expert builds a graph of states and transitions, based on the place-cells activity provided by a simplified model of the hippocampus (states), and the experience of which action allows to go from one place to another. Places where rewards are encountered are learned, so that path planning can then be computed.

The taxon expert implements a standard neural implementation of a Q-learning algorithm. It takes as input state space a representation of the perceived landmarks configuration (namely, which landmark is visible and at which distance) and learns in each state to choose a landmark towards which the robot should orient. The following transformation of this choice into an egocentric direction of movement using the camera information is hardwired (see 2.1 below for more details).

Finally, the gating network also implements a Q-learning algorithm, which takes as input the activity of the place cells (hence the current estimation of position), and learns to choose the strategy which is the most efficient in each position for maximizing future reward.

A specificity of the [17] model is that the adaptive navigation experts (here taxon and planning) receive reward signals even when they did not generate the movement, so that they can learn from each other: for the taxon, the Q-value of the landmark whose direction is the closest to the direction of the real movement (no matter if it was generated by the taxon itself, by the planning or by the exploration expert) is updated; for the planning the position of the reward is recorded whichever expert led the robot there.

The planning expert (place cells + planning graph) has been extensively described in [17, 18]. In following sections, we provide the equations for the newly implemented taxon expert and for its coordination with other experts by the gating network.

2.1   Taxon Strategy

The taxon learns to choose the best landmark to guide the orientation behavior, based on the current distance configuration of the landmarks, using a Q-learning algorithm [20]. This means that the taxon selects a landmark (action) based on the landmark configuration (state) that the robot sees and then proposes the direction towards this landmark to the gating network (section 2.2).

Indeed, the visual system automatically detects the N_L visual landmarks, based on the color of contrasted patches and on their shape, and evaluates their distance based on the binocular disparity. The distance (in meters) of each landmark is discretized in five possible ranges ($[0, 0.5],]0.5, 1],]1, 2],]2, 3],]3, +\infty[$), and for each landmark l a corresponding 5-component vector I_l with a one on the component corresponding to the detected distance is produced, and zeros elsewhere. The input I of the taxon is the concatenation of all these landmark distance vectors. The output $O = W^{taxon}I$ is a N_L-long vector, attributing a value to each of the possible landmark choice. The final selection of the landmarks is simply greedy, the chosen landmark L is:

$$L = \arg\max_{i\in[0,N_L]} O_i \tag{1}$$

as the exploration necessary for the convergence of such an algorithm is handled by the exploration expert, and the regulation of the amount of exploration results directly from the gating network choices (see below).

A standard neural implementation of a Q-learning algorithm [20] is then used to learn the $[N_L \times 5] \times [N_L]$ weight matrix W^{taxon} associating landmark distance configurations to landmarks. The prediction error δ is computed and the matrix W^{taxon} updated accordingly:

$$\delta(t+1) = r_{t+1} + \gamma \max_{i\in[0,N_L]}(W^{taxon}I(t+1))_i - W^{taxon}I(t) \tag{2}$$

$$W^{taxon}_{i,L} \leftarrow W^{taxon}_{i,L} + \alpha\delta(t+1) \tag{3}$$

When the taxon is not chosen by the gating network, and thus has not chosen the last direction of movement, the landmark $\hat{L}$ whose direction $d_{\hat{L}}$ is the closest to the chosen direction d is updated as follows:

$$\hat{L} = \arg\max_{i\in[0,N_L]} d_i \cdot d \tag{4}$$

$$W^{taxon}_{i,\hat{L}} \leftarrow W^{taxon}_{i,\hat{L}} + \alpha\delta(t+1) \tag{5}$$

2.2   Gating Network

The gating network learns the Q-value $g^k(t)$ of each expert k (called gating-values), based on a matrix of weights $z^k_j(t)$:

$$g^k(t) = \sum_{j}^{N_{PC}} z^k_j(t)n^{PC}_j(t) \tag{6}$$

The selection probability of an expert is then computed as follows:

$$P(\Phi^*(t) = \Phi^k(t)) = \frac{g^k(t)}{\sum_i g^i(t)} \tag{7}$$

Here $\Phi^k(t)$ is the action proposed by expert k at time t. $\Phi^*(t)$ is the final action proposed by the gating network. The gating network is a strategy selection mechanism instead of an action selection mechanism. It selects a winning strategy ($*$) at each action step and the action (a new heading direction) proposed by this strategy will be executed unless a higher priority mechanism (e.g. hardwired wall avoidance) is activated. This is an imporant part of the system, as the gating network is independent of the actions proposed by the strategies when it selects a strategy. If the executed action was not produced by any of the strategies feeding into the gating network (e.g. wall avoidance), the gating network and the strategies themselves can still learn as the global reward and executed action are shared between all strategies and the gating network.

Learning is sped up by using action generalization and eligibility traces, the detailed equations for these techniques are to be found in [18]. To modify the Q-values, a modified Q-learning algorithm [20] is applied:

$$\Delta z_j^k(t) = \xi \delta(t) e_j^k(t+1) \tag{8}$$

$$\delta(t) = R(t+1) + \gamma \max_k (g^k(t+1)) - g^{k^*}(t) \tag{9}$$

where ξ is the learning rate of the algorithm and δ the reward prediction error, and $e_j^k(t+1)$ the eligibility trace.

The reward prediction error δ is based on the observed reward when performing action $\Phi*$ and the future expected reward (g^{k^*} is the activation of the winning output neuron and γ the discount factor). The eligibility trace e_j^k allows the reinforcement of previously selected strategies and the strategies proposing a direction close to the one proposed by the winning strategy [20]:

$$e_j^k(t+1) = \nu(t)\Psi(\Phi^*(t) - \Phi^k(t))r_j^{PC}(t) + \lambda e_j^k(t). \tag{10}$$

The eligibility traces depend on a time-varying value $\nu(t)$ which is a measure of the quality of the sensory input as estimated by the robot, $r_j^{PC}(t)$, the place cells' activations and a decay factor λ [18].

The gating network is a simple but effective way to combine competition and cooperation between strategies, as experimentally observed in rat behavior [15]. While the gating network itself only directly provides competition, the strategies cooperate by sharing rewards and their actions (e.g. the taxon uses the executed action for learning, instead of its proposed action (Eq. 4)). Hence the gating network is advantageous for strategies as they can learn from each other, while at the global level the performance can also increase because the best performing strategy can be used in each situation.

3   Experimental Setup

The robot is a mobile platform equipped with two motorized wheels and a rotating head on which sensors are fixed (see Fig. 2-left). We use here the two frontal cameras, and given their limited aperture (60 degrees), the robot regularly makes head rotation movements so as to acquire panoramas. The combination of these panoramas with a memory of the previously seen landmarks allows to generate estimations of the landmark configuration over 360 degrees (see [18] for a detailed description of the different layers of the visual system of the model). The

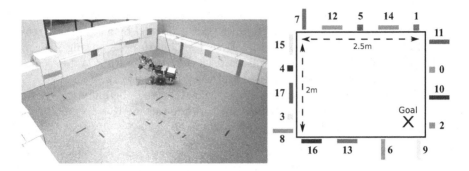

Fig. 2. Open field arena used for the experiments with the Psikharpax rat robot. The arena is surrounded by 18 landmarks differentiated by their color and shape. The goal is located in the south-east part of the arena.

landmarks are identified based on their color and shape (only unique combinations were used so as to avoid aliasing, see Fig. 2-right), and their distance was roughly estimated using binocular discrepancy.

The robot makes discrete movements, moving 10 cm at each timestep in an open 2 m by 2.5 m environment (Fig. 2). 18 different landmark cues are distributed around the arena. An invisible 20 cm diameter zone (314 cm^2, or 1/160th of the environment) is defined as the goal location that the robot has to learn to efficiently reach by trial-and-error. When the robot reaches this zone, the reward is set to one, then the robot is moved to a different starting location to begin a new trial. Anywhere else the reward is 0. The egocentric reference frame has the neck of the robot as its origin and the orientation is defined by the direction of the head.

All experiments presented here follow an exploration phase where the robot moves randomly without getting reward and builds place cells and a cognitive map (topological links between places) of the environment based on vision and odometry. This process and its robustness to the number of landmarks, to noise and to the model's parameters have been extensively described in [18]. Here we focus on the learning processes that both allow the taxon strategy to progressively select appropriate landmarks to orient toward, and simultaneously allow the gating network to progressively select the most appropriate strategy (among taxon, planning and exploration strategies) in each part of the arena.

4   Results

In all experiments, the robot is initially positioned in one of the three corners, far from the goal. If after 5 meters of movement the robot has still not found the goal, it is guided directly towards the goal as in rat experiments and as in our previous robotic work [18]. During guiding, the strategies (taxon/planning) can still learn, hence guiding can speed up learning. Guiding is also used to lead the robot back to one of the corners after receiving a reward to start a new trial.

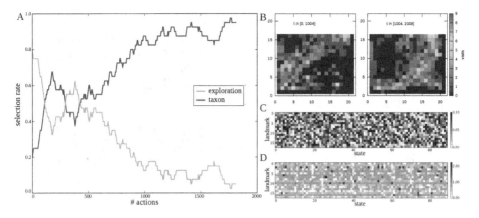

Fig. 3. Behavioral adaptation in the taxon experiment. A: the selection rate (averaged over 200 actions) illustrates that the model has successfully learned to prefer the taxon strategy over time. B: histogram of the number of times the taxon strategy is selected by the gating network at each position in the environment, during the first (left) and the second (right) half of the experiment. C and D: the taxon's weight matrices (associating the landmark distance configuration (state) with the landmark to aim for (action)) before and after learning (note the change of scale).

4.1 Evaluation of the Taxon

As we modified the taxon from our previous work, we first test the efficiency of this new expert, by running the model without planning. An experiment is run for 2008 timesteps, corresponding to 32 goal-reaching sequences (*i.e.* 32 trials).

The results confirm that the taxon has learned and has become more and more efficient, as it is predominantly selected by the gating network (Fig. 3, A). The histogram locations occupied by the robot when the taxon strategy is selected (Fig. 3, B) shows a pattern that switches from random locations in the first half of the experiment (left), to locations on trajectories leading to the goal for the second half (right). Finally, the matrices of weights at the beginning of the experiment (Fig. 3, C) and at the end (Fig. 3, D) have evolved from a random pattern to the selection of a limited number of landmarks, most of them close to the goal (mainly landmarks #2, #6, #9 and #10).

4.2 Full Model (Taxon+Planning+Exploration)

We then test the full model, in an experiment lasting 2664 timesteps, corresponding to 43 goal-reaching sequences. Here, the strategy selection evolution (Fig. 4-A) shows that exploration is rapidly replaced by the taxon and planning strategies as they become more and more efficient. Unfortunately, this experiment is not long enough for convergence of the gating network's learning mechanism, so that exploration is still selected 20% of the time. Nevertheless, the histograms of positions where the taxon (Fig. 4-B) is selected show that the taxon has learned very similarly to the previous experiment: during the second

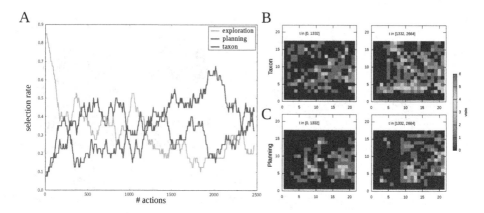

Fig. 4. Behavioral adaptation in the full model experiment. A: evolution over time of the selection rate (averaged over 200 actions) of the three stategies, illustrating the progressive learning of the taxon and planning strategies. B and C: histograms in the first (left) and the second (right) half of the experiment for the taxon (B) and planning (C) strategies indicating how many times the strategy was selected by the gating network. Similarly to the previous experiment, the taxon learns direct trajectories to the goal, while the planning is used in areas where the taxon is less efficient.

part of the experiment ($t \in [1332 : 2664]$) the taxon is mainly recruited along direct paths to the goal (*i.e.* along the southern wall and along the diagonal from the northwestern corner until the goal). Moreover, the planning is selected in places where the taxon is less efficient (Fig. 4-C), thus showing a complementary recruitment of the two strategies by the gating network. Quantitatively, excepting moments when the robot is exploring, the taxon and planning strategies are both selected during a substantial proportion of time during the second part of the experiment (respectively 62% and 38% of the time). More precisely, there are particular regions within the arena where the taxon is not sufficiently efficient and where the planning is thus preferred (Fig. 5). This is especially clear along the eastern wall, where landmark #2 is often not seen. This is also the case in the south central part of the arena, which falls outside the paths followed by the taxon. At the end of the experiment, exploration is mainly used far away from the goal, where neither taxon nor planning are yet fully efficient.

5   Discussion

We have presented the integration of a multiple landmark taxon in our strategy selection model allowing an autonomous robot to navigate in an initially unknown environment. The model selects among two parallelly learned navigation strategies: a response strategy learning to orient towards relevant cues in the visual field; a place strategy building a map of place cells and planning trajectories between different locations in the arena. This model constitutes an extension to a previously published model of multiple navigation strategies [17] which was

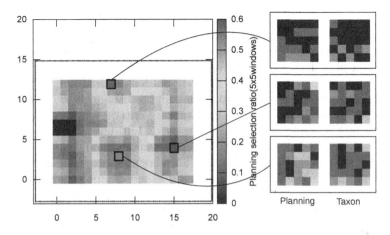

Fig. 5. Relative proportion of selection of the planning strategy over the taxon strategy during the second part of the full model experiment, averaged over $50cm \times 50cm$ sliding windows. Note the 0 to 60% scale. The planning strategy is overtly preferred in three different regions. The right insets show, for three points (black squares) in these regions, the corresponding windows of Planning and Taxon selection extracted from figure 4. The upper region seems to be an artefact, given the low number of measurements; however in the two others, the taxon is clearly less chosen, being locally less efficient or reliable.

tested in simulation to replicate a series of rat behavioral experimental results [15, 16], and which was previously applied to robotics in tasks involving a single intramaze cue [18].

Here we first show that the new taxon can succesfully learn to orient towards relevant landmarks to reach the goal, and thus that the gating network can parallely learn to prefer the taxon strategy over the exploration strategy. Then we combine this taxon strategy with the whole model, thus competing with the planning and exploration strategies. We find that the robot successfully learns to rely less and less on the exploration strategy through learning. Moreover, at the end of learning the robot has learned to prefer the taxon strategy in subparts of the environment where the robot can robustly perceive salient landmarks near the goal, while it has learned to prefer the planning strategy in parts of the environment where perceived individual landmarks are less reliably associated with goal reaching.

These results show that the model generalizes well with a taxon that takes into account all visual landmarks in the environment. They also validate in a new experiment the ability of the model to use the specific advantages of each strategy in each subpart of the environment (*e.g.* a local taxon strategy combined with a global but coarse path planning strategy). The complete validation of the model will require to test it in non-stationary environments (*e.g.* changes in goal location), as shown with the previous simplified taxon expert [18]. Future work will also test the ability of the model to achieve more complex robotic tasks involving a larger environment and the apparition/vanishing of new objects that can constitute obstacles for the robot.

Besides the interest of such modelling approach to contribute to a better formalization of rodent navigation behavior, this work has also the potential of contributing to mobile robotics. Indeed, the bio-inspired ability to rapidly switch between several behavioral strategies, and to memorize which strategy is the most efficient and appropriate in each subzone of the environment could help improve current control architectures for robots. Multi-layered control architectures with different levels of decisions have become more and more popular in robotics and are now widely used [21, 22]. Such architectures raise issues such as managing the interactions between submodules, coordinating multiple competing learning processes and providing alternative solutions to motion planning in situations where such strategy is limited [23]. Indeed the planning strategy can be approximate when coping with uncertainties, *e.g.* when there is perceptual aliasing as we illustrated in [18], and can also require high computational costs and long times to propagate possible trajectories through mental maps [21]. In contrast, in situations where animals have developed habits under the form of cue-guided taxon or response strategies to solve a particular task, they can perform quick and accurate decision-making. Taking inspiration from computational models of how mammals progressively shift from costly decision-making to habits as a function of a speed-accuracy trade-off may constitute the basis of great future advances in robotics [24].

Acknowledgments. This research was funded by the EC FP6 IST 027819 ICEA Project and a Ph.D. fellowship of the Research Foundation - Flanders (FWO).

References

[1] Pfeifer, R., Lungarella, M., Iida, F.: Self-organization, embodiment, and biologically inspired robotics. Science 318, 1088–1093 (2007)

[2] Arbib, M., Metta, G., van der Smagt, P.: Neurorobotics: From vision to action. In: Handbook of Robotics, pp. 1453–1480. Springer, Berlin (2008)

[3] Meyer, J.A., Guillot, A.: Biologically-inspired robots. In: Handbook of Robotics, pp. 1395–1422. Springer, Berlin (2008)

[4] Arleo, A., Gerstner, W.: Spatial cognition and neuro-mimetic navigation: a model of hippocampal place cell activity. Biological Cybernetics 83(3), 287–299 (2000)

[5] Krichmar, J., Seth, A., Nitz, D., Fleischer, J., Edelman, G.: Spatial navigation and causal analysis in a brain-based device modeling cortical hippocampal interactions. Neuroinformatics 3(3), 147–169 (2005)

[6] Meyer, J.-A., Guillot, A., Girard, B., Khamassi, M., Pirim, P., Berthoz, A.: The Psikharpax project: towards building an artificial rat. Robotics and Autonomous Systems 50(4), 211–223 (2005)

[7] Barrera, A., Weitzenfeld, A.: Biologically-inspired robot spatial cognition based on rat neurophysiological studies. Autonomous Robots 25, 147–169 (2008)

[8] Giovannangeli, C., Gaussier, P.: Autonomous vision-based navigation: Goal-oriented action planning by transient states prediction, cognitive map building, and sensory-motor learning. In: Proceedings of the International Conference on Intelligent Robots and Systems, vol. 1, pp. 281–297. University of California Press (2008)

[9] Milford, M., Wyeth, G.: Persistent navigation and mapping using a biologically inspired slam system. The International Journal of Robotics Research 29(9), 1131–1153 (2010)

[10] Arleo, A., Rondi-Reig, L.: Multimodal sensory integration and concurrent navigation strategies for spatial cognition in real and artificial organisms. Journal of Integrative Neuroscience 6(3), 327–366 (2007)

[11] Packard, M.G., Knowlton, B.J.: Learning and memory functions of the basal ganglia. Annual Review of Neuroscience 25, 563–593 (2002)

[12] Burgess, N.: Spatial cognition and the brain. Year In Cognitive Neuroscience 2008 1124, 77–97 (2008)

[13] O'Keefe, J., Nadel, L.: The Hippocampus as a Cognitive Map. Clarendon Press, Oxford (1978)

[14] Johnson, A., Redish, A.D.: Neural ensembles in CA3 transiently encode paths forward of the animal at a decision point. Journal of Neuroscience 27(45), 12176–12189 (2007)

[15] Pearce, J.M., Roberts, A.D., Good, M.: Hippocampal lesions disrupt navigation based on cognitive maps but not heading vectors. Nature 396(6706), 75–77 (1998)

[16] Devan, B.D., White, N.M.: Parallel information processing in the dorsal striatum: Relation to hippocampal function. Journal of Neuroscience 19(7), 2789–2798 (1999)

[17] Dollé, L., Sheynikhovich, D., Girard, B., Chavarriaga, R., Guillot, A.: Path planning versus cue responding: a bioinspired model of switching between navigation strategies. Biological Cybernetics 103(4), 299–317 (2010)

[18] Caluwaerts, K., Staffa, M., N'Guyen, S., Grand, C., Dollé, L., Favre-Félix, A., Girard, B., Khamassi, M.: A biologically inspired meta-control navigation system for the psikharpax rat robot. Bioinspiration and Biomimetics (to appear, 2012)

[19] Stein, B.E., Meredith, M.A.: The merging of the senses. The MIT Press, Cambridge (1993)

[20] Sutton, R., Barto, A.: Reinforcement Learning: An Introduction. MIT Press (1998)

[21] Gat, E.: On three-layer architectures. In: Kortenkamp, D., Bonnasso, R.P., Murphy, R. (eds.) Artificial Intelligence and Mobile Robots: Case Studies of Successful Robot Systems, pp. 195–210. AAAI Press (1998)

[22] Kortenkamp, D., Simmons, R.: Robotic systems architectures and programming. In: Siciliano, B., Khatib, O. (eds.) Handbook of Robotics, pp. 187–206. Springer (2008)

[23] Minguez, J., Lamiraux, F., Laumond, J.: Motion planning and obstacle avoidance. In: Siciliano, B., Khatib, O. (eds.) Handbook of Robotics, pp. 827–852. Springer (2008)

[24] Keramati, M., Dezfouli, A., Piray, P.: Speed/accuracy trade-off between the habitual and goal-directed processes. PLoS Computational Biology 7(5), 1–25 (2011)

Bioinspired Tunable Lens Driven by Electroactive Polymer Artificial Muscles

Federico Carpi[1,2,*], Gabriele Frediani[1], and Danilo De Rossi[1,2]

[1] University of Pisa, Interdepartmental Research Centre 'E. Piaggio', School of Engineering,
56100 Pisa, Italy
f.carpi@centropiaggio.unipi.it, frediani.gabriele@gmail.com
[2] Technology & Life Institute, 56122 Pisa, Italy
d.derossi@centropiaggio.unipi.it

Abstract. Electrical control of optical focalisation is important in several fields, such as consumer electronics, medical diagnostics and optical communications. As an alternative to complex, bulky and expensive current solutions based on shifting constant-focus lenses, here we report on an electrically tunable lens made of dielectric elastomers as 'artificial muscle' materials. The device is inspired to the architecture of the crystalline lens and ciliary muscle of the human eye. A fluid-filled elastomeric lens is integrated with an annular elastomeric actuator that works as an artificial muscle. Electrical activation of the artificial muscle deforms the lens, with a relative variation of focal length comparable to that of the human lens. Optical performance is achieved with compact size, low weight, fast and silent operation, shock tolerance, no overheating, low power consumption, and inexpensive off-the-shelf materials. Results show that combing bio-inspired design with dielectric elastomer artificial muscles can open new perspectives on tunable optics.

Keywords: Actuator, artificial, bioinspired, biomimetic, electroactive, lens, muscle, optical, polymer, tunable.

1 Introduction

Electrical control of optical focalisation is an important feature for a number of devices and systems in several fields, such as consumer electronics, medical diagnostics and optical communications. In those and other fields, focal length tuning is usually obtained by displacing one or more constant-focus lenses [1,2]. Owing to the need for moving parts, miniaturization of such systems is typically complex, and results in bulky, ineffective and expensive structures.

Aimed at overcoming such drawbacks and addressing the need for radically new solutions, we recently reported on a new technology for electrically tunable lenses inspired to the architecture of the crystalline lens and ciliary muscle of the human eye [3]. Here, we describe salient features of that technology and discuss future developments towards a specific application.

T.J. Prescott et al. (Eds.): Living Machines 2012, LNAI 7375, pp. 74–82, 2012.
© Springer-Verlag Berlin Heidelberg 2012

The human lens is a biconvex structure connected to a ciliary muscle via the zonule ligament, as sketched in Fig. 1. The ligament fibres are stretched or relaxed by the ciliary muscle. The combined action of the muscle and zonule changes the radius of curvature of the lens and thus its focus.

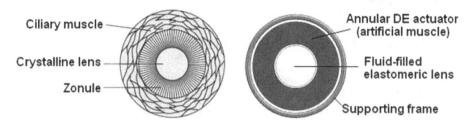

Fig. 1. Schematic representations of the human crystalline lens and the bioinspired lens. The former is surrounded by the annular ciliary muscle and connected to it via the annular zonule. The latter consists of a fluid-filled lens surrounded by an annular artificial muscle actuator.

This natural architecture was a source of inspiration to conceive a variable-focus device as a fluid-filled elastomeric lens integrated with and surrounded by an artificial ciliary muscle. The enabling artificial muscle technology is described below.

2   Enabling Artificial Muscle Technology

The artificial muscle technology that was used in this work is one of the most recent and most promising for polymer-based electromechanical transduction. It is known as dielectric elastomer (DE) actuation [4-7]. Within the family of electroactive polymers (EAP) as actuators [8,9], DEs are quickly emerging as a top choice of 'smart' materials for new kinds of soft actuators capable of high strains, high energy density, high efficiency, fast response, noise- and heat-free operation, high resilience and light weight [4-7]. These properties provide DE actuators with a widely recognized potential as 'artificial muscle' transducers [4-7].

   DE actuators take advantage of mechanical and electrical properties exhibited by soft insulating elastomers. Thin layers of such materials coated with compliant electrodes can be electrostatically deformed upon electrical charging, as a result of a Maxwell stress [7]. For a planar layer with thickness d and relative dielectric constant ε_r, an applied voltage V generates the following compressive stress along the thickness direction [7]:

$$p = \varepsilon_0 \varepsilon_r (V/d)^2 = \varepsilon_0 \varepsilon_r E^2 \qquad (1)$$

where ε_0 is the dielectric permittivity of vacuum and E is the electric field. For an elastic material with Young's modulus Y, the stress p gives the following thickness strain:

$$S = -p/Y = -\varepsilon_0\varepsilon_r(V/d)^2/Y = -\varepsilon_0\varepsilon_r E^2/Y \tag{2}$$

This field of science and technology at present is experiencing a fast growth owing to a widely recognized potential [5]. In this work, DE actuation was used to design and implement a bioinspired electrically tunable optical lens, as described below.

3   Bioinspired Design and Implementation

Inspired to the architecture of the crystalline lens and ciliary muscle of the human eye, we conceived a variable-focus device as a fluid-filled elastomeric lens integrated with and surrounded by an annular DE actuator working as an artificial muscle (Fig. 1). The structure is as follows.

The lens and the annular actuator form a single body, consisting of two membranes of an optically transparent dielectric elastomer that are bi-axially pre-stretched, coupled together and fixed to a circular plastic frame. To obtain a symmetrical biconvex lens, a transparent fluid is confined in a central region of the membranes, creating a closed chamber. The remaining part of the membranes, corresponding to an annular planar region, is coated with a compliant-electrode material so as to obtain an annular DE actuator.

Upon electrical activation or deactivation, the actuator radially relaxes or stretches the lens so as to change its radius of curvature and thus its focal length, mimicking exactly the same functional operation of the natural system (Fig. 1). Fig. 2 shows the exact correspondence between the human crystalline lens and the bioinspired device, and describes the functional analogy between them. The working principle of the bioinspired lens can be detailed as follows. Upon electrical charging of the compliant electrodes, the annular region of the DE membranes tends to be squeezed in thickness and expanded in surface, at a constant volume (Poisson's ratio equal to 0.5). The membrane surface expansion in the active state is due to a release of radial tension, and is made possible by pre-stretching in the passive state. The membrane deformation induces a reduction of the lens diameter that corresponds to an increase of the lens thickness, since the fluid preserves its volume constant. Therefore, both the radius of curvature and the focal length of the lens are reduced (Fig. 2a,a'). This working principle is functionally analogous to the accommodation process that takes place in the human eye. In fact, in the rest state, the ciliary muscle is relaxed and the zonule fibers are stretched, so that the crystalline lens is 'flattened'. To focus an object at a closer distance, the ciliary muscle contracts, releasing tension of the zonule fibers and causing a decrease of both the radius of curvature and the diameter of the crystalline lens. This leads to a reduction of the focal length (Fig. 2b,b'). Therefore, from a purely functional standpoint, the annular DE actuator behaves as an analogue of the combined ciliary muscle and zonule, while the fluid-filled elastomeric lens is an analogue of the crystalline lens.

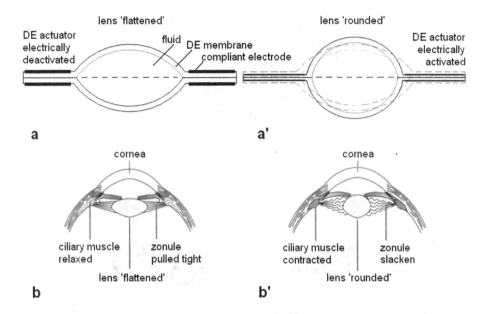

Fig. 2. Functional analogy between the bioinspired lens and the human lens. (a,b) Schematic sectional views (not in scale) of the two systems in the rest state. (a',b') Corresponding views in an activation state. The annular DE actuator works as an artificial muscle, functionally analogous to the combined ciliary muscle and zonule (adapted from [3]. Copyright Wiley-VCH Verlag GmbH & Co. KGaA. Reproduced with permission).

A prototype implementation of the lens was obtained by using as elastomeric membranes an acrylic-polymer-based film (VHB™ 4905, 3M, USA). The lens-filling fluid was a silicone pre-polymer (Sylgard 184, Down Corning, USA), while the electrode material consisted of a carbon grease (Carbon Conductive Grease 846, M.G. Chemicals, Canada). Details about materials and construction are reported in [3]. The following section presents results of the prototype device.

4 Results

Fig. 3 shows a picture of the prototype bioinspired device. The central lens had a rest diameter of 7.6 mm, a rest thickness of 0.75 mm and a refractive index of 1.43. The focalisation ability of the tunable device is presented in Fig. 4.

Fig. 5 reports the experimentally assessed dependence of the lens diameter D and focal length f on electrical driving. With respect to the human lens, the bioinspired lens showed comparable performance in terms of relative variation of focal length: − 29.1% versus −26.4%, respectively [3].

Fig. 3. Picture of a prototype bioinspired device. The lens is surrounded by an annular artificial muscle actuator.

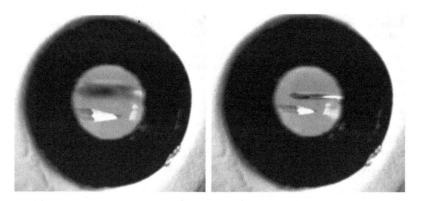

Fig. 4. Bioinspired tunable lens in action. Electrically controlled focalisation of a pencil and a syringe needle, located 10 and 3 cm far from the lens (adapted from [3]. Copyright Wiley-VCH Verlag GmbH & Co. KGaA. Reproduced with permission).

This comparable performance of the bioinspired lens (not optimized) with respect to the human lens is an interesting achievement. It shows the suitability of the described approach to develop a new type of tunable optical devices.

Also, the possibility of achieving high performance with a limited overall encumbrance (Fig. 3) is an attractive feature, especially for those potential applications wherein miniaturization is strategic and wherein other technologies requiring bulky (although powerful) driving units, such as hydraulic, pneumatic and electromagnetic drives, might introduce limitations. Furthermore, as compared to those technologies, a fully elastomeric lens is prone to exhibit a higher tolerance to mechanical shocks.

In terms of response speed, the lens showed a maximum frequency of operation of 1-10 Hz. This is a relatively fast response that, in comparison with alternative technologies, such as thermally, electrothermally, pH- or electrochemically driven lenses, as well as some hydraulic or pneumatic based mechanisms, is an advantage for applications wherein fast tuning ability is important (e.g. focusing of targets in rapid motion). Even faster tuning could be achieved with less viscous elastomers, such as silicones or polyurethanes [4-6]. Extensive discussions about material choice are reported in [3].

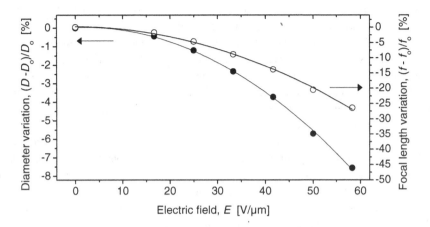

Fig. 5. Performance of the bioinspired tunable lens in terms of electrically induced variations of its diameter (•) and focal length (o). Quadratic fitting curves are inserted as a guide to the eye.

It is worth stressing that, in general, the low driving currents needed to charge DE devices, that are capacitive loads, allow for low-power driving. The prototype lens described in this work absorbed about 40 μA as a peak current during charging at 4 kV, corresponding to a peak power of 160 mW. Notably, electric power is mostly consumed during charging, i.e. during focal changes, while in static conditions the lens does not consume energy (losses across the elastomer are negligible). Low power consumption has two evident advantages: it allows for energy saving, and it favours miniaturization of the driving circuitry. Additionally, low power consumption is associated to low heating (and high efficiency), as compared to the typical behaviour of other technologies, such as thermal, electrothermal and electromagnetic drives. While power consumption increases at high rates, as for any other device, a lower exposure to overheating is advantageous for applications. All of these features make DE-based systems suitable even for portable applications.

5   Future Developments

Future developments of this work should include, first of all, optimizations. For a given actuation performance, i.e. a given radial strain of the actuator, thinner lenses are expected to exhibit a higher relative variation of focal length [3]. Therefore, in order to widen the focalisation range, thinner lenses driven at higher strains are needed.

On a longer time scale, we are working in the perspective of applying this technology to a specific application of interest: new systems for artificial vision. Within such a framework, increasing the accommodation ability would be helpful. To this end, tunable crystalline-like lenses might be combined with tunable corneal-like lenses. In fact, while in the human eye the corneal lens is static, in some animal eyes (e.g. eagle eye) it is tunable. We are currently developing a technology similar to that described above, in order to obtain effective tunable corneal-like lenses. Combinations of bioinspired crystalline and corneal lenses are particularly attractive to mimic some of the unique features of the huge variety of eyes available in the animal world, with the aim of designing improved artificial vision systems. While the opportunity for new optical systems with unprecedented properties is evident, this field remains completely unexplored so far, owing to the lack of enabling technologies. DE transducers might represent a solution for such a need.

Furthermore, bioinspired tunable lenses might effectively be combined with smart photo-detectors to develop innovative systems for artificial vision. In a recent paper by Jung et al. [10], tunable lenses were combined with conformable arrays of photo-electric transducers arranged on surfaces with tunable curvature. The conformability of the detector surface and the control of its curvature (Fig. 6-left) were shown to allow for significant improvements in image quality, overcoming typical limitations introduced by flat and static surfaces [10]. The important implications of this result motivate future investigations. However, in [10] tuning was achieved by means of hydraulic driving, which is not a viable approach for practical usage, especially for miniaturized and portable applications. Therefore, we aim to combine our DE-based tunable lens technology with a tunable photo-detecting technology still driven by DE actuation.

A combination of the technologies described above is expected to allow for a radically new generation of performing artificial vision systems, with a number of possible advantages:

1) Unique pre-retinal hardware processing, enabled by the bioinspired design of corneal-like and crystalline-like tunable lenses and artificial muscles (as opposed to rigid lenses, linearly displaced by motors): versatile control of the field of view and accommodation;

2) Unique structural and driving properties, enabled by the use of smart functional materials (as opposed to the use of motors and mechanical transmissions): compact size, simple structure, low weight, fast and silent operation, shock tolerance, low power consumption, no overheating;

3) Unique retinal hardware processing, enabled by the use of hemispherical and conformable photo-detectors (as opposed to flat and static detectors): high image quality, thanks to optimal image detection and reconstruction.

We plan to demonstrate this, by developing an advanced electronic/robotic eye, as sketched in Fig. 6-right.

We hope that such an attention for hardware processing of the optical information before and during its transduction into electrical information might effectively enhance the well consolidated efforts from many groups on post-detection software processing, so as to overcome the inherent limitations of approaches purely based on image elaboration.

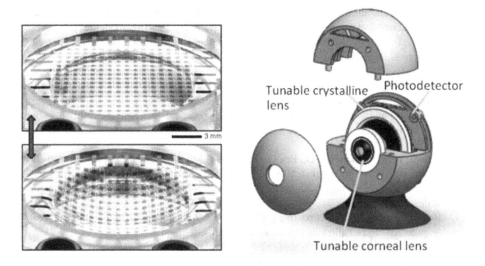

Fig. 6. Left: stretchable photo-detecting array manufactured on a membrane with tunable curvature (adapted from [10]). Right: exploded schematic view of a designed electronic/robotic eye that should integrate tunable crystalline-like and corneal-like lenses, as well as a tunable photo-detecting array.

6   Conclusion

The bioinspired design and technology were shown to be suitable to achieve useful optical tuning in devices that have compact size, low weight, fast and silent operation, shock tolerance, no overheating and low power consumption, and that can be assembled with inexpensive off-the-shelf materials.

Results clearly show the potential of combing dielectric elastomers as 'artificial muscle' materials with bioinspired design, in order to enable a new generation of electrically tunable optical lenses. Much of this potential is completely unexplored and unexploited, leaving opportunities for expected significant developments in the near future.

Acknowledgements. This work was supported in part by the European Commission, within the project "CEEDS: The Collective Experience of Empathic Data Systems" (FP7-ICT-2009.8.4, Grant 258749), and "ESNAM: European Scientific Network for Artificial Muscles" (COST Action MP1003), and in part by 'Fondazione Cassa di Risparmio di Pisa', within the project "POLOPTEL" (Grant 167/09).

References

[1] Friese, C., Werber, A., Krogmann, F., Mönch, M., Zappe, H.: Materials, effects and components for tunable micro-optics. IEEJ Trans. Electr. Electron. Eng. 2, 232–248 (2007)

[2] Eldada, L.: Optical communication components. Rev. Sci. Instrum. 75, 575–593 (2004)

[3] Carpi, F., Frediani, G., Turco, S., De Rossi, D.: Bioinspired tunable lens with muscle-like electroactive elastomers. Advanced Functional Materials 21, 4152–4158 (2011)

[4] Brochu, P., Pei, Q.: Advances in dielectric elastomers for actuators and artificial muscles. Macromol. Rapid Comm. 31, 10–36 (2010)

[5] Carpi, F., Bauer, S., De Rossi, D.: Stretching dielectric elastomer performance. Science 330, 1759–1761 (2010)

[6] Carpi, F., De Rossi, D., Kornbluh, R., Pelrine, R., Sommer-Larsen, P. (eds.): Dielectric Elastomers as Electromechanical Transducers: Fundamentals, materials, devices, models and applications of an emerging electroactive polymer technology. Elsevier, Oxford (2008)

[7] Pelrine, R.E., Kornbluh, R.D., Pei, Q., Joseph, J.P.: High-speed electrically actuated elastomers with strain greater than 100%. Science 287, 836–839 (2000)

[8] Carpi, F., Smela, E. (eds.): Biomedical Applications of Electroactive Polymer Actuators. Wiley, Chichester (2009)

[9] Bar-Cohen, Y. (ed.): Electroactive polymer (EAP) Actuators as artificial muscles. Reality, potential, and Challenges, 2nd edn. SPIE Press, Bellingham (2004)

[10] Jung, I., et al.: Dynamically tunable hemispherical electronic eye camera system with adjustable zoom capability. P. Natl. Acad. Sci. USA 108, 1788–1793 (2011)

A Pilot Study on Saccadic Adaptation Experiments with Robots

Eris Chinellato[1], Marco Antontelli[2], and Angel P. del Pobil[2]

[1] Imperial College London, South Kensington Campus, London SW7 2AZ, UK
`e.chinellato@imperial.ac.uk`
[2] Jaume I University, Campus Riu Sec, 12071, Castellón de la Plana, Spain
`{antonell,pobil}@icc.uji.es`

Abstract. Despite the increasing mutual interest, robotics and cognitive sciences are still lacking common research grounds and comparison methodologies, for a more efficient use of modern technologies in aid of neuroscience research. We employed our humanoid robot for reproducing experiments on saccadic adaptation, on the same experimental setup used for human studies. The behavior of the robot, endowed with advanced sensorimotor skills and high autonomy in its interaction with the surrounding environment, is based on a model of cortical sensorimotor functions. We show how the comparison of robot experimental results with human and computational modeling data allows researchers to validate and assess alternative models of psychophysical phenomena.

1 Introduction

Artificial intelligence has been from its very foundation a meeting-place for scientists of seemingly unrelated disciplines. The impact of interdisciplinary research involving high technology fields and life sciences is now growing faster and faster. Two disciplines which met thanks to artificial intelligence are robotics and neuroscience, and their encounter is producing mutually beneficial developments. For example, bio-mimetic robotics constitutes an important way for technological improvement, while representing a scientific advancement toward the understanding of how biological systems work. The research of common grounds on which to pursue an effective cross-fertilization is not new [9,10,14], and concrete proposals have been put forth in order to facilitate communication and interchange between roboticists and neuroscientists [3,5,17]. Still, fundamental differences in research goals, methodologies and language prevent a more proficuous collaboration between the fields.

With this work we provide a new contribution to the above goals, by training a humanoid robot to perform simulated psychophysical experiments. The behavior of the robot is based on a model designed with the purpose of achieving visuomotor awareness of the environment by using eye and arm movements [4]. The implementation of the model on the humanoid robot provides it with the capability of performing concurrent or decoupled gazing and reaching movements toward visual or memorized targets placed in its peripersonal space [1]. In this

T.J. Prescott et al. (Eds.): Living Machines 2012, LNAI 7375, pp. 83–94, 2012.

paper we check the robot skills on an experimental setup resembling those used in human psychophysical studies. On the one hand, these tests allow to validate the underlying properties of the computational model on which the robot behavioral abilities are built upon. On the other hand, we wish to verify if our robotic system can represent a research tool able to emulate human experiments, and thus constitute a potential aid in the design and analysis of actual experimental protocols. Simulating psychophysical experiments on the robot can be useful to check in advance the appropriateness of experimental protocols, reducing the expensive and complicated preliminary tests with human subjects [8].

We present here a benchmark test for our proposal, which also constitutes the first theoretical contribution of our system towards the study of eye gazing mechanisms in humans. To achieve such a goal we present our robot with a cognitive science experimental setup similar to those used for saccadic adaptation experiments in humans [7,16]. The comparison of human experimental data with a computational simulation and with the robot tests provide insights on theoretical aspects related to visuomotor cognitive aspects, and contextually allows us to validate our model.

In the next Section 2 we describe how saccadic adaptation experiments are normally executed with human subjects and their typical outcome, and how we built our robotic system in order to be able to replicate such experiments with a humanoid robot. Section 3 describes the experimental setup and the outcome of both simulated and actual robot experiments, which are compared and discussed in Section 4.

2   Saccadic Adaptation in Humans and Robots

2.1   Saccadic Adaptation in Humans

The phenomenon of saccadic adaptation can be observed when the visual feedback provided to a subject after an eye movement is inconsistent with the location of the target before the movement (see Fig. 1(a), which will be described with more detail in Section 3). In these conditions, the subject gradually learns to perform a saccadic movement that allows her to fixate the expected final position of the stimulus, even though this is displaced with respect to its initial position [12]. Saccadic adaptation experiments can be either inward, when the target is displaced toward the starting point, thus causing a reduction of the saccadic movement amplitude, or outward, when the target is displaced further away from the starting point, thus causing larger saccadic movements. Saccadic adaptation is not limited to the point of adaptation, but transfers to the nearby visual space. Amplitude of post-adaptation saccades on stimuli around the adapted target varies according to the distance and the relative position with respect to the target point, a phenomenon called adaptation transfer.

The reference graphs for human experiments are reproduced in Fig. 2, above for the inward protocol [7], below for the outward protocol [16]. For both cases three fundamental aspects are analyzed, the same that we will explore in our experiments. Fig. 2(a) and 2(d) show the adaptation trend, i.e. the time course of

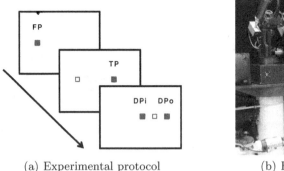

(a) Experimental protocol

(b) Humanoid robot

Fig. 1. Experimental protocol (filled squares represent currently visual stimuli, empty squares are disappeared stimuli) and humanoid robot with pan/tilt/vergence head

the gradual shift of the subject response from the initial movement amplitude to the displaced target, for inward and outward adaptation respectively. Saccadic adaptation fields, displayed in Fig. 2(b) and 2(e), are a local representation of adaptation transfer, assessing how the mis-trained movement affects saccades directed towards different targets in space. Black dots represent movement average endpoint before adaptation, while the end of the segments represent the average endpoint after adaptation. Finally, adaptation transfer accumulated along the fundamental axes is visualized in Fig. 2(c) and 2(f), where the horizontal (x) and vertical (y) components of the displacements of the adaptation field are represented separately as a function of the x coordinate.

2.2 The Model and the Humanoid Robot Implementation

Our humanoid robot system was designed with the goal of achieving advanced capabilities in the interaction of an autonomous system with its nearby environment. To this purpose, we conceived and implemented a sensorimotor framework composed of three Radial Basis Function Networks (see Fig. 3, details can be found in [4]): one for converting the visual position of a stimulus into an oculomotor position (left transformation block in the diagram), and the other two for hand-eye movement coordination (right transformation block). Networks of suitable basis functions are able to naturally reproduce the gain-field effects often observed in parietal neurons and are particularly suitable for maintaining sensorimotor associations [15]. We have designed our computational schema and the neural networks which compose it directly from insights drawn from the analysis of monkey single-cell data on reaching and gazing experiments registered from posterior parietal area V6A [2,6].

Similarly to the way primates explore their environment, in our framework the robot incrementally builds a sensorimotor representation of the peripersonal space, through subsequent, increasingly complex interactions composed by sequences of saccades and reaching movements [1,4]. As a first step, the system

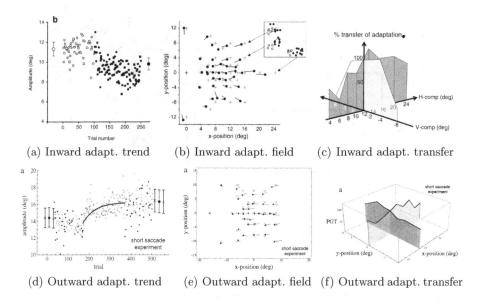

(a) Inward adapt. trend   (b) Inward adapt. field   (c) Inward adapt. transfer

(d) Outward adapt. trend   (e) Outward adapt. field   (f) Outward adapt. transfer

Fig. 2. Human saccadic adaptation results for inward (above, from [7]) and outward adaptation experiments (below, from [16])

learns to associate retinal information and gaze direction (i.e. proprioceptive eye position), using successive foveation movements to salient points of the visual space. This allows to create a mapping, performed by the first neural network, between visual information and oculomotor (vergence - version) coordinates. Gaze direction is then associated to arm position, by moving the arm randomly and following it with the gaze, so that each motor configuration of the arm joints is associated to a corresponding configuration of the eye motor control, and vice versa. This process allows the robot to learn the bidirectional link between different sensorimotor systems, so that it can look where its hand is but also reach the point in space it is looking at. Hence, the representation of the peripersonal space is maintained contextually by both limb sensorimotor signals on the one hand and by visual and oculomotor signals on the other hand.

The described learning framework endows the robot with the ability of exploring and building an egocentric representation of its surrounding environment, through the execution of coupled or decoupled gazing and arm reaching actions, on either visible or memorized targets. Whilst the above skills have been described in previous works [1,4], we introduce in this paper a novel scenario, in which the robot visual environment is constituted by a computer screen placed right in front of it, on which small geometrical shapes are visualized. This scenario corresponds to the typical setup of psychophysical experiments on humans and other primates, and we present it here to the robot in order to engage it in saccadic adaptation experiments similar to those typically performed on human subjects.

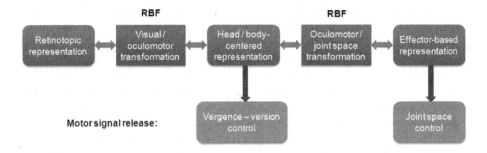

Fig. 3. Computational model for building a visuomotor awareness of the environment through gazing and reaching movements

3   Experimental Evaluation

The typical psychophysics experimental setup on which the skills of our humanoid robot are tested consists of a computer screen placed within reaching distance, visualizing different visual stimuli associated to action signals. It is worthwhile to clarify that, although vergence varies very little in this setup, all transformations are fully tridimensional, and the robot keeps acting as in the usual 3D configuration. On the other hand, as these experiments require no arm movements, only the visual to oculomotor neural network is involved in this study. Details on the experimental setup are provided below.

3.1   Experimental Setup

Our robot is a humanoid torso endowed with a pan-tilt-vergence stereo head and two multi-joint arms (Fig. 1(b)). The head mounts two cameras separated by a 270mm baseline, and having a resolution of 1024x768 pixels that can acquire color images at 30 Hz. After the RBFNs underlying the robot behavior have been trained, saccadic adaptation experiments are performed according to the protocol described below (see Fig. 1(a)). A computer monitor (1440x900, 19") is put in front of the robot at a distance of 720mm, which allows to obtain version angles similar to human experiments without getting too close to image periphery. The experiment program is designed to display at required positions small red squares (5x5 pixels), unambiguously identified by the robot module for blob detection.

The task begins with the robot looking straight ahead at a fixation point stimulus (FP) corresponding to null version angles (frame 1 of Fig. 1(a)). A second visual stimulus is then showed at the target position (TP), having the same vertical but different horizontal coordinate, increased by a certain amount Δx, parameter of the experiment, while FP disappears (frame 2 of Fig. 1(a)). The robot is required to perform a saccade toward this new stimulus. When the saccade movement signal is released and the robot starts moving, the stimulus is displaced toward a third point (DP), either closer (DPi) or further (DPo) on the x axis with respect to TP (for *inward* and *outward* saccadic adaptation protocols, respectively, frame 3 of Fig. 1(a)). At the end of the movement, the robot perceives a visual error between its final position, corresponding to TP if the saccade is correctly executed, and the visible target DP. Such visual difference is used to adapt the weights of the network performing the transformation from retinal to oculomotor coordinates. The starting stimulus is then shown again and the robot saccades back toward it. The whole sequence is repeated 100 times.

Before performing the saccadic adaptation tests with the real robot, we simulated them using the robot model on a corresponding virtual setup. This simulation is useful for predicting the sort of results that experiments with the real robot are expected to provide. On the one hand, this is done to avoid keeping the robot busy with the execution of irrelevant experiments. On the other hand, it allows to assess the impact that real world tests have on the purely theoretical insights provided by the simulation.

3.2   Simulated Experiments

We tested each of the two experimental protocols, inward and outward adaptation, with two different configurations of the visual to oculomotor radial basis function network. The *uniform* configuration has the centers of the basis functions distributed evenly on the input space (x and y cyclopean coordinates and horizontal disparity). In the *logarithmic* distribution neurons are placed closer to each other at the center of the visual field and for small disparities. While for the uniform distribution all neurons have the same spread, in the logarithmic case radii vary according to the distance of a neuron from its neighbors. For what concerns the parameters of the experimental setup, the target was fixed for all experiments at 11.89° on the right of the starting point, while the displaced point was set at 7.96° for inward displacements and at 15.73° for outward displacements. These values were set considering the robotic setup, in order to generate eye version movements comparable to those of human tests. The results we obtained with our simulation are shown in Fig. 4. Adaptation trend, adaptation field and adaptation transfer (columns) are depicted for: uniform inward, uniform outward, logarithmic inward and logarithmic outward tests (rows).

Adaptation trend graphs (Fig. 4(a), 4(d), 4(g) and 4(j)) show a plausible learning curve which reduces (in the inward case) or increases (in the outward case) movement amplitude according to the deceiving feedback provided by the displaced target stimulus. In the uniform distribution tests, movement amplitude gets at trial 100 to 8.79° in the inward protocol and to 14.92° in the outward protocol, from the initial 11.89°, for a final adaptation of 79.5% and 77% respectively. The average adaptation over all trials is of 2.1° in both cases, about 54% of the target step. Slightly higher values (faster adaptation) have been obtained in the logarithmic case. In general, average adaptation values for humans are smaller than what we found in our simulation. For the inward case, only a 13% adaptation was observed [7], whilst 33%-45% adaptations were registered for outward experiments [16], depending on the initial saccade amplitude.

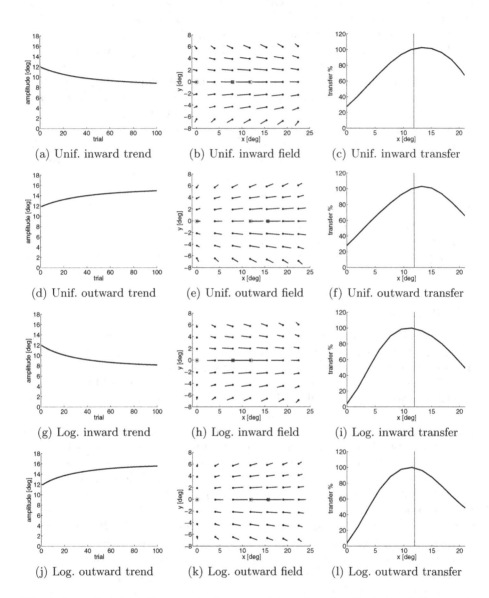

Fig. 4. Saccadic adaptation results for computational model, for both Uniform and Logarithmic distribution of RBFs

Adaptation fields (Fig. 4(b), 4(e), 4(h) and 4(k)) have in all cases a perceptible radial trend, with a y component indicating wider movements toward the top or the bottom of the screen for both protocols and net configurations. This is rather consistent with Schnier and colleagues outward tests (see Fig. 2(e)), but much less apparent for what concerns inward experiments (Fig. 2(b)).

The overall trend of the adaptation over the horizontal (x) component can be observed in the adaptation transfer graphs of Fig. 4(c), 4(f), 4(i) and 4(l). Human experiments suggest that, both for input and outward adaptation (see Fig. 2(c) and Fig. 2(f)), the difference between pre- and post-adaptation movements peaks just after the abscissa of the target. Also, while transfer decreases with the distance from the peak, such decrease is slower for larger saccades than for shorter ones (gentler slopes on the right side of the peak). It seems that the uniform configuration captures the first of this phenomena, showing a peak in transfer for movement amplitudes slightly after the target. Still, the transfer is symmetrical with respect to the peak. The opposite occurs for the logarithmic distribution, which transfer peak appears slightly before the target abscissa. The transfer trend is though asymmetrical, showing a less pronounced decrease on the right of the peak. As observed in both inward and outward studies on humans, the adaptation vertical (y) component had a very small error rather homogeneous for different movement amplitudes, with no clear trend worth visualization.

3.3   Robot Experiments

The same two configurations of the visual to oculomotor RBFN employed in the simulation were used also in the real robot experiments. The adopted configurations were chosen through an exhaustive search of the center locations and spreads providing the highest precision in approximating the goal function.

The network weights found on the model, and used in the simulated saccadic adaptation experiments described above, were transferred to the real robot. A short training phase with on-screen visual stimuli was then executed in order to adapt the network to possible distortions and unavoidable differences between the model and the real robot setup. This was performed by randomly showing a point on the screen, which the robot had to saccade to. The possible residual error of the movement was employed to train the network. As in the simulation, four saccadic adaptation experiments were conducted with the robot, characterized by the basic structure of the visual to oculomotor network (uniform or logarithmic distribution of the centers) and by the direction of the displacement (inward or outward). Target and displaced point were the same as above: initial target 11.89°, inward displaced point 7.96°, outward displaced point 15.73°.

All results are depicted in Fig. 5, matching the correspondent graphs of Fig. 4, obtained in the simulation for the same conditions. The adaptation trend is shown in Fig. 5(a), 5(d), 5(g) and 5(j) for the four different tests. The only noticeable difference with the simulation is a short initial sequence of trials in which the system seems to be resilient to learning the new movement amplitude, but after that the trend is very similar to what observed for the model and in human experiments. The final and average movement amplitudes are 66.5%

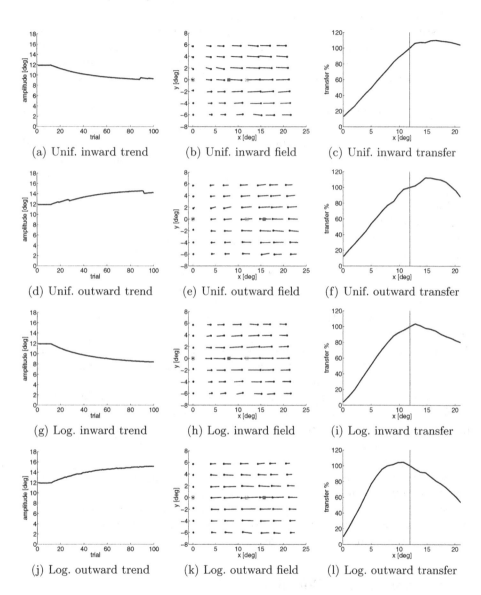

Fig. 5. Saccadic adaptation results for robot experiments, for both Uniform and Logarithmic distribution of RBFs

and 43.0% for inward and 60.5% and 40.7% for outward adaptation, respectively. These smaller values suggest that, employing the same learning rate, the robot achieves a better approximation of the human data with respect to the simulation. Again, the logarithmic network provides higher adaptation values.

To study adaptation transfer, an adaptation field was created by defining a 20x25 lattice on the screen. All points on the lattice were shown one at time, and the robot was required to perform a saccade toward each stimulus starting from FP. At the end of the movement, the visual position of the stimulus and the oculomotor angles were compared. This process was performed before and after saccadic adaptation, but could be executed at any one of the 100 steps of the experiment, in order to monitor the progress of adaptation transfer. It is important to clarify that, during this evaluation task, learning is suspended and the network is frozen in its current state. This solution allows to monitor precisely the evolution of the saccadic adaptation learning process, and constitutes thus an advantage with respect to human experiments, where such freezing is clearly not possible. Adaptation fields are shown in Fig. 5(b), 5(e), 5(h) and 5(k), in which, for clarity reasons, only a subset of the lattice points have been visualized. The radial effect, when present, is very light and not consistent across different positions, showing a pattern more similar to the human data than to the simulation results, which present a stronger radial effect.

Interesting insights can be drawn by observing the horizontal (x) component of the movement change (Fig. 5(c), 5(f), 5(i) and 5(l)). A late peak can be observed in both experiments for the uniform configuration (more pronounced than in the simulation), and for the logarithmic distribution in the inward adaptation test. Moreover, practically all cases show an asymmetry of the transfer, with curves descending more slowly for larger saccades, as it happens in the human case, whilst simulation curves are clearly symmetrical. This effect is again stronger for the logarithmic network configuration. Such slightly more plausible results achieved by the logarithmic networks is consistent with the non homogeneous distribution of the neural receptive fields in the retina and primary visual cortex. Once more, no relevant effects were observed for what concerns the vertical adaptation transfer component.

As a general consideration, it can be observed that the robot results approximate the human data better than the simulated results. The reduced radial aspect of the adaptation field and the trend of the horizontal component in peak position and slope asymmetry are more consistent between human and robot than the correspondent simulated results.

4   Discussion and Conclusion

Different properties observed in human saccadic adaptation studies were captured by our tests. Both the simulation and the robot experiments showed plausible adaptation trends, slightly radial adaptation fields and typical features of the adaptation transfer on the horizontal component, such as asymmetry and late peak. Reminding that our model is based on a direct mapping between visual stimuli and oculomotor signals, the above results support the view that saccadic adaptation is a multifold phenomenon related not only to motor aspects, but also to more complex cortical sensorimotor processes [11]. On this regard, the role of posterior parietal areas, and especially of area V6A on which our model was based upon, seems to be of particular interest and worth of further exploration.

From a computational point of view, it is worth highlighting that real robot experiments provided a better approximation of human data with respect to the model. This is especially interesting considering that exactly the same parameters were employed in the two cases. This phenomenon might reflect implicit properties of the embodiment that affect the way untrained movements are biased by learning processes applied to similar movements. Unmodeled aspects, such as the noise in the identification of stimulus location, might also contribute to this effect, and we are conducting further studies in order to clarify this issue. In any case, the phenomenon represents a rationale supporting robot emulation of psychophysical experiments as preferable to computational simulations.

Abstracting from the study case of saccadic adaptation analyzed in this work, having humanoid robots performing psychophysical experiments represents an important common tool for robotics and cognitive sciences. We can use robot experiments for comparing, and validate, possible explanations of experimental results deriving from different models. Moreover, the rigorous model design required by actual robotic implementation is likely to represent an ideal environment on which to devise more detailed and plausible models, and can be an inspiration source for planning completely new experiments.

There is an additional important advantage that we plan to achieve through our system, which is, the possibility of performing "impossible experiments". In fact, many neuroscientific theories derive from observations done on neurally-impaired people. Such experiments cannot be reproduced, neither it is possible to decide beforehand the type of brain damage on which to investigate, but they can be tested on a robot system on which some modules have been damaged or important parameters altered. In this regard, we are planning to assess the effect of saccadic adaptation on arm movements, to verify how a visuomotor phenomenon transfers to other motor domains. We can do this by artificially altering the parameters of the model on which the oculomotor to arm-motor transformation is performed, and verifying if and how this affect reaching movements in the saccadic adaptation paradigm (see e.g. [13]).

Acknowledgments. This work was supported by the National Research Foundation of Korea (WCU program) funded by the Ministry of Education, Science and Technology (Grant R31-2008-000-10062-0), by the Spanish Ministerio de Ciencia e Innovación (FPI Grant BES-2009-027151 and DPI2011-27846), by Generalitat Valenciana (PRO-METEO/2009/052) and by Fundació Caixa Castelló-Bancaixa (P1-1B2011-54).

The authors would like to thank Markus Lappe for many helpful discussions and advice regarding the experiments and earlier drafts of this paper.

References

1. Antonelli, M., Chinellato, E., del Pobil, A.P.: Implicit mapping of the peripersonal space of a humanoid robot. In: IEEE Symposium Series on Computational Intelligence, SSCI (2011)
2. Bosco, A., Breveglieri, R., Chinellato, E., Galletti, C., Fattori, P.: Reaching activity in the medial posterior parietal cortex of monkeys is modulated by visual feedback. J. Neurosci 30(44), 14773–14785 (2010)
3. Cheng, G., Hyon, S.-H., Morimoto, J., Ude, A., Colvin, G., Scroggin, W., Jacobsen, S.C.: Cb: A humanoid research platform for exploring neuroscience. In: Proc. 6th IEEE-RAS Int. Humanoid Robots Conf., pp. 182–187 (2006)
4. Chinellato, E., Antonelli, M., Grzyb, B.J., del Pobil, A.P.: Implicit sensorimotor mapping of the peripersonal space by gazing and reaching. IEEE Transactions on Autonomous Mental Development 3(1), 43–53 (2011)
5. Chinellato, E., del Pobil, A.P.: fRI, functional robotic imaging: Visualizing a robot brain. In: IEEE International Conference on Distributed Human-Machine Systems, DHMS, Athens, Greece (March 2008)
6. Chinellato, E., Grzyb, B.J., Marzocchi, N., Bosco, A., Fattori, P., del Pobil, A.P.: The dorso-medial visual stream: from neural activation to sensorimotor interaction. Neurocomputing 74(8), 1203–1212 (2010)
7. Collins, T., Dore-Mazars, K., Lappe, M.: Motor space structures perceptual space: Evidence from human saccadic adaptation. Brain Res. 1172, 32–39 (2007)
8. Culham, J.C.: Functional neuroimaging: Experimental design and analysis. In: Cabeza, R., Kingstone, A. (eds.) Handbook of Functional Neuroimaging of Cognition, pp. 53–82. MIT Press, Cambridge (2006)
9. Dario, P., Carrozza, M.C., Guglielmelli, E., Laschi, C., Menciassi, A., Micera, S., Vecchi, F.: Robotics as a future and emerging technology: biomimetics, cybernetics, and neuro-robotics in European projects. IEEE Robotics & Automation Magazine 12(2), 29–45 (2005)
10. Kawato, M.: Robotics as a tool for neuroscience - cerebellar internal models for robotics and cognition. In: Proceedings of the 9th International Symposium on Robotics Research, pp. 321–328 (2000)
11. Lappe, M.: What is adapted in saccadic adaptation? J. Physiol. 587(pt. 1), 5 (2009)
12. McLaughlin, S.: Parametric adjustment in saccadic eye movements. Percept. Psychophys. 2, 359–362 (1967)
13. Nanayakkara, T., Shadmehr, R.: Saccade adaptation in response to altered arm dynamics. Journal of Neurophysiology 90(6), 4016–4021 (2003)
14. Prescott, T.J., Redgrave, P., Gurney, K.: Layered control architectures in robots and vertebrates. Adaptive Behavior 7(1), 99–127 (1999)
15. Salinas, E., Thier, P.: Gain modulation: a major computational principle of the central nervous system. Neuron 27(1), 15–21 (2000)
16. Schnier, F., Zimmermann, E., Lappe, M.: Adaptation and mislocalization fields for saccadic outward adaptation in humans. Journal of Eye Movement Research 3(3), 1–18 (2010)
17. Tsagarakis, N.G., Metta, G., Sandini, G., Vernon, D., Beira, R., Becchi, F., Righetti, L., Santos-Victor, J., Ijspeert, A.J., Carrozza, M.C., Caldwell, D.G.: icub: the design and realization of an open humanoid platform for cognitive and neuroscience research. Advanced Robotics 21(10), 1151–1175 (2007)

Jumping Robot with a Tunable Suspension Based on Artificial Muscles

Sanjay Dastoor, Sam Weiss, Hannah Stuart, and Mark Cutkosky

Stanford University, Stanford CA 94305, USA
sanjayd@stanford.edu
http://bdml.stanford.edu

Abstract. This paper describes the design and control of a suspension based on electroactive polymers for controlling the landing dynamics of a jumping robot. Tunable suspension elements can electrically change their stiffness up to a factor of 10 in less than 0.01 seconds. We discuss design parameters and performance relevant to bio-inspired systems and demonstrate the ability to operate in positive (actuator), neutral (spring-like), or negative (damping or braking) workloops. When applied to a single-legged robot, positive workloops allow sustained periodic hopping while negative workloops can be used to rapidly achieve equilibrium during a landing event, acting in a similar manner to muscle in jumping animals. Extended bio-inspired applications are discussed.

Keywords: bio-inspired robotics, variable stiffness.

1 Introduction

The ability of animals to successfully and efficiently perform motion tasks, from posture control to grasping to locomotion, is thought to rely greatly on the inherent and controllable compliance in their musculoskeletal systems. For example, studies on animal and human running suggest that correctly tuned leg stiffness, dependent on mass, number of legs, and substrate stiffness, minimizes specific cost of transport and may be used to control running speed[1]. Several studies have also examined the ability of a tuned compliant system to reject external disturbances and recover quickly without neural feedback, termed "preflexes" [2][3][4]. In addition, running robots designed with carefully chosen leg stiffness demonstrated robust and efficient running [5].

Transient locomotion events, such as perching or landing from a jump, also benefit from this ability, where the right compliance profile can limit ground reaction forces while staying within joint compression limits [6][7]. Robots that perch require carefully tuned suspension systems to successfully engage claws or spines while dissipating excess kinetic energy [8]. Finally, non-locomotory tasks, such as posture control and arm movement, have also been shown to implement controlled muscle and joint stiffness[9].

T.J. Prescott et al. (Eds.): Living Machines 2012, LNAI 7375, pp. 95–106, 2012.

1.1 Related Work

Tuning the performance of a robot or vehicle suspension is usually a matter of physically swapping or reconfiguring elements with fixed properties. The ability to change stiffness and damping properties *in situ* would permit these platforms to operate efficiently over a wide range of tasks and environments. With this goal in mind, several researchers have attempted to create biomimetic platforms with tunable stiffness elements. A review of variable stiffness mechanisms and actuators is provided in [10].

Among the earliest hopping robots to exploit variable stiffness, Raibert's planar biped used an air spring to vary leg stiffness [11]. The controller could change pressure, increasing both spring preload and stiffness, thereby increasing stride frequency, reducing stance duration, and increasing running speed. Pneumatics were also used in a bipedal robot, where pneumatic McKibben muscles were arranged in antagonistic pairs and co-contracted to change joint stiffness [12].

Hurst demonstrated a tunable stiffness leg with his AMASC design for a bipedal robot [13]. This antagonistic system uses nonlinear leaf springs that are preloaded and actuated using two motors, allowing independent control of joint position and stiffness. The leg showed an increase in efficiency and the ability to tune the running gait. Other systems have also been designed to change the arrangement of nonlinear spring elements [10].

Changing effective spring length [14] was implemented on the Edubot robot, based on the RHex platform. This structure-controlled system allowed the robot to adapt its system dynamics to different gaits, changes in payload, and changes in terrain, or to optimize for efficiency and speed.

The foregoing designs have limitations that affect their suitability for small bio-inspired systems. First, many of them are powered by electromagnetic actuators designed for prime mover tasks. These actuators require transmissions and mechanisms to effect changes in configuration, adding weight and mechanical complexity. There is also a tradeoff among actuator size, transmission ratio, and the speed at which changes can accomplished, with the fastest systems changing stiffness on the order of 0.1 seconds [15]. The pneumatic solutions require a source of compressed gas, and their response time is limited by actuator volume, gas pressure and valve selection. As a result, most previous designs are limited to changing stiffness properties from stride to stride or by applying feedforward control strategies, but have low bandwidths for closed-loop control. In addition, their weight and complexity are obstacles to their use on small, bio-inspired platforms.

2   Tunable Stiffness Modules

2.1 Electroactive Polymers

Electroactive polymer actuators [16], also called dielectric elastomers, EPAMs or EAP actuators, have been investigated extensively as actuators for robotics [17].

For recent comprehensive surveys, see [18][19]. Although they are relatively efficient and have can have a high power/weight ratio, common difficulties include modest numbers of cycles before failure, either due to mechanical tearing or dielectric breakdown. In addition, they are inherently compliant and viscoelastic, which may be a disadvantage when trying to use them as an ideal "force source." However, when used in a suspension, the passive compliance and damping can be an advantage.

The tunable suspension modules used here are based on a diaphragm configuration, as shown in Figure 1. Operation is based on an effective Maxwell pressure applied to a soft dielectric film via compliant electrodes, where the pressure is given by [16]:

$$p = \epsilon_0 \epsilon_r E^2 = \frac{\epsilon_0 \epsilon_r V^2}{t^2} \tag{1}$$

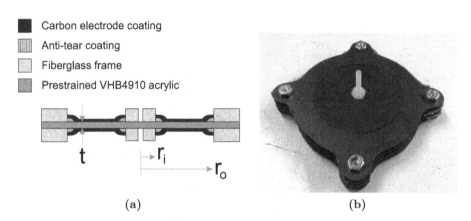

Fig. 1. Variable stiffness diaphragm: (a) cross section of single diaphragm (b) photo of 4-diaphragm stack; each (D1) unit has prestrain=400%, t=62.5μm, $r_i = 10mm$, and $r_o = 45mm$

To extend lifetime, a silicone rubber anti-tear coating is applied at the edges of the electrode area. This prevents mechanical tearing due to stress concentrations at the film-frame interface and prevents shorting due to dielectric breakdown [20]. The diaphragms are manufactured by masking and spray coating a prestrained acrylic film with successive layers of anti-tear coatings and compliant electrodes and then gluing on a rigid fiberglass frame. Further details on the design and manufacturing process are provided in [21].

2.2 Variable Stiffness Performance

The diaphragm geometry used here has been discussed in previous work, usually where two are biased in an antagonistic configuration and displacement is proportional to the difference in activation between units [22]. The implementation here uses a single unbiased diaphragm so that activation voltage induces a stiffness change instead of displacement. This change occurs because the effective Maxwell pressure induces a stress that counteracts the internal membrane stresses due to the initial manufacturing prestrain. The maximum stiffness is achieved at zero voltage, while the minimum is limited by buckling of the film. Figure 2 shows the relationships among force, displacement, and voltage, showing different effective slopes for different activation voltages. For this particular dielectric material and prestrain, 6kV is approximately the maximum voltage possible before dielectric breakdown occurs [23]. Note that while there is some nonlinearity due to material properties and due to the axisymmetric geometry [21], the stiffness at each voltage is approximately constant for displacements below 2.5 mm.

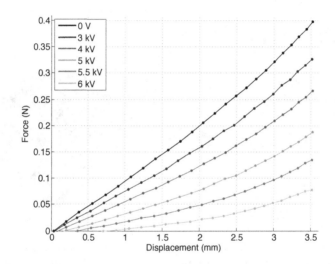

Fig. 2. Relationship between voltage and stiffness profiles. Data are from a single diaphragm (D2: prestrain=400%, t=62.5μm, $r_i = 5mm$, and $r_o = 25mm$), somewhat smaller than the unit shown in Figure 1.

2.3   Workloops and Energy Dissipation

The workloop technique pioneered by Josephson [24] can be used to show the ability of muscle to act as a motor, brake, or spring, depending on the relationship between forced cycling and activation phase [25]. To investigate the possibility of using a tunable spring in a similar manner, we subjected diaphragms to workloop experiments using a muscle lever (Cambridge 305B, Aurora Scientific, Canada). A feedforward voltage activation using a Trek 610B power supply (Trek, Inc., USA) was used to induce stiffness changes and the phase between activation and forced cyclic oscillation was varied.

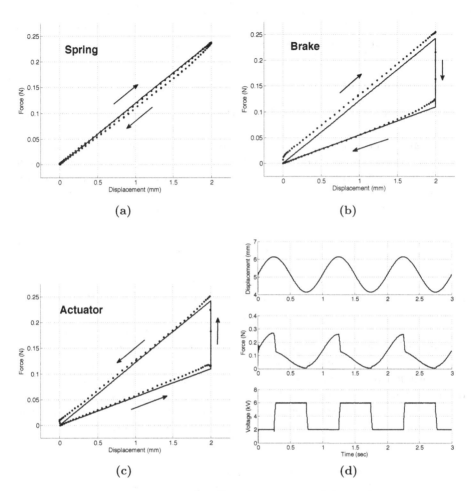

Fig. 3. Ideal linear spring model (solid) and experimental (dotted) workloops from diaphragm (D1). (a) Passive workloop shows slight energy loss due to damping (b) Negative workloop shows energy removed (c) Positive workloop shows energy added (d) Forced displacement, measured force, and forced voltage are shown for a negative workloop.

As seen in Figure 3, changing stiffness during forced cyclic oscillation can change the device from a passive spring to either a brake or a motor. Positive area inside the workloop indicates energy added to the system, while negative area indicates energy removed. The figure also shows the forced sinusoidal displacement, measured force, and applied voltage for the negative workloop. The differences between the simulated and measured cycles arise from approximating the unit as a pure linear spring without damping in this example.

It is important to note that the ability to change stiffness quickly and repeatably, relative to the frequency of the dynamic system, is critical to this multi-functionality. Also, while these data were obtained using a feedforward voltage signal, a feedback control scheme could allow more complex behaviors and the addition or subtraction of energy at different points during the cycle.

2.4 Response Time

Since the stiffness change occurs due to the Maxwell pressure from the applied electric field, it does not require the material to strain or mechanically change configuration. The time constant for the stiffness change is dependent only on the ability of the power supply to supply enough charge to produce the required electric field. Since the stiffness-voltage relationship is quadratic [21], we can determine the stiffness as a function of time and the maximum current capability of the power supply. Voltage on a capacitor is given by

$$V = \frac{Q}{C} = \frac{1}{C} \int_0^t i(t)dt = \frac{d}{\epsilon A} \int_0^t i(t)dt \tag{2}$$

The bench-top power supply used for the workloop experiments, a Trek 610B, can source 2mA. A more portable solution, the AH60 (Emco High Voltage, USA), weighs 8.4g, is 33.8 x 11.4 x 6.4 mm, and can supply 250 μA. Another portable solution uses this small converter to charge a buffer capacitor with greater capacitance than the diaphragm and a high voltage reed relay, which can handle 2-3 A of current and has a 2-3 ms operation time. The converter can continuously source 250 μA to the buffer while the buffer relay periodically opens to charge the diaphragm. Table 1 shows the stiffness change in diaphragm D2, which has a capacitance of 1.24 nF, as a function of time for these different supply ideas, assuming negligible ramp-up time for maximum current output.

Table 1. Response Time ($k_{max} = 100N/m$ to $k_{min} = 10N/m$)

Power Supply	Max Current (mA)	Time (ms)
Trek 610B	2	3.9
Emco AH60	0.25	31.1
Buffer Cap and Relay	2000	3

2.5 Power Consumption

The power consumption of the diaphragm is solely based on the electrical energy sourced or sunk by the power supply. When the diaphragm switches from stiff to soft, the power supply adds charge, and vice versa. Figure 4 shows the power supply output for a step input in commanded stiffness from k_{min} to k_{max}. The power consumption peaks at 11.6 W and the total energy consumption is approximately 14.3 mJ. This consumption scales with the number of diaphragms and the capacitance of each diaphragm.

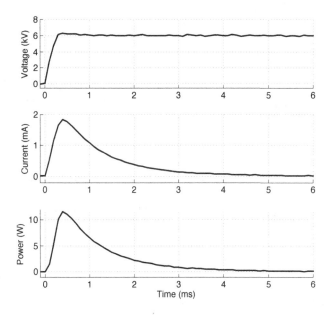

Fig. 4. Voltage and current step response of diaphragm D2, measured at the power supply, and calculated power output as diaphragm stiffness changes from 100 N/m to 10 N/m

3   Hopping Robot

3.1   Physical Design

A small 1 DOF hopper was created by attaching a D1 diaphragm to a long, partially counter-balanced boom (Figure 5). The resulting system has an equivalent mass of 27 g and weight of 87.3 mN. When in contact with the ground, the hopper has a resonant frequency of 10.6 Hz if no voltage is applied. Displacement is sensed using a laser micrometer (Omron Z4W, Omron, USA), velocity is sensed using a three-axis gyroscope (Sparkfun, USA), and force using a six-axis force-torque sensor (ATI, USA).

3.2   Control Law

To illustrate the effect of changing stiffness dynamically, a simple discontinuous control strategy was implemented, using the velocity signal from the gyro sensor to change the stiffness at each velocity reversal:

$$k = \begin{cases} k_1 & \text{if } \dot{x} < 0 \\ k_2 & \text{if } \dot{x} > 0 \end{cases} \tag{3}$$

If $k_1 = k_{max}$ and $k_2 = k_{min}$, this control achieves the negative workloop of Figure 3b; if $k_1 = k_{min}$ and $k_2 = k_{max}$ it achieves the positive workloop of

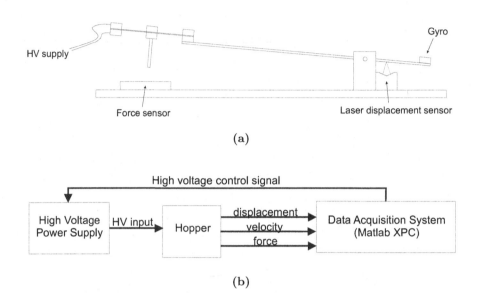

(a)

(b)

Fig. 5. Diagram of hopper and block diagram of control system

Figure 3c. The stiffness range for the D1 diaphragm is $k_{max} = 121N/m$ at 0 Volts and $k_{min} = 55N/m$ at 6kV.

3.3   Results

Figure 6 shows the displacement of the hopper during different experiments. In Figure 6a, the passive landing response is compared to an actively damped landing response when the negative workloop controller is applied. Figure 6b shows the positive workloop controller causing the hopper to increase its kinetic energy from rest until it achieves flight and enters a sustained hopping trajectory. The stance period during landings was approximately 50-60ms.

4   Extended Applications

The suspension design could be implemented on a variety of robotic, haptic, prosthetic, and rehabilitation devices. A platform of particular interest is a small UAV that relies on a tuned suspension system to achieve stable perching on vertical walls [8]. The suspension system's stiffness and damping are chosen to maximize the range of initial conditions that result in a successful landing, and are heavily dependent on variables such as mass, initial velocity, landing substrate stiffness, and the presence of external disturbances such as wind. Several constraints exist that cause landing failure, including spine attachment failure due to incorrect loading vectors, spine detachment due to force overload or tail rebound, or propeller damage due to the wrong pitch angle at contact. To determine the effects of a variable stiffness suspension on the perching maneuver, the simulation described in [26] was run with different values for suspension stiffness.

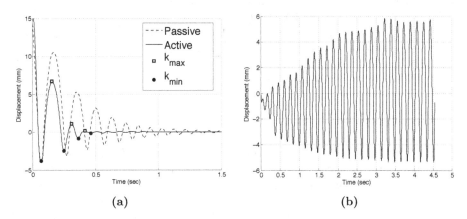

Fig. 6. Experimental data showing hopper vertical displacement. (a) Passive (dashed) and actively damped (solid) landings of the hopper (b) Positive workloop controller causing hopper to resonate into aerial phase and sustained hopping.

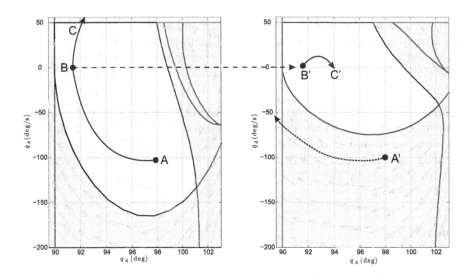

Fig. 7. Simulated region of attraction for a perching UAV (mass=0.3kg, I = 0.0164kgm^2, b = .0344Nms/rad) with two different suspension stiffnesses (left = 1.41 Nm/rad, right = 0.235 Nm/rad). Boundary curves show constraints on landing; trajectories crossing shaded areas eventually lead to failure. Initial conditions at A or A', without changes in stiffness, lead to failure. High initial stiffness (A to B) and a switch from high to low stiffness (B to B') result in a successful landing (C').

Figure 7 shows regions of attraction for different suspension stiffnesses, where the shaded areas represent different failure modes. Initial conditions at point A or A' would result in an unsuccessful landing, but a stiffness change, represented by a shift in plots from B to B', allows a successful landing. This scenario uses a discontinuous control signal to the suspension via a feedback loop, but other strategies are possible using feedforward techniques.

The total system weight for the UAV platform is approximately 300g, and the combined weight of a 10-element stack of D2 modules (with combined stiffness ranging from 100 to 1000 N/m) and compact high voltage power supply is 25g. This tunable suspension system, even with the high voltage requirement, is compact enough to be used on small mobile platforms.

5   Conclusions

We have demonstrated the application of a tunable suspension system to a one-legged hopping robot in a manner resembling the multifunctional nature of biological muscle. Simple dynamic models can be used to derive control strategies and substantially change the behavior of the passive system. Several key properties, such as the low power consumption, light weight, fast response time, and scalable design, make this suspension relevant to many other mobile bio-inspired platforms.

Future work will be split between applications on cyclic locomotion and on transient events, such as landing or perching. This will include both feedforward controllers, such as transient input shaping [27] and feedback strategies. We will also address manufacturing and assembly of multi-layer stacks, and further examine control circuits that fulfill high-voltage power and switching requirements while staying within volume and mass constraints.

Acknowledgements. The authors thank David Christensen and Alexis Lussier Desbiens for valuable advice during discussions. They also thank Tom Libby and Robert Full at UC Berkeley for generous use of test equipment, and Roy Kornbluh and Harsha Prahlad at SRI for fabrication advice. This work was supported in part by the ARL RCTA, W911NF-10-2-0016, and by a Stanford Bio-X Fellowship.

References

1. McMahon, T.A.: The role of compliance in mammalian running gaits. Journal of Experimental Biology 115, 263–282 (1985)
2. Brown, I., Loeb, G.: A reductionist approach to creating and using neuromusculoskeletal models. Biomechanics and Neural Control of Posture and Movement (2000)
3. Blickhan, R., Seyfarth, A., Geyer, H., Grimmer, S., Wagner, H., Gnther, M.: Intelligence by mechanics. Philosophical Transactions of the Royal Society A: Mathematical, Physical and Engineering Sciences 365(2007), 199–220 (1850)

4. Full, R.J., Koditschek, D.E.: Templates and anchors: neuromechanical hypotheses of legged locomotion on land. Journal of Experimental Biology 202(23), 3325–3332 (1999)
5. Kim, S., Clark, J.E., Cutkosky, M.R.: iSprawl: Design and tuning for high-speed autonomous open-loop running. The International Journal of Robotics Research 25(9), 903–912 (2006)
6. Dyhre-Poulsen, P., Simonsen, E., Voigt, M.: Dynamic control of muscle stiffness and h reflex modulation during hopping and jumping in man. Journal of Physiology, 287–304 (1991)
7. Farley, C.T., Houdijk, H.H.P., Van Strien, C., Louie, M.: Mechanism of leg stiffness adjustment for hopping on surfaces of different stiffnesses. Journal of Applied Physiology 85(3), 1044–1055 (1998)
8. Lussier Desbiens, A., Asbeck, A., Cutkosky, M.R.: Landing, perching and taking off from vertical surfaces. The International Journal of Robotics Research 30(3), 355–370 (2011)
9. Bizzi, E., Accornero, N., Chapple, W., Hogan, N.: Posture control and trajectory formation during arm movement. Journal of Neuroscience 4, 2738–2744 (1984)
10. Van Ham, R., Sugar, T., Vanderborght, B., Hollander, K., Lefeber, D.: Compliant actuator designs. IEEE Robotics and Automation Magazine, 81–94 (September 2009)
11. Raibert, M.H., Brown, H.B., Chepponis, M.: Experiments in balance with a 3D one-legged hopping machine. The International Journal of Robotics Research 3(2), 75–92 (1984)
12. Vanderborght, B., Verrelst, B., Van Ham, R., Van Damme, M., Lefeber, D., Duran, B.M.Y., Beyl, P.: Exploiting natural dynamics to reduce energy consumption by controlling the compliance of soft actuators. International Journal of Robotics Research 25(4), 343–358 (2006)
13. Hurst, J.W., Chestnutt, J.E., Rizzi, A.: Design and philosophy of the BiMASC, a highly dynamic biped. In: Proc of Intl. Conf. on Robotics and Automation (IEEE-ICRA), pp. 1863–1868 (April 2007)
14. Galloway, K., Clark, J., Koditschek, D.: Design of a tunable stiffness composite leg for dynamic locomotion. In: Proc. of Intl. Design Engineering Technical Conf, ASME-IDETC/CIE (2009)
15. Tsagarakis, N.G., Sardellitti, I., Caldwell, D.: A new variable stiffness actuator (compact-vsa): Design and modeling. In: Proc. of Intl. Conf. on Intelligent Robots and Systems (2011)
16. Pelrine, R., Kornbluh, R., Pei, Q., Joseph, J.: High-speed electrically actuated elastomers with strain greater than 100%. Science 287(5454), 836–839 (2000)
17. Cianchetti, M., Mattoli, V., Mazzolai, B., Laschi, C., Dario, P.: A new design methodology of electrostrictive actuators for bio-inspired robotics. Sensors and Actuators B: Chemical 142, 288–297 (2009)
18. Brochu, P., Pei, Q.: Advances in dielectric elastomers for actuators and articial muscles. Macromolecular Rapid Communications 31 (2009)
19. Carpi, F., De Rossi, D., Kornbluh, R., Pelrine, R., Sommer-Larsen, P. (eds.): Dielectric Elastomers as Electromechanical Transducers. Elsevier (2008)
20. Pelrine, R., Kornbluh, R., Prahlad, H., Sharma, S., Chavez, B., Czyzyk, D., Wong-Foy, A., Stanford, S.: Tear resistant electroactive polymer transducers. U.S. Patent (7,804,227) (2010)
21. Dastoor, S., Cutkosky, M.R.: Design of dielectric electroactive polymers for a compact and scalable variable stiffness device. In: Proc. of Intl. Conf. on Robotics and Automation, IEEE-ICRA (2012)

22. Kornbluh, R.: Fundamental configurations for dielectric elastomer actuators. In: Carpi, F., De Rossi, D., Kornbluh, R., Pelrine, R., Sommer-Larsen, P. (eds.) Dielectric Elastomers as Electromechanical Transducers. Elsevier (2008)
23. Kofod, G., Sommer-Larsen, P., Kornbluh, R., Pelrine, R.: Actuation response of polyacrylate dielectric elastomers. Journal of Intelligent Material Systems and Structures 13, 787–793 (2003)
24. Josephson, R.K.: Mechanical power output from striated muscle during cyclic contraction. Journal of Experimental Biology 114, 493–512 (1985)
25. Dickinson, M.H., Farley, C.T., Full, R.J., Koehl, M.A.R., Kram, R., Lehman, S.: How animals move: An integrative view. Science 288(5463), 100–106 (2000)
26. Glassman, E.L., Desbiens, A.L., Tobenkin, M., Cutkosky, M., Tedrake, R.: Region of attraction estimation for a perching aircraft: A lyapunov method exploiting barrier certificates. In: Proc. of Intl. Conf. on Robotics and Automation, IEEE-ICRA (2012)
27. Hyde, J.M., Cutkosky, M.R.: Controlling contact transition. IEEE Control Systems 14, 25–30 (1994)

Static versus Adaptive Gain Control Strategy for Visuo-motor Stabilization

Naveed Ejaz, Reiko J. Tanaka, and Holger G. Krapp

Department of Bioengineering, Imperial College London, London SW72AZ, UK
nejaz@imperial.ac.uk

Abstract. Biological principles of closed-loop motor control have gained much interest over the last years for their potential applications in robotic system. Although some progress has been made in understanding of how biological systems use sensory signals to control reflex and voluntary behaviour, experimental platforms are still missing which allow us to study sensorimotor integration under closed-loop conditions. We developed a fly-robot interface (FRI) to investigate the dynamics of a 1-DoF image stabilization task. Neural signals recorded from an identified visual interneuron were used to control a two-wheeled robot which compensated for wide-field visual image shifts caused by externally induced rotations. We compared the frequency responses of two different controllers with static and adaptive feedback gains and their performance and found that they offer competing benefits for visual stabilization. In future research will use the FRI to study how different sensor systems contribute towards robust closed-loop motor control.

Keywords: fly-robot interface, closed-loop, visuo-motor control, adaptive gain control, frequency response.

1 Introduction

The blowfly *Calliphora* belongs to the most sophisticated fliers amongst the orders of insects. During semi-free flight where its flight dynamics are limited, *Calliphora* performs angular rotations up to 2000 deg/s and reaches top speeds up to 1.2 m/s and horizontal acceleration of around 2g [1]. To maintain flight stability during such rapid manoeuvres, the fly must constantly acquire sensory information about changes of its flight trajectory relative to its environment. In addition to sensory signals obtained from other modalities, flies rely heavily on vision for flight stability and control. A variety of visuo-motor tasks including chasing, escape, and equilibrium reflexes, are achieved by a relatively simple nervous system and are amenable to quantitative behavioural and electrophysiology studies. This makes flies an attractive model system to investigate mechanisms underlying visuo-motor control.

The optomotor response, for instance, is a visual stabilization reflex that has been extensively studied in flies [2] and other arthropods (review: [3]). Walking and flying flies placed inside a rotating cylinder, the walls of which are lined

T.J. Prescott et al. (Eds.): Living Machines 2012, LNAI 7375, pp. 107–119, 2012.

with a visual pattern, turn in the direction of pattern motion to reduce retinal slip-speed and stabilize the visual input. Key neuronal elements supporting this behaviour are identified to be visual interneurons in the lobula plate which analyse optic flow and estimate the animal's self-motion components ([4,5], review: [6]). The signals of theses cells are then used for motor control in order to compensate for image shifts as a result of external perturbations.

Typically, optomotor responses in flies have been measured at the behavioural level using a flight simulator with 1-DoF [7,8]. In such studies, a tethered fly was placed in front of a visual display that generates a vertically oriented grating pattern. By coupling the yaw torque of the fly to the horizontal motion of the pattern, the performance of the optomotor responses in stabilizing visual motion could be investigated. However, Heisenberg & Wolf (1993), distinguished physical torque from neural torque in the control of the optomotor response. The latter refers to the neuronal signal sent to the flight motor while physical torque gives the angular momentum the fly motor generates. An injury to one or both of the wings might require the fly to increase/decrease its neural torque in order to maintain its desired angular momentum. While the physical torque generated during the optomotor response has been extensively studied, the control strategies by which the neural torque is used are poorly understood. The only study that took into account the neural activity involved in the optomotor response used visual stimuli with constant image velocities and focused on the analysis of neuronal steady-state responses [9]. Such steady-state analysis, however, fails to capture the dynamic properties of the optomotor control system, for example the speed at which a visual perturbation is corrected for [10].

In this paper, we use a fly-robot interface (FRI) to investigate the dynamics of the optomotor response along 1 DoF. In our experimental setup, the neural responses of an identified visual interneuron were used to control the rotation of a mobile robot as it compensated for oscillating image motion inside a rotating cylinder. We investigated the frequency response of the FRI and compared the performance when using a proportional (static) and an adaptive feedback gain controller. Our results suggest that both controller types offer competing benefits for visual stabilization. Specifically, the bandwidth of the optomotor frequency response was higher for the adaptive gain feedback controller, whereas the corresponding pass-band system gain was slightly higher for some parameters for the static gain feedback controller. We found that no one controller (static versus adaptive gain) works well for all stimulus conditions and that the best proposed strategy for the fly would be to adjust its control law and/or parameters depending on the dynamics of the visual input.

2   Methods

The experimental system used consisted of a fly-robot interface (FRI) as shown in Fig 1. A detailed description of FRI can be found in [11]. A fly was prepared for electrophysiology and placed into a 1-DoF visual flight simulator setup. A mobile robot was fixed on a turn-table placed inside a cylindrical arena whose walls were lined with high contrast black and white stripes. The neural activity of the H1 cell - an identified fly visual interneuron responding to horizontal image motion [12] - was recorded and the signals were used to control the angular velocity, ω_r, of the mobile robot. External visual perturbations were introduced into the system by rotating a turn-table the robot was placed on, at an angular velocity ω_p defined as a sine-wave with a DC-offset: $\omega_p = 72\left[\sin(2\pi f_i t) + 1\right]$, where the input frequency f_i ranged from $0.03 \leq f_i \leq 3.0$ Hz. The responses of the H1 cell induced as a result of the relative motion, $\omega_p - \omega_r$, between the turn-table and the robot were used to control the rotation speed of the robot to minimize visual motion under closed-loop conditions.

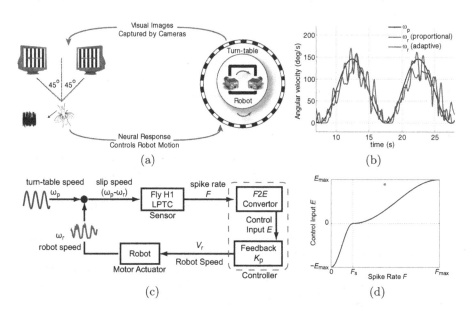

Fig. 1. Experimental setup. (a) A fly was placed in front of a visual display consisting of two high-speed CRT computer monitors. Input to the two monitors were provided by two high-speed video cameras mounted on a mobile robot. The robot was positioned on a turn-table placed inside a cylindrical arena with pattern. Robot and turn-table movements were limited to rotations around a fixed vertical axis. Any visual motion as a result of the rotation of the turn-table was captured by two cameras. Electrophysiology recordings were made from the H1 cell and the signals were used to control the rotation of the robot. (b) The sample responses of the robot with a proportional ($K_p = 0.1$) and an adaptive ($\Delta T_{ws} = 0.10$ ms) controller. (c) Block diagram of the closed-loop FRI. Relative motion between the turn-table and the robot, $\omega_p - \omega_r$, causes spiking in the H1 cell. The H1 responses (instantaneous spike rate F) were used by a controller to compensate for externally generated turn-table movements by driving the robot in the opposite direction. The $F2E$ convertor maps F onto the control input E based on the piece-wise sigmoid functions in (d). E was then used to update the robot speed V_r.

2.1 Fly Preparation and Electrophysiology

A 2-3 day old female blowfly *Calliphora* was cooled on ice. Its legs, wings, and proboscis were removed and the holes were sealed with beeswax. The head of the fly was aligned with the horizontal plane of the visual stimulation setup by using the deep pseudo-pupil technique [13]. Two small windows were cut into the cuticle of the left and right rear head capsules to gain access to the lobula plate. Tungsten electrodes (signal & ground, fh-co.com, Item code - UEW SHG SE 3P1M) were used to record the activity of the H1 cell from the left hemisphere extracellularly. Ringer solution was supplied to the brain to keep it from drying out. The H1 signals were amplified (NPI EXT 10-2F, gain - x 10k), band-pass filtered (300-2k Hz) and sampled using a data acquisition card (National Instruments USB 6215, 16-bit) at 10 kHz. An amplitude threshold was applied to discriminate H1 spikes from the background activity of other neurons. The H1 spikes were convolved with a causal half-gaussian kernel ($\sigma_{sp} = 0.05$ s) to obtain the instantaneous spike rate, F, which was then used by a closed-loop controller to modulate the speed of the robot. The kernel width, σ_{sp}, was chosen such that the time interval was long enough to reliably estimate the H1 spiking rate yet also short enough not to induce a significant time delay for closed-loop control.

2.2 Visual Stimulation of the H1 Cell

The prepared fly was placed in front of two high-speed cathode ray tube (CRT) computer monitors as shown in Fig 1(a). Input to the CRT monitors was provided by two high-speed video cameras (Prosilica, GC640, mono, 10-bit) at 200 fps. The cameras were mounted on a small, two-wheeled customized mobile robot (Arexx Engineering, ASURO Robot Kit) fixed on the turn-table.

In the absence of visual motion (pattern is stationary), the H1 cell responds with a spontaneous spiking rate F_s. The back-to-front visual motion (PD: preferred direction) with $\omega_p - \omega_r > 0$ increases the spiking rate F of the H1 cell, up to a maximum spiking rate F_{max}, which is achieved for a visual pattern with contrast frequency $\left(\frac{\text{pattern velocity}}{\text{pattern spatial wavelength}}\right)$ in the range of 2-4 Hz [14]. Front-to-back motion decreases spiking activity, F, in the H1 cell. Prior to the actual closed-loop experiments, both F_s (Mean ± SE : 19.67 ± 2.3) and F_{max} (Mean ± SE : 78 ± 4.27) were estimated in open-loop for each fly using 3 repeats of 5 second stimulation with no pattern motion and motion in the PD (contrast frequency - 3 Hz), respectively.

2.3 Closed-Loop Controllers

A feedback controller (shown in the dotted red box in Fig 1(c)) specifies how the H1 spike rate, F, is converted into a motor signal, V_r, that drives the wheels of the robot. We used two different controllers, a proportional (static) and adaptive feedback gain controller. For the proportional controller, a feedback gain was chosen prior to the start of the experiment and remained constant for the duration of the run. A detailed description of the proportional controller is found in [15]. In contrast, the adaptive controller used a variable feedback gain that depended on the recent activity of the H1 cell (cf. adaptive gain controller). The components and structure of the closed-loop FRI for both controllers remained the same.

The spiking rate, F, of the H1 cell was considered to be a measure of the visual slip-speed or compensation error, $\omega_p - \omega_r$, under closed loop conditions. F had to be mapped onto an 8-bit value E that represents a control input used to update the robot angular velocity. The 8-bit restriction on the value E was imposed by the 256 possible range of input values that could be used to control the two motors on the robot. A controller contained a signal converter, "F2E converter", to map F onto E and "Feedback" to update the V_r, which is also an 8-bit value to control the angular velocity of the robot: $\omega_r = V_r \cdot \frac{1.73}{s^2+1.87s+0.86}$. The transfer function for ω_r was identified by open-loop experiments with sinusoidal inputs V_r.

Proportional Controller. For the proportional controller, we assumed that F is mapped to E within the range described by $-E_{\max} \leq E \leq E_{\max}$ (Fig 1(d)). Accordingly, $F2E$ converter is defined by

$$
E \begin{cases}
= -\frac{E_{\max}}{2}\left\{\cos\left(2\pi\phi\right) + 1\right\}\left\{\frac{1}{4}(V_r(t) - \hat{V}_r)\right\}, \ \phi = -\frac{1}{2}\left(\frac{F}{F_s}\right) & \text{, for } F < F_s, \\
= 0 & \text{, for } F = F_s, \quad (1)\\
= \frac{E_{\max}}{2}\left\{\cos\left(2\pi\phi\right) + 1\right\}, \ \phi = -\frac{1}{2}\left(\frac{F-F_{\max}}{F_s-F_{\max}}\right), & \text{, for } F > F_s,
\end{cases}
$$

with $\hat{V}_r$ being a speed signal offset that was required to initiate robot rotation i.e. the minimum speed signal to overcome the robot's inertia. E was then used to update the robot speed, based on the fixed controller gain K_p by: $V_r(t+1) = E + K_p \cdot V_r(t)$. We used $K_p = [0.01, 0.1, 0.5, 1.0, 3.0]$ in our experiments. A sample response of the robot with a proportional controller for $K_p = 0.1$ is shown in Fig 1(b).

Adaptive Gain Controller. The adaptive gain controller used the same $F2E$ converter as described by Eq (1) and the updating rule: $V_r(t + 1) = E + K_p \cdot V_r(t)$ with $K_p = 1$. However, $F_{\max}$ was repeatedly estimated by: $F_{\max} = \max\{F_s, F(\tau)\}$, over the time window $t - \Delta T_{ws} \leq \tau \leq t$ every 50 ms (discrete control loop delay). We used $\Delta T_{ws} = [0.05, 0.10, 0.15, 0.50]$ ms in our experiments. A sample response of the robot with an adaptive gain controller ($\Delta T_{ws} = 0.10\,\text{ms}$) is shown in Fig 1(b). Continuously updating $F_{\max}$ enabled us to dynamically re-scale the sigmoid mapping of F to E in the preferred direction of the H1 cell $(F_s, F_{\max})$ (Fig 1(d)). This scaling method was adopted from an adaptive neural coding strategy proposed by Laughlin (1994).

3   Results

3.1   Bode Magnitude and Phase Plots

Fig 2 shows the bode magnitude and phase plots for both the adaptive and the proportional controllers. The plot for the adaptive gain controller (Fig 2(a))

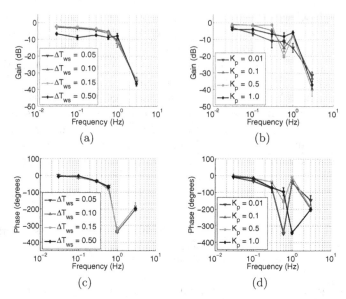

Fig. 2. Bode magnitude and phase plots for the adaptive and proportional controllers. The responses for both controllers were probed over the frequency range $0.03 \leq f_i \leq 3$ Hz. The Bode magnitude plot of the (a) adaptive and (b) proportional controllers both show low-pass filter characteristics, except for the resonator/anti-resonator response at 1.0 and 0.6 Hz (see text) for the proportional controller with $K_p = 0.1, 0.5$ & 1.0. The phase profiles of the (c) adaptive gain controller for all ΔT_{ws} are nearly identical with those of the (d) proportional controller for a higher gain $K_p = 1.0$.

shows a low-pass filter behaviour of the system for all values of ΔT_{ws} investigated, with a smoother roll-off for $\Delta T_{ws} = [0.05, 0.10, 0.15]$ ms compared to $\Delta T_{ws} = 0.50$ ms. The system gain (magnitude) for the adaptive gain controller with $\Delta T_{ws} = [0.05, 0.10, 0.15]$ ms are higher than those for $\Delta T_{ws} = 0.50$ ms across the entire frequency range except for 1 Hz. For the proportional controller, different values of K_p leads to an increase in the frequency dependant system gain (Fig 2(b)). Between the range $0.03 \leq f_i \leq 0.3$ Hz, the system gain is lowest for $K_p = 0.01$. Larger values of $K_p = 0.1$&0.5 within this range of f_i increase the system gain, while it decreases when K_p is very large (=1.0). For larger $K_p = [0.1, 0.5, 1.0]$, a resonator and anti-resonator response phenomenon is observed at 1.0 Hz and 0.6 Hz, respectively. A 1 Hz natural frequency of the closed-loop FRI has previously been identified [15]. Its impact on the system gain is considerably smaller for the adaptive gain controller than it is for the proportional controller over the range $0.6 \leq f_i \leq 1.0$.

Similarly for the phase, a smaller K_p for the proportional controller leads to an increased phase shift between turn-table motion and the compensatory response of the robot. In comparison, the adaptive gain controller for all values of ΔT_{ws} has a phase profile similar to that of the proportional controller with larger $K_p = 1.0$. This allows the adaptive gain controller to respond faster to

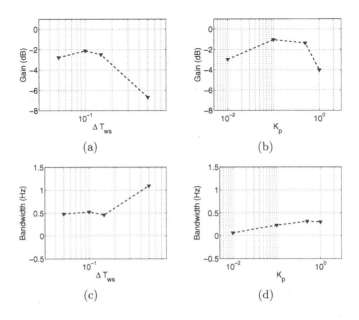

Fig. 3. Pass-band gain and bandwidth of the adaptive (a,c) and the proportional controllers (b,d). The pass-band gains (calculated as the system gain at $f_i = 0.03$ Hz) are shown for the (a) adaptive and (b) proportional controllers. The bandwidth for the (c) adaptive and (d) proportional controllers are calculated for the -3 dB cut-off frequency level.

any changes of the visual input which modulates the spiking rate of the H1 cell, while maintaining a high system gain over the entire range of input frequencies.

3.2   Motor Pass-Band Gain and Bandwidth

The system gains of both the adaptive and the proportional controller had low-pass filter characteristics. Low-pass filters can be characterised by a combination of the pass-band gain, G_{dc} dB, and the cut-off frequency, f_c Hz [17]. G_{dc} is the system gain in the flat pass-band region of the filter, while f_c is the frequency at which the system gain first drops below -3 dB relative to G_{dc}, or half the maximum power in the filter pass-band. In the case of the frequency response curves obtained for the adaptive and proportional controllers (Fig 2), the bandwidth of the system is equivalent to f_c.

Fig 3 shows G_{dc} and f_c for both the adaptive and proportional controllers. For all controller parameters, the adaptive gain controller had larger bandwidth than the proportional controller (Figs 3(c) & 3(d)). For the adaptive gain controller with $\Delta T_{ws} = [0.05, 0.10, 0.15]$ ms, the bandwidth (0.5 Hz) was the maximum acheivable given the limitations imposed by the robot dynamics while the pass-band gain was above the -3 dB absolute cut-off value (Fig 3(a)). For

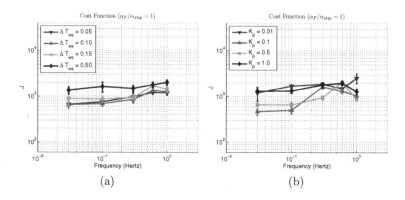

Fig. 4. Proposed performance indexes for the (a) adaptive and (b) proportional controllers

$\Delta T_{ws} = 0.50$ ms, the bandwidth ($\approx$ 1Hz) was artificially large due to the very low corresponding pass-band gain ($< \frac{1}{4}$ the peak power measured for any controller configuration tested). The bandwidth for the proportional controller for all K_p was less than 0.5 Hz (Fig 3(d)). However, for $K_p = [0.1, 0.5]$, the proportional controller had a higher pass-band gain than the adaptive gain controller for all ΔT_{ws} (Fig 3(b)). On the other hand, distortions in the frequency response due to the system's natural frequency were more pronounced for the proportional controller with $K_p = [0.1, 0.5]$. To summarize, characterization of pass-band gain and bandwidth based on the frequency response shows that the adaptive gain controller maximized the output bandwidth at the cost of system gain.

3.3   Performance Index

To compare the performance of the two controllers from an optimal control perspective, we introduce a performance index J_i, to assess the efficiency of the two controllers in stabilizing the visual motion observed by the fly. The performance of the system for the input frequency f_i is defined here by a performance index:

$$J_i = \frac{1}{T_i} \int_0^{T_i} (\omega_p^i - \omega_r^i)^2 + (F_i)^2 \, dt \tag{2}$$

Here, T_i is the total time of the closed-loop measurements, ω_p^i and ω_r^i are the angular velocities of the turn-table and robot and F_i is the spiking rate of the H1 cell, all measured at the input frequency f_i, respectively.

$\omega_p^i - \omega_r^i$ in the first term corresponds to the visual slip-speed ovserved by the fly under closed-loop conditions and indicates the compensation error. The larger it is, the worse the compensation by the controller. F_i in the second term indicates the energy spent by the H1 cell. An optimal control therefore should minimizes J_i over the desired operating range f_i, by achieving the smaller error $(\omega_p^i - \omega_r^i)$ with smaller input energy. J_i for both the adaptive and the proportional controller are shown in Fig 4. J_i at $f_i = 3$ Hz was disregarded since the system gains for the adaptive and the proportional controllers were very low.

For the adaptive gain controller, increasing the input frequency increased J_i for all ΔT_{ws}. The best performance of the adaptive gain controller was achieved with $\Delta T_{ws} = 0.10$ ms for $0.03 \leq f_i \leq 0.3$ and $\Delta T_{ws} = 0.05$ ms for $0.6 \leq f_i \leq 1.0$. The difference in the performance of the adaptive gain controller for different values of $\Delta T_{ws} = [0.05, 0.10, 0.15]$ ms was negligible compared to the changes in performance observed for the proportional controller with different K_p. Furthermore, the peak performance (minimization of J_i) of the adaptive gain controller compared favourably to that of the proportional controller over all input frequencies f_i and all combinations of the controller parameters ΔT_{ws} and K_p (except for $f_i = 0.03 - 0.1$ Hz for $K_p = 0.1$).

4   Discussion

In this paper, we used a FRI to characterize the sensorimotor integration for optomotor yaw control in the blowfly. While there have been many studies on neural coding of motion information in the blowfly (for review see [18,19]), the mechanisms that convert sensory signals into motor commands under closed-loop conditions, are still poorly understood (see [20]). This is primarily due to methodological limitations with obtaining neural recordings in freely behaving animals. Here, we characterized the frequency response dynamics of a visuo-motor control task in a FRI using different feedback controllers which transform neural signals into motor commands.

The use of a robotic controller to understand animal behaviour provides real-world physical interactions typically missing from modeling studies where a low-pass filter is used to model the dynamics of the fly flight motor system. As argued by Webb (2006), this lack of physical interaction would mean that complex motion dynamics such as slipping due to friction cannot be accounted for in the computer model. Indeed, recent work by Dickinson (2010) showed that both body-inertia and -damping play a significant role in the dynamics of saccadic turns in *Drosophila*.

The FRI in this paper used spiking activity of the H1 cell to control the compensatory rotation of a robot emulating an optomotor response. The H1 cell belongs to a specific class of hetero-lateral neurons that connect the two hemispheres of the fly brain with each other [23,24]. This connection enables the integration of binocular motion information which increased the detection of specific self-motion components. The H1 cell provides excitatory inputs to the contra-lateral horizontal equitorial cell (HSE), a directionally selective output neuron that has been suggested to encode yaw rotations around the fly's vertical axis [25,26]. While optomotor control mechanisms were observed to be linear at the behavioural level [20,27], the responses of the H1 cell are highly non-linear [28]. The mechanisms by which nonlinear neuronal signals are combined to result in an overall linear performance of the system at the behavioural level are still not fully understood.

A possible control strategy for the yaw optomotor response of the fly would be to employ a proportional controller with a static feedback gain K_p that is based on the activity of neurons sensitive to horizontal motion, including those of the H1 cell. The only study to use neuronal activity to stabilize visual motion under closed-loop conditions was carried out by Warzecha and Egelhaaf (1996) using a proportional controller with a static feedback gain. They used constant velocity motion as an input to the closed-loop system and found that the intrinsic properties of the elementary movement detectors prevent the optomotor system from becoming unstable even for high feedback gains. However, here we showed that the optimal K_p differs depending on the input frequency and that no single value of K_p was able to minimize the retinal slip speed over the range of the input frequencies investigated under closed-loop conditions (Fig 2(b)). This is likely to be due to the relationship between K_p and the overall systems behaviour [15]. A lower K_p leads to longer rise time but a smaller steady state error and overshoot. This makes it ideal to compensate for externally forced perturbations which result in slow accelerations of the visual input. However, to compensate for high accelerations a higher K_p is required to achieve fast rise times of the controller. Fast response times are particularly important for reflexes such as the optomotor response so that the fly is able to compensate for slip-speeds with high acceleration. However, although higher K_p decrease the response times for the system, they may cause oscillations, leading to a resonator/anti-resonator phenomenon in the frequency response (Fig 2(b)).

An alternative visuo-motor control strategy might be to dynamically scale the feedback gain at which H1 cell activity is converted into optomotor yaw torque. The adaptive gain controller presented in this paper is such an example. While many variations of an adaptive gain controller exist, our choice was inspired by the matched coding strategy proposed by Laughlin (1994) to maximize information transmission from fly photoreceptors to large monopolar cells. Laughlin identified two conditions necessary for an adaptive coding strategy: (i) the stimulus-response function should have a sigmoid shape and (ii) the slope at the midpoint of the curve can be altered dynamically to scale the input-output relationship. For the proportional controller, the function used to map the instantaneous firing rate, F, onto the control input, E, was static. The initial scaling of the curve was determined by the maximum firing rate, F_{max}, calculated under open-loop conditions. With the proportional controller, the function transforming F onto E only partly fulfilled the second condition above. While the mapping function had a sigmoid shape, the slope at the curve's midpoint in both the preferred and null directions were static. The adaptive gain controller complies with Laughlin's second condition by continually estimating and updating F_{max} and re-scaling the sigmoid relationship between F and E.

In behavioural studies, Egelhaaf (1987) visually stimulated *Musca domestica* with oscillating patterns and observed that the yaw torque had low-pass filter characteristics. For low input frequencies, $f_i \leq 0.0625$ Hz, the measured yaw torque was approximately constant. However, increasing the input frequency beyond 0.0625 Hz resulted in the yaw torque declining approximately linearly.

Beyond $f_i = 4$ Hz, the measured yaw torque response was nearly zero. In the closely related species *Calliphora*, Borst (2003) observed similar low-pass filter charactersitics in the response of the H1 cell to white-noise stimuli but at a significantly higher cut-off frequency ($f_c=20$ Hz). From the two studies, it can be concluded that the bandwidth of the visual sensor, the H1 cell, is considerably higher than that of the motor system. Ultimately, in a freely moving fly, the frequency response of visuo-motor behaviours are limited by the dynamics of the motor systems [27]. For the optomotor yaw response the limits would depend on a combination of the fly's flight motor system and its aerodynamic properties. Consequently, the higher priority for the fly would possibly be to maximize the bandwidth of the motor system in comparison to the sensory systems, in this case the visual system. Any control strategy that does not maximize the motor system bandwidth would be operating sub-optimally.

Maximizing the pass-band gain and bandwidth offer functional advantages for the fly in an attempt to stabilize its body attitude based on visual motion information under closed-loop conditions. A higher pass-band gain would more efficiently reduce any retinal slip speed over the fly's eyes and would therefore increase the stabilization performance. Similarly, a higher cut-off frequency or bandwidth means that the fly would be able to stabilize visual motion over a larger dynamic input range. In the case of the FRI, the frequency response properties are limited by the dynamics of the robot. For the robot, the maximum achievable frequency response bandwidth is 0.5 Hz. While the proportional controller is able to achieve near perfect compensation in the pass-band region for $K_p = 0.1$ & 0.5 (Fig 3(b)), the corresponding bandwidth of the motor actuator i.e. the robot is sub-optimal (Fig 3(d)). In contrast, the bandwidth of the motor actuator for the adaptive gain controller is approximately equal to the maximum possible with the robot dynamics. However, while the adaptive gain controller is able to maximize the bandwidth of the motor actuator, it does so at the slight expense of the pass-band gain. This is consistent with the proposition that sensory systems of the fly encode differences rather than absolute values [31] and that perfect compensation might not be required as long as the visual slip-speed remains within the sensory bandwidth limits. The adaptive gain controller also manages to keep the phase differences lower than the proportional controller does (Fig 2(c)). This is extremely important for stable control as an increased phase difference can lead to unwanted oscillations that increase the visual slip-speed observed by the fly. Additionally, the overall performance of the adaptive gain controller in minimizing the visual slip-speed with less energy compares favourably to that of the proportional controller as measured in the time domain by the performance index (Fig 4).

Two conclusions can be drawn from the work presented with the FRI in this paper. Firstly, no single control strategy works equally well for all input frequencies; i.e. both static and adaptive gain feedback controllers have benefits and drawbacks. The best proposed strategy for the flies would be to adjust the control laws and/or parameters depending on the stimulus dynamics. Secondly, to fully understand the closed-loop sensorimotor performance in the fly, all component systems must be fully characterized; i.e. sensory and motor systems as well as the feedback control laws and delays. Under closed-loop conditions, the bandwidth of sensory and motor systems will have a bearing on one another and must therefore be considered as a cohesive system and not looked at in isolation. In the future, we will use the FRI to study multisensory control strategies in the fly, and specifically how sensory signals from different modalities are integrated in order to increase their robustness for motor control and to reduce the response delays.

Acknowledgments. This work was supported by the Higher Education Commission Pakistan, EPSRC Career Acceleration Fellowship and the US Airforce Research Labs [grant FA 8655-09-1-3022]. We would like to thank Kris Peterson and Kit Longden for helping with the electrophysiology experiments and to Martina Wicklein for helpful discussion on the work presented in the manuscript.

References

1. Schilstra, C., van Hateren, J.H.: Blowfly flight and optic flow: I. thorax kinematics and flight dynamics. J. Exp. Biol. 202, 1481–1490 (1999)
2. Heisenberg, M., Wolf, R.: The sensory-motor link in motion dependent flight control of flies. In: Miles, F.A., Wallman, J. (eds.) Visual Motion and its Role in the Stabilization of Gaze. Rev. of Oculomotor Res., vol. 5, pp. 265–283. Elsevier (1993)
3. Wehner, R.: Spatial vision in arthropods. In: Autrum, H. (ed.) Handbook of Sensory Phys. VII/6C, pp. 287–616. Springer (1981)
4. Krapp, H., Hengstenberg, R.: Estimation of self-motion by optic flow processing in single visual interneurons. Nat. 348, 463–466 (1996)
5. Krapp, H., Hengstenberg, B., Hengstenberg, R.: Dendritic structure and receptive-field organization of optic flow processing interneurons in the fly. J. Neurophys. 348, 1902–1917 (1998)
6. Krapp, H.G.: Neuronal matched filters for optic flow processing in flying insects. Int. Rev. Neurob. 44, 93–120 (2000)
7. Gotz, K.G.: Hirnforschung am navigationssystem der fligen. Naturwis 62, 468–475 (1975)
8. Reichardt, W., Wenking, H.: Optical detection and fixation of objects by fixed flying flies. Naturwis 56, 674–689 (1969)
9. Warzecha, A.K., Egelhaaf, M.: Intrinsic properties of biological motion detectors prevent the optomotor control system from getting unstable. Phil. Trans.: Bio. Sci. 351, 1579–1591 (1996)
10. Collet, T., Nalbach, H., Wagner, H.: Visual stabilization in arthropods. In: Miles, F.A., Wallman, J. (eds.) Visual Motion and Its Role in the Stabilization of Gaze. Rev. of Oculomotor Res., vol. 5, pp. 239–263. Elsevier (1993)
11. Ejaz, N., Peterson, K., Krapp, H.: An experimental platform to study the closed-loop performance of brain-machine interfaces. J. Vis. Exp. 10(3791) (2011)
12. Krapp, H.G., Hengstenberg, R., Egelhaaf, M.: Binocular contributions to optic flow processing in the fly visual system. J. Neurophys. 85, 724–734 (2001)
13. Franceschini, N.: Sampling of the visual environment by the compound eye of the fly: fundamentals and applications. In: Snyder, A.W., Menzel, R. (eds.) Photoreceptor Optics, pp. 98–125. Springer (1975)

14. Warzecha, A.K., Horstmann, W., Egelhaaf, M.: Temperature-dependence of neuronal performance in the motion pathway of the blowfly *Calliphora erythrocephala*. J. Exp. Biol. 202, 3161–3170 (1999)

15. Ejaz, N., Tanaka, R.J., Krapp, H.G.: Closed-loop performance of a proportional controller for visual stabilization using a fly-robot interface. In: IEEE Int. Conf. Rob. Biomim., pp. 1509–1515 (2011)

16. Laughlin, S.B.: Matching coding, circuits, cells, and molecules to signals - general principles of retinal design in the fly's eye. In: Prog. in Ret. and Eye Res., vol. 13, pp. 165–196. Elsevier (1994)

17. Ogata, K.: Modern control engineering, 3rd edn. Prentice Hall PTR (1997)

18. Borst, A., Haag, J.: Neural networks in the cockpit of the fly. J. Comp. Phys. A 188, 419–437 (2002)

19. Krapp, H.G., Wicklein, M.: Central processing of visual information in insects. In: Masland, R., Albright, T.D. (eds.) The Senses: a Comprehensive Reference, vol. 1, pp. 131–204. Academic Press (2008)

20. Theobald, J.C., Ringach, D.L., Frye, M.A.: Dynamics of optomotor responses in Drosophila to perturbations in optic flow. J. Exp. Biol. 213, 1366–1375 (2009)

21. Webb, B.: Validating biorobotic models. J. Neur. Eng. 3, 25–35 (2006)

22. Dickson, W.B., Polidoro, P., Tanner, M.M., Dickinson, M.H.: A linear systems analysis of the yaw dynamics of a dynamically scaled insect model. J. Exp. Biol. 213, 3047–3061 (2010)

23. Hausen, K.: Functional characterisation and anatomical identification of motion sensitive neurones in the lobula plate of the blowfly *Calliphora erythrocephala*. Z. Naturforsch 31, 629–633 (1976)

24. Krapp, H.G.: Estimation and control of self-motion and gaze in flying insects. In: ION 63 Ann. Meet. (2007)

25. van Hateren, J.H., Kern, R., Schwerdtfeger, G., Egelhaaf, M.: Function and coding in the blowfly h1 neuron during naturalistic optic flow. J. Neuro. 25, 4343–4352 (2005)

26. Kern, R., van Hateren, J.H., Michaelis, C., Lindemann, J.P., Egelhaaf, M.: Function of a fly motion-sensitive neuron matches eye movements during free flight. PLoS Biol. 3, 1130–1138 (2005)

27. Graetzel, C.F., Nelson, B.J., Fry, S.N.: Frequency response of lift control in drosophila. J. Royal Soc. Int. 10 (2010)

28. Borst, A., Reisenman, C., Haag, J.: Adaptation of response transients in fly motion vision ii: model studies. Vis. Res. 43, 1309–1322 (2003)

29. Egelhaaf, M.: Dynamic properties of two control systems underlying visually guided turning in house-flies. J. Comp. Phys. A. 161, 777–783 (1987)

30. Borst, A.: Noise, not stimulus entropy, determines neural information rate. J. Comp. Neuro. 14, 23–31 (2003)

31. Taylor, G.K., Krapp, H.G.: Sensory systems and flight stability: What do insects measure and why? Adv. Ins. Phys. 34, 231–316 (2007)

Learning and Retrieval of Memory Elements in a Navigation Task

Thierry Hoinville[1,2], Rüdiger Wehner[3], and Holk Cruse[1,2]

[1] Center of Excellence Cognitive Interaction Technology (CITEC)
[2] Department of Biological Cybernetics, University of Bielefeld, Germany
[3] Brain Research Institute, University of Zürich, Switzerland

Abstract. Desert ants when foraging for food, navigate by performing path integration and exploiting landmarks. In an earlier paper, we proposed a decentralized neurocontroller that describes this navigation behavior. As by real ants, landmarks are recognized depending on the context, i.e. only when landmarks belong to the path towards the current goal (food source, home). In this earlier version, neither position nor quality of the food sources can be learnt, the memory is preset. In this article, we present a new version, whose memory elements allow for learning food place vectors and quality. When the agent meets a food source, it updates the quality value, if this source is already known, or stores position and quality, if the source is new. Quality values are used to select food sources to be visited. When one source has a too low quality, the agent also finds a shortcut to another known food source.

Keywords: Ant, Navigation, Learning, Decentralized Memory, Navinet.

1 Introduction

The capability to navigate, i.e., the faculty to find distant (not directly receivable) locations as food places or some home site, and the ability to exploit learned landmarks is an excellent paradigm to study the architecture of biological memories, in this case procedural memory [5]. Crucial questions concern how, in an autonomous agent, the memory content is learnt, in which way it is stored and how it can be retrieved in a context-dependent manner.

Intensively studied examples are various insects, in particular honey bees and desert ants. These animals are able to navigate using path integration and landmarks. Although a huge amount of experimental data is available [3,6,10,2], the neural and computational mechanisms underlying how food sites and landmarks are learnt, stored and retrieved are still unknown. A quantitative and therefore testable hypothesis on how this information may be represented in long term memory and how retrieval is possible has been proposed by [4]. However, the way learning and storing of this information may be performed has not been considered in this study. Here we extend and specify the earlier hypothesis by proposing a neuronal architecture that with respect to the general structure called Navinet I, follows the basic ideas of the earlier model, but now shows how both learning and retrieval may be realized using this neural architecture.

T.J. Prescott et al. (Eds.): Living Machines 2012, LNAI 7375, pp. 120–131, 2012.

In this earlier version, a food source memory simply consisted of two numbers, the vector describing the position of the goal. Neither position nor quality of the food source can be learnt. In this article we present a new version of Navinet, called Navinet II, whose memory elements are much more complex to allow the agent to learn food place memory vectors and food quality. During foraging, when the agent meets a food source, it updates the quality value, if this source is already known, or stores position and quality, if the source is a new one. Quality values are used to select food sources to be visited. When one source has a too low quality, the agent also finds a shortcut to another known food source.

2   Navinet I: Navigation through Predefined Places

The earlier model [4] consists of four parts (see Fig. 1) as does the new version.

2.1   Path Integrator and Area Concentrated Search

The first part is composed of two control modules, one being a Path Integrator (PI) and the other implementing an Area Concentrated Search (ACS). The PI continuously integrates displacements to maintain the so-called "current vector" pointing from the nest (null vector) to the actual position of the ant, and used to guide it back home. ACS consists in a randomized spiral-like exploration walk (see [4]) and is performed whenever the ant should have reached the nest according to the PI output, but arrives elsewhere in the neighborhood, due to the inherent PI error accumulation. Note that PI and ACS, combined together, repeatedly lead the agent back to the assumed nest position as shown in real ants [11]. Both PI and ACS capabilities are plausibly assumed to be innate since those are already observed when ants leave the nest for the first time [11].

2.2   Motivation Units

The second part of Navinet I (and II) is a motivation network consisting of so-called motivation units that represent and control different internal states, like "stay-in-nest" or "forage". Motivation units are self-exciting, nonlinear summation units (saturing activation function) connected in such a way that related states excite each other, whereas conflicting states inhibit reciprocally in a winner-take-all fashion. Thus, "forage" state involves activating either "outbound" state, for walking from the nest to a food site, or "inbound" state, for walking back to the nest when satiated. One motivation unit is also associated with each food source, in order to decide which one should be visited. Here again, the whole motivation network is assumed to be an innate structure.

2.3   Predefined Food Places

Navinet I and Navinet II differ with respect to the organization of food place memories and landmark memories. In Navinet I, three types of memories have

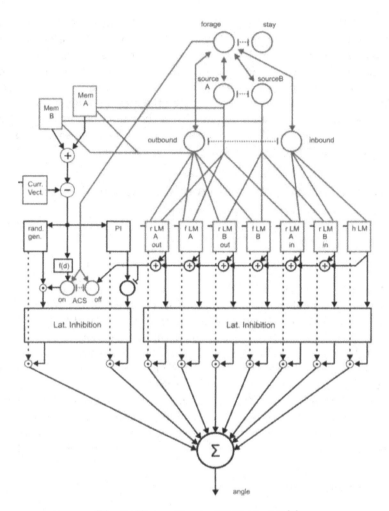

Fig. 1. The earlier model Navinet I [4]

been implemented (food place memories, food/home landmark memories, and route landmark memories). Concerning the food place memories, In Navinet I it is assumed that the agent has already encountered every food site and, for each, stored the "current vector" provided by the PI system in a food place memory as a so-called food place vector, which will be used later as the Reference vector. When activated during "outbound" walks, these memories are used to head the agent directly to the food source. As will be explained below, in Navinet II there are only food place memories whose structure will be explained in Sect. 4 in detail. The simple food place memory of Navinet I containing just two numbers will be replaced by the circuit shown in Fig. 3. Landmark memories will not be considered in this article as will be justified in Sect. 3.4.

2.4   Output Selection

The fourth part of both models, Navinet I and II, concerns how to determine the actual walking direction from the outputs of PI, ACS and active memory elements. Indeed, all provide a vector that represents a walking direction, and a "salience value" given by the length of the vector. A competition to select one or a mixture of walking direction occurs thanks to a lateral inhibition network between saliences. The partially asymmetric connectivity of this network, also considered innate, is explained in more detail in [4].

3   Navinet II: Navigating and Learning Places

In the new version, several changes have been introduced. The overall architecture of Navinet II is depicted in Fig. 2.

3.1   Default Food Search

In our earlier model [4], we did not explicitly introduce a procedure that controls the search behavior that ants exhibit when they leave the nest without heading for specific, already known food sources [11]. In Navinet II, "default food search" is performed when the motivation unit D is active, that is when "outbound" unit is active and no interesting food source (Fig. 2, units A, B, C) is currently available. As far as we know, there are not many data from biological studies concerning the mechanisms controlling this food search path (but see [8,12]). As a simple hypothesis, we use an ACS controller parameterized for exploring larger distances with normally distributed random walk (i.e. no spiral tendency).

3.2   Motivation Units

A second, minor, change concerns the embedding of the food place motivation units into the motivation unit network. Whereas in the earlier network they were directly connected with the unit "forage", they are now connected to the unit "outbound". This solution is simpler and is adopted because there is no clear experimental evidence to support one of these assumptions.

As an example, Fig. 2 shows how three food place memories (in blue) with their respective motivation units (A, B, C in red) are embedded in Navinet II. More have to be used if the agent should store more than three food sources. Like in Navinet I, compatible states excite each other, whereas uncompatible states form winner-take-all inhibitory nets.

3.3   Area Concentrated Search

Third, as a result of this rearrangement, two ACS procedures have been introduced, one for the search for the home site (ACSh, depending on "outbound" deactivation), the other for the search of a food place (ACSf, depending on "inbound" deactivation).

3.4   Landmark Memory

An important change refers to the landmark memories. Whereas in the earlier model landmark elements where considered as to be represented by separate, discrete memory elements, recently an exciting alternative concept has
been proposed by [1]. Thus, one three-layered feed-forward network involving
an infomax-based learning rule can be used to implement robust path following
with respect to, in practice, an unlimited number of landmarks. Therefore, in
Navinet II, there are two types of memories, a bank of food place memories and a
unique landmark network (Fig. 2). This landmark memory receives visual input
stream, as well as inputs from the "inbound" and food place motivation units,
accounting for the current context internal to the agent. However, here we adopt
this idea about landmark memory only conceptually. In fact, in this article we
do not apply the landmark memory but just mention how it could be integrated
into the architecture proposed here.

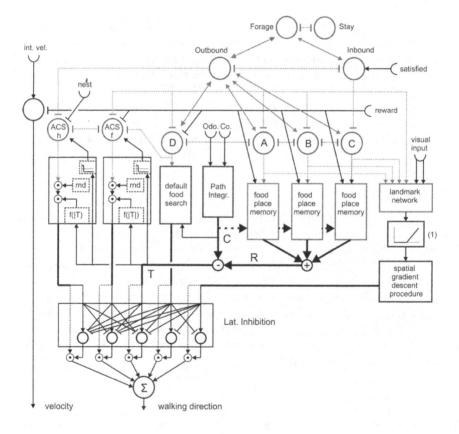

Fig. 2. Navinet II. In contrast to the earlier version [4] this architecture allows for
learning and retrieval of the memory contents. Memory elements are shown by blue
rectangles, motivation units in red. For further explanation, see section 3.

3.5   Inputs/Outputs

Compared to the previous model, we have designed Navinet II to be embedded in actual agents by making explicit several sensory input signals which trigger learning and switching between motivations. A "satisfied" signal indicates having received enough food and thus activates "inbound" unit to turn back to the nest. To compute the "current vector", the PI procedure is provided wih the distance traveled at each move given by an odometer ("Odo."), and the agent's heading direction according to a compass ("Co."). The landmark network receives a visual stream. As described in the following, all food place memories receive a reward signal used to trigger learning of both the location and quality of all food sites. This signal moreover suppresses units D, ACSf and a unit transmitting a signal controling the walking velocity (ie. it stops the agent). The ACSh unit is suppressed by a "nest" signal indicating when the agent reaches the nest.

   As in the earlier model, all components of Navinet II output different possible walking directions (five in total). In particular, the so-called "travel vector" T is obtained from substracting the current vector C, maintained by the PI system, to the reference vector R, retrieved from the food place memories. As before, a lateral inhibition network is used to select a unique output vector based on its length, called salience. The inhibitory influences are partly asymmetric. In accordance with experimental results [13], the landmark salience inhibits all other output vectors. The same is true for the output of the Default Search procedure (in particular, the travel vector has to be inhibited, because during the Default Search, the PI system points to the home place as no food place vectors are activated; note that nonetheless the PI system is continuously running). Travel vector and both ACS outputs show mutual inhibition, but do not influence the other two vectors.

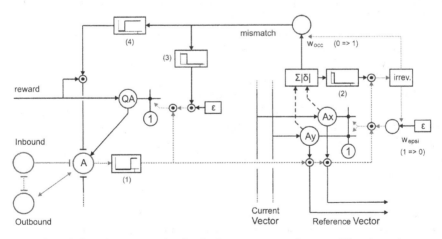

Fig. 3. Detailed architecture of a food place memory element. The three important parts are: (i) motivation unit A, (ii) quality unit QA, and (iii) the food place net consisting of units Ax and Ay. Inputs are: reward and Current Vector, output is Reference Vector. ϵ: learning rate. For further details see the text.

4   Learning and Retrieving Food Places

Fig. 3 shows, as an example, the circuit of food place memory A. The food place memories have to be structured in a way that they can cope with the following three problems. First, when the agent has found an interesting food place, the vector actually provided by the PI system, the Current Vector, has to be stored in the LTM. Second, when the agent decides to visit this food source again, this vector should be retrieved from LTM and used as a reference vector for the PI-based controller. There is a third problem to be considered. Assume that the agent has already stored one or several food source places and now, while searching for a selected source, meets a food source. How does the agent know whether this is indeed the stored source it is searching for or another source, possibly not yet stored in the LTM? How is, in the latter case, the new memory established, a problem also occurring when a food source is found during a Default Search. To begin with, we will discuss how the first two questions can be solved by this circuit.

4.1   Structure of Food Place Memories

The network required to store and retrieve one food place memory consists of three sections, (i) a motivation unit, (ii) a one-unit network to store the quality of the food source (i.e. the value of the reward signal), called "quality net", and (iii) a two-unit network to store the food place vector itself, called "food place net" (Fig. 3, from left to right). Both networks are equipped with a bias unit of activation one (neuronal elements depicted by ① in Fig. 3). The food place vector is given in Cartesian coordinates which is (i) computationally simpler than using polar coordinates, for example, and (ii) also biologically more plausible [9]. The structure of each food place memory element is given (or "innate"). Only the single weight in the quality net and the two weights in the food place net, marked by black dots in Fig. 3, as well as the "occupied" weight w_{occ} and w_{epsi} explained below have to be learnt in each memory element.

Both networks, the "quality net" and the "food place net", are constructed using so called Input Compensation (IC) units, as they allow for easy training [7]. In Fig. 3, these IC units are marked by QA and Ax, Ay, respectively. An IC unit receives two kinds of input signals, the external input (arrow entering the unit from the left hand side) and the internal input (entering the unit from the right hand side). As in this simple case there is only one weight per unit, the internal input of neuron i represents the value of the weight (note that bias unit activation is one)

$$s_{i,t} = w_{i,t} \tag{1}$$

The external input $I_{i,t}$ is not added to $s_{i,t}$ and therefore not influencing the output (arrow pointing downwards from each IC unit), but only used to compute an error $\delta_{i,t} = I_{i,t} - s_{i,t}$. This error signal will be used in two ways. First, it is applied to change the weights of the unit according to

$$\Delta w_{i,t} = \epsilon \, \delta_{i,t} = \epsilon \left(I_{i,t} - s_{i,t} \right) \tag{2}$$

which formally corresponds to the delta rule [14]. In case of a quasistatic external input, application of this learning rule changes the weight in such a way that the recurrent signal $s_{i,t}$ asymptotically approaches the mean value I_i. Using a learning rate of 0.5, learning takes 15 iterations to reach an error $< 0.1\%$. After learning is finished, the unit forms a memory representing the input value I_i.

4.2   Learning of Food Place Memory Information

Let us begin with the first problem mentioned above. Assume that the agent has been staying in the nest (activation of unit Stay) but the motivation unit Forage is now activated. This activation suppresses unit Stay and elicits a competition between motivation units Inbound and Outbound. Motivation unit Outbound wins the competition due to a small asymmetry between both motivation units indicated by the slightly stronger coupling with unit Forage (see larger arrows in Fig. 2). In turn, activation of motivation unit Outbound elicits a competition between default food search unit D and all food place motivation units. Let us assume that the agent has not yet learnt a food place vector before, as for instance when leaving the home for the first time. The Default Food Search unit will win the competition, due to a slightly stronger connection from unit Outbound to unit D than to units A–C.

When the agent then finds a food source being rich enough, sensory input (Fig. 2, "reward") will inhibit unit D. Also, as already mentioned, the reward signal sets walking velocity to zero. If this were not the case, the Reference Vector being zero would immediately attempt to lead the agent to the home site instead of staying at the food place and continue learning. Due to the WTA connection between units A–D, unit D will lose the competition and instead anyone of the units A–C will win. Say source A has reached the threshold value (Fig. 3, (1)) meaning it has been selected by the WTA, only this food place net A will learn the actual value of the Current Vector (although the latter is given to all food place nets). During learning, if the total learning error $\sum |\delta|$ becomes smaller than a given threshold (see Fig. 3, (2)) and the motivation unit is above threshold (Fig. 3, (1)), learning is finished irreversibly by setting w_{epsi} from one to zero (Fig. 3, box "irrev."), i.e. for this food place net the learning rate ϵ is effectively set to zero (see blue dashed arrow). Thereby, from that moment on, the values of the weights are kept fixed. This end signal in addition changes a weight, called "occupied" weight w_{occ}, from zero to one, which opens a channel to transmit this error value, now called mismatch error, to the quality network and to the motivation unit. The reward value will be learnt if the motivation unit is active above a threshold (Fig. 3, (1)) and the mismatch error is small enough (Fig. 3, (3)). When enough food has been collected so that the animal is satisfied (Fig. 2, input to Inbound), the motivation network switches to inbound state. The Inbound unit inhibits the motivation units A–D and thereby sets the reference vector to zero which leads the agent back to its home site.

4.3   Retrieval and Updating Food Place Memory Information

How can our second problem be solved? When the next time the agent decides to travel to a food source (unit Outbound being active), the WTA net connecting the motivation units A–D selects one of these based on the food quality values previously stored. In fact, as depicted in Fig. 3, the food place motivation units are excited back by their corresponding quality net (unit Q). Therefore, the motivation unit of the memory element with the highest quality will win (of course, a more complex quality network might also contain further aspects of the food source). If no food source with positive quality is given the Default Search unit will win. Depending on the strength of the asymmetry between unit D and units A–C, the default unit may already win, when the quality values of elements A–C are positive, but below a given threshold value. When then one of these sources is found again by chance and its quality has increased in the meantime, this memory element will be "reactivated" and therefore this source will again have the possibility to win later competitions and to be visited.

Let us assume that unit A has won the WTA competition. This opens the output of the corresponding food place net (a further effect controlled by the motivation unit), which becomes the new Reference Vector the agent will travel to. When the animal then meets the food source again, the quality network will be updated (as reward is on, motivation unit A is on, and mismatch error is zero), but the weights of the net storing the food place vector are not changed. Note that, although the reward signal is projected to all motivation units, the already actually activated motivation unit, in our example motivation unit A, suppresses all other food place motivation units, which means that the reward signal will not influence the other food source memory elements. Thus, the first two problems mentioned above, storing as well as retrieval of a food source vector are solved by this network.

4.4   Recognition of a Food Source

There is however the following already mentioned third problem. If the agent arrives at the food source, three cases have to be distinguished. One possibility (case 1) is that the agent has found the correct food source, i.e. the one it has searched for. If, however, the agent accidentally came across another source, there are yet two further possibilities. Either (case 2) this source is one of the other food sources learned earlier, say source B, or (case 3) the food source found might be an unvisited one and a new food source memory element should be activated to store the corresponding food place vector and its quality value.

First of all, to cope with these questions, a distinction is required between case 1 (the actually found source is the correct one) or not (cases 2, 3). To this end, the mismatch error is exploited in the following way: In any case, motivation unit A is active due to the competition that took place when deciding on which source should be visited. As the Current Vector is given to all food place nets, the mismatch error is used in each of the memory elements that have been learnt. If this mismatch error is larger than a given threshold (Fig. 3, (4)), it inhibits its own motivation unit. In case 1, if the correct food source is found, the mismatch error for memory element A is zero and therefore no inhibition takes place. This means that this motivation unit will continue to win the competition. Therefore the quality net of this memory element will learn the corresponding reward signal. If another source is found, that has also been learnt earlier (case 2), the motivation unit of this memory element will win the competition (whereas all the others will be suppressed) and this memory element will learn the correct reward signal. If, case 3, a new food source has been detected, i.e., a food source for which no memory element exists yet, all already learned memory elements are inhibited because all of them show a large mismatch error. In addition, the default search unit is suppressed by the reward signal to avoid the unwanted activation of a default search in this situation. As a consequence, one of the motivation units belonging to a "naive" memory element will win the competition and this element will then learn the properties (food place vector, quality) of this new source.

To give a naive memory element a chance to win the competition, we add small noise to all food place motivation units, the already "occupied" ones and the still naïve ones. Another problem has to be considered. Of course, a naive network produces a large mismatch error. This signal must not inhibit its own motivation unit already before learning can take place. To this end, we have above introduced a circuit allowing that the channel transmitting the error signal to the motivation unit is only opened (i.e., $w_{occ} = 1$) after learning of the food place net has been finished. Therefore, only non-naive elements can inhibit their own motivation unit (and only if reward and large total error occur at the same time). Therefore, also in case 3 learning takes place and the new source can be stored in a separate memory element.

5   Results

The network depicted in Figs. 2 and 3 and explained above comprises a complex system the properties of which may not exhaustively be understood after this verbal description. However, a compact mathematical description is also not possible. Therefore, the only way to test and illustrate the properties of this architecture is to resort to a quantitative simulation. As a preliminary validation, before testing it on a physical robot, we used a software simulation and considered the following four scenarios:

1. The naive network starts with a default search until it finds a source and stores its location. Then the agent returns to the nest. As unit Forage is still on, the agent will search for this location again.
2. Following scenario 1, the agent does not find the selected location, but another one not yet stored. The latter should then be stored in a new memory.
3. As scenario 2, but the unexpectedly found source has already been stored. The memory concerning this source should be activated, the quality value updated and, after reward, the agent should turn home.
4. Two sources with somewhat different quality have been stored earlier. The agent is heading for the one with the best quality, finds it, but detects that

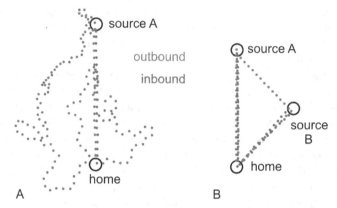

Fig. 4. (A) While searching, the agent finds a food source A. It learns position and quality, returns home and again visits the food source several times. (B) An example for finding a shortcut between source A and source B. See text for further explanation. Top view; outbound travels marked by red dots, inbound travels by blue dots.

the quality is zero. The agent then directly heads to the second source via a new shortcut path. The quality memories have to be updated accordingly.

In various simulations we applied different arrangements of home and food place positions. In any situation and in all four tests the agent did solve the problem without error. As representative examples, we show two figures. Fig. 4(A) demonstrates a Default Search path. After having found a food source, this is then visited repeatedly. Fig. 4(B) shows an example for a shortcut (scenario 4). To simplify the figure, the agent has already learnt the position of sources A and B, before the beginning of the experiment. Now the agent looks for source A two times. The second time quality of source A is reduced by the experimenter. Then the agent heads to source B, returns home and visits source B repeatedly.

6   Discussion

Navinet II is based on the same decentralized architecture as the earlier Navinet I, and therefore has the same properties with respect to memory retrieval. Via the motivation unit network the agent can adopt different internal states and select procedural memories depending on the actual context. In particular, like the earlier Navinet, Navinet II is able to find new shortcuts too. As Navinet II is equipped with the capability to learn new memory contents, it can deal with all four scenarios mentioned in the Result section. For example, the agent realizes whether a food place found is the one it was looking for, or another already known one, or a new food source that then will be learnt. In all cases, the memory concerning the quality of the food source will be updated.

Navinet II is also able to be expanded with a network that exploits learned visual input for landmark navigation as illustrated in Fig. 2. This procedure proposed by [1] has however not yet been tested. Introduction of a vision system and implementation on a robot will be our next steps to be taken.

As a possible disadvantage of Navinet II, one might consider the fact that the neuronal structure for storing a food place vector cannot be used later (when for instance the food would disappear or its quality would completely drop) to store another food place. In case no free memory would be available, either new ones could be created online or the discovered food place could be simply ignored. We are actually working on an alternative solution based on forgetting places whose food quality gets poor. However, the current version has the advantage that relearning a food place that would later provide good quality food again, is not required. In this case, the memory element used earlier can be exploited again and only the quality has to be updated.

Acknowledgments. This work has been supported by the "Center of Excellence Cognitive Interaction Technology" (CITEC EXC 277) and the EU project EMICAB FP7 – 270182 (T.H, H.C.) and by the Humboldt Foundation (R.W.).

References

1. Baddeley, B., Graham, P., Husbands, P., Philippides, A.: Model of ant route navigation driven by scene familiarity. PLoS Comput. Biol. 8(1), e1002336 (2012)
2. Cheng, K., Narendra, A., Sommer, S., Wehner, R.: Traveling in clutter: navigation in the central australian desert ant Melophorus bagoti. Behav. Proc. 80, 261–268 (2009)
3. Collett, T.S., Graham, P., Harris, R.A., Hempel-de-Ibara, N.: Navigational memories in ants and bees: memory retrieval when selecting and following routes. Adv. Stud. Behav. 36, 123–172 (2006)
4. Cruse, H., Wehner, R.: No need for a cognitive map: Decentralized memory for insect navigation. PLoS Comput. Biol. 7(3), e1002009 (2011)
5. Fuster, J.M.: Memory in the Cerebral Cortex: an Empirical Approach to Neural Networks in the Human and Nonhuman Primate. MIT Press, Cambridge (1995)
6. Graham, P.: Insect navigation. In: Breed, M.D., Moore, J. (eds.) Encyclopedia of Animal Behavior, vol. 2, pp. 167–175. Academic Press, Oxford (2010)
7. Makarov, V.A., Song, Y., Velarde, M.G., Hübner, D., Cruse, H.: Elements for a general memory structure: Properties of recurrent neural networks used to form situation models. Biol. Cybern. 98, 371–395 (2008)
8. Schmid-Hempel, P.: Foraging characteristics of the desert ant Cataglyphis. Experientia. Suppl. 54, 43–61 (1987)
9. Vickerstaff, R., Cheung, A.: Which coordinate system for modelling path integration. Journal of Theoretical Biology 263, 242–261 (2010)
10. Wehner, R.: The desert ant's navigational toolkit: procedural rather than positional knowledge. J. Navigation 55, 101–114 (2008)
11. Wehner, R.: The architecture of the desert ant's navigational toolkit (Hymenoptera: Formicidae). Myrm News 12, 85–96 (2009)
12. Wehner, R., Meier, C., Zollikofer, C.: The ontogeny of foraging behaviour in desert ants, Cataglyphis bicolor. Ecol. Entomol. 29, 240–250 (2004)
13. Wehner, R., Michel, B., Antonsen, P.: Visual navigation in insects: coupling egocentric and geocentric information. J. Exp. Biol. 199, 129–140 (1996)
14. Widrow, B., Hoff, M.: Adaptive switching circuits. In: Anderson, J., Rosenfeld, E. (eds.) Neurocomputing: Foundations of Research, pp. 96–104. MIT Press, Cambridge (1960)

Imitation of the Honeybee Dance Communication System by Means of a Biomimetic Robot

Tim Landgraf[1], Michael Oertel[1], Andreas Kirbach[2],
Randolf Menzel[2], and Raúl Rojas[1]

[1] Freie Universität Berlin, Institut für Informatik, Arnimallee 7, 14195 Berlin,
Germany
[2] Freie Universität Berlin, Institut für Neurobiologie,
Königin-Luise-Str. 28/30, 14195 Berlin, Germany

Abstract. The honeybee dance communication system is one of the
most intriguing examples of information transfer in the animal kingdom.
After returning from a valuable food source honeybee foragers move vig-
orously, in a highly stereotypical pattern, on the comb surface conveying
polar coordinates of the field site to a human observer. After 60 years of
intense research it remains still unknown how the bees decode the dance.
To resolve this question we have built a robotic honeybee that is able
to reproduce all stimuli found to be generated in the dance ([12]). By
imitating single stimuli or combinations and tracking the bees' ensuing
behavior we are able to identify essential signals in the communication
process. In this paper we describe the design of our current prototype,
show how we validated the function of the robotic wing buzzes and propose a reactive behavior control on the basis of relative body configurations of nearby bees measured by custom smart camera modules. We will
conclude by showing first promising result of field experiments within a
live honeybee colony.

Keywords: honeybee robot, dance communication, biomimetic robots,
biomimetics.

1 Introduction

The brain of a honeybee is not bigger than a pinhead. Nonetheless, it implements
amazing cognitive capabilities. Honeybees are quick learners - they explore unknown terrain and form detailed neuronal representations of the environment.
Intriguingly, honeybees use their location memory of valuable field spots to translate them into body movements, a behavior called the waggle dance. Nestmates,
in turn, translate their percept of those movements back to field coordinates
and are able to find that location themselves only based on the communicated
information. This amazing communication system has been discovered by Karl
von Frisch more than 60 years ago and still is only partially understood. A
tail-wagging honeybee forager moves forward in an almost straight line on the

T.J. Prescott et al. (Eds.): Living Machines 2012, LNAI 7375, pp. 132–143, 2012.
© Springer-Verlag Berlin Heidelberg 2012

vertical comb surface, throwing her body from side to side in a pendulum-like motion at a frequency of about 13 Hz. This so called waggle-phase is followed by a return-phase, in which the dancer circles back to the approximate starting point of the previous waggle, alternatingly performed clockwise and counter-clockwise such that the path of a dancer resembles the figure 8. Von Frisch found that certain dance parameters reflect properties of the food source ([5]). In the waggle phase, the body's angle with respect to gravity approximates the direction to the food relative to the sun's azimuth. The length and duration of the waggle run correlate highly with the distance to the target location ([6],[18]). In addition to information regarding the location, the dance communicates also the profitability or quality of the food source with respect to the current hive's needs: Foragers tend to dance more lively and perform longer dances when feeding on a highly profitable source ([8],[18]).

An amazing amount of knowledge on navigation, memory and communication in honeybees has been gathered ([2]). We can rely on compelling evidence indicating that honeybees actually evaluate and use the information encoded in the dance, rendering this communication system unique in the insect world. However, today, more than 60 years after its discovery, it still remains unknown how follower bees decode the information contained in the dance.

Follower bees, the bees standing in a close proximity to and showing interest in the movements of the dancer, are most likely to be recruited after attending several dance periods. In that process they actively pursue the dancer in order to remain in, or establish a, close contact with her. They detect a variety of stimuli. Mechanical cues like antenna and head contacts to the body of the dancer are frequently observable and likely transmit information about the dancer's body orientation ([1],[17], [7]). Wing bursts in the waggle run produce complex patterns of laminar air flows, three-dimensional fields of short-ranged air particle oscillations and comb vibrations that might as well deliver meaningful multisensory input ([3],[10],[20], [15]). The body temperature of dancers is significantly higher than of non-dancing foragers ([19]). Recently, a dance-specific scent has been reported ([21]) as yet another possible signal. Floral odors and regurgitated food samples are associated cues. However, after more than 60 years of intense research it is still unknown how exactly information is decoded by the followers. Which of the many stimuli carry information? Can we assign specific meanings to single stimuli? How do the followers use that complex mosaic of stimuli they perceive? Do bees extract such abstract concepts as angles to integrate them in and read them from the dance? Or is the encoding and the decoding process more of a memory playback and recording, respectively?

In order to investigate the characteristics of the communication process we built a robotic honeybee that is able to reproduce all known stimuli, that can blend into a honeybee society. This allows us to directly observe the effect of different stimuli or stimulus combinations on the foraging behavior. By tracking newly recruited bees with a customized radar system we are even able to record and analyze the complete flight trajectory from the time they exit the hive until they come back. This will hopefully enable us to finally resolve questions that have remained unanswered for a long time. Building biomimetic robots to investigate the dynamics of animal groups has been shown previously to be a feasible approach ([9], [4]). Using a robot the researcher has full and reproducible control over morphology and behavior but introduces an additional problem of making the robot an accepted member of the animal group. The idea of using a honeybee robot to investigate honeybee dance communication is also not entirely new ([16]). Although using a similar mechanical setup, we propose substantial improvements to the robot's design. Current experimental results suggest that these changes improve the performace of the robot, measured by behavioral parameters of the nestmates in the hive. The robotic dance excites stereotypical following movements, a behavior bees display before hurrying out the hive to forage. This behavior has never been shown before to be displayed in association with robotic dances.

In this contribution we provide a detailed description of the important parts of the system. The very dance motion, the base for all other stimuli, has been modeled using a parameter set identified through a statistical variance analysis of hundreds of dance trajectories that were captured from highspeed dance video recordings via an automatic tracking program ([11]). This analysis was used as a conceptual support for the hardware design. The robot, essentially a plotter-based positioning system with a small, life-sized replica of a bee, can move in a two dimensional plane. The positioning system carries a rod on whose end the body of the small artificial honeybee body is affixed. It can be inserted into the hive such that it "hovers" 1-2 mm over the surface of the densely populated comb. The body can be heated and can deliver small drops of sugar solution through a tiny syringe. The robot exhibits a single wing that is vibrated with an electro-magnetic driver and therewith produces the above mentioned oscillating air currents and laminar air flows. Those "jet streams" have been hypothesized to convey meaningful stimulation of the follower bees ([15]). We show an inexpensive alternative to hot wire anenometry to measure laminar air currents at the abdomen of the robot.

As a further contribution we describe an embedded computer vision system that we use to obtain a 360° obstacle map used for evading collisions with bees. The bee robot can recognize two crucial situations in which different behavioral programs are executed. Therewith a closed feedback loop is established. We have also begun to test the robot under natural conditions and whether it has an influence on foraging. Here, we will show preliminary results obtained from several field tests and discuss their implications in the remainder of this paper.

2   Mechanical Description and Hive Setup

The robot is based on a plotter (Roland DXY-1300). We fully replaced the proprietary electronics, added a third stepper motor to the pen carriage for rotational motion and cut out most of the plot surface. The plotter is encased into an aluminum frame that is used to stabilize the system once it is stationed in front of an observation hive. The bee replica is life sized and integrates two

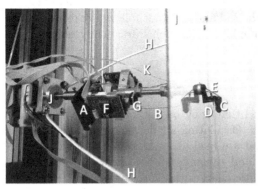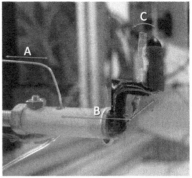

Fig. 1. Left: Top view of the threaded bar that holds all essential parts of RoboBee. A small loudspeaker (A) produces oscillations that are transmitted by a stiff wire (B) down to the wing (D) of the bee replica (C). Details on the mechanics are depicted in the photograph to the right. Through a tube (K) we can deliver small amounts of sugar water to the head part of RoboBee (E). Two camera modules (F) are affixed to the central rod such that the sensors (G) observe a perimeter around the replica. A polycarbonate sheet (I) held by two rods (H) is used to cover the hole through which the replica is inserted into the hive. The central rod can be rotated by a motor (J) that is mounted to the carriage of a plotter. Right: Closeup view of the replica at the end of the central rod. This body is inserted through a glass window of an observation hive. The picture shows the wing mechanism. Vibrations from a loudspeaker are transmitted through a stiff wire (A) and conducted through a metal pipe (B). The end of the wire is glued to the tip of a pivoted plastic wing. Vibrations are transformed to radial movements of the wing (C). The body's temperature can be regulated using a resistor to heat and a thermosensor to measure the temperature. Tiny drops of sugar water can be presented at the head of the replica by pressing it through a syringe near the head part.

resistors and a thermosensor used for temperature regulation. It is moved by a central rod attached to the third motor. It can be moved in the x/y plane parallel to the comb surface using the plotter's motors and mechanics. Besides the bee model, the central rod carries a loudspeaker used for the wing vibrations, two embedded camera modules for the behavior control, the wiring and a small tube used to deliver small drops of sugar water. To reduce mechnical noise in the waggle portion of the dance we have attached the bee replica excentrically. The 13 Hz waggle motion is produced by the rotation motor only; very similar to the motion of real waggle dancing bees. To introduce the robot into the hive a small window to the surface of the comb has to be opened. This introduces a great amount of disturbance making it at some point impossible to interact with the bees naturally. Opening a window leads to a change of hive temperature and air movement that alters the hive's micro-climate. As a reaction one observes a growing "curtain" of bees holding tightly to one another to seal the hole through which we insert the robot. This constrained our first experiments to short dance durations of 5-10 minutes. To overcome this problem we have added a light-weight transparent plastic sheet to cover the opening when inserting the robot. The sheet is not coupled to the rod and does not follow any rotations like the fast waggle movements. This simple solution helps drastically reducing the "clumping" of bees.

3   Wing Vibrations

Michelsen ([15]) previously indicated that wing oscillations are the source of a rich and complex mixture of stimuli. Oscillating air flows as well as continuous ("jet") streams can be detected and are likely to convey meaningful informa-tion to the nestmates. Also static and dynamic electric fields were shown to be produced ([23]) by motion of static charges that are carried by the bees on their wings and body parts. To replicate the wing vibrations and the resulting stimuli we utilize a small speaker that is fixed to the central rod. In the waggle portion of the dance bees buzz their wings in short bursts of a 280 Hz carrier. The robotic bee can produce a wide range of frequencies. We drive the speaker with rectangular bursts of +- 5V of a frequency and duration that are adjustable parametrically. A stiff wire is attached to the speaker's diaphragm and serves to transmit its vibration to the replica where it is attached to a wing model pivoted near the thorax. The wing model is composed of 10 layers of a plastic sheet of decreasing lengths such that a stiffness gradient from the base to the tip is created. We have used highspeed video microscopy to validate the ampli-tude and motion of the wing oscillations. By now, the mean amplitude (tip to

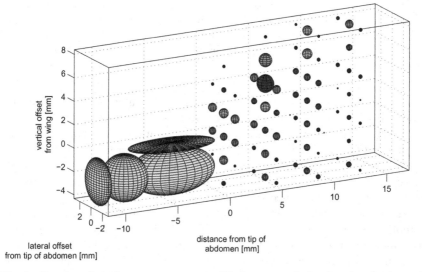

Fig. 2. Results of air flow measurements. We have sampled pointwise the intensity of air flow at distances of 1mm, 5mm, 9mm and 13mm for various heights and lateral displacements. Dorso-ventrally the "jet" is rather broad but remains in the limits of approximately 4 mm, i.e. the width of the wing.

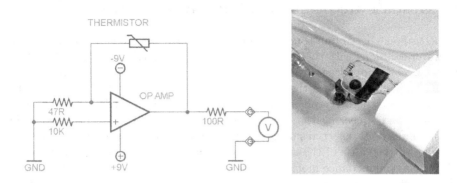

Fig. 3. Left: Wiring diagram of the circuit used to measure airflows produced by the wings of the robot. We use a trait of operational amplifiers producing a small voltage even though no input voltage is given. We feed back the offset voltage into the amplifier over a thermo-resistor. The amplification of the OP AMP is determined by the ratio of the two resistances to the left. Right: The thermoresistor is positioned near the wing tips coupled to a heating resistor 1 mm on top of it.

tip displacement) with 5 V driving voltage is about 1 mm. This is even more than real bees produce (personal communication with Axel Michelsen). We have built a custom anemometer to detect and measure the range of air flows around the robot's abdomen. Figure 3 depicts the circuit of the anemometer. We use a simple operational amplifier coupled to a temperature sensor whose resistance is regulating the voltage entering the amplifier. We use an interesting property of operational amplifiers that by themselves exhibit a kind of leakage output of some mV that in our circuit is fed into the amplifier's inverted input. By using a second circuit we heat a resistor close to the measuring device. When air is moved by the flapping wings heat is carried away from the sensor and its resistance is decreased. The output voltage in effect is increased and can be measured using an AD converter (CED MICRO 1401 mkII). Using a micromanipulator we measured for a set of sampling points behind the bee replica the difference of the output voltage with wing movements to the output when inoperative at a constant room temperature (23°C). Because a feasible calibration method is lacking, we cannot convert this information into the actual amount of air that is moved. However, the relative magnitude of the air flow can be measured and moreover one is able to learn about the length of the jet streams that typically reach distances of 10 mm ([15]). Figure 5 shows a false color coded intensity plot of the air motion around the robot body. Each sphere depicts a sample point. Its radius and color code the intensity of airflow measured. The jet is notably narrow in width but exhibits measurable intensities of air flow from the floor level up to 8 mm above the wings, typically the height where the glass window is positioned. This matches previous observations of jetstreams produces in natural dances [15].

4   Obstacle Recognition and Behavior Control

Dancing honeybees often touch nestmates. Frequent collisions can be observed. However collisions with the robotic bee induce high energy impacts and furthermore, if not stopped, the robot might squeeze bees when running over them. This can happen because the comb surface is not perfectly planar. Hence the space between the bee replica and the comb surface varies. It is crucial to avoid collisions and squeezing the bees. The release of an alarm pheromone would be the immediate effect resulting in the attraction of bees to the dance floor seeking for the cause of the alarm. Aside from the effect on recruitment probability this renders dance motion almost impossible due to the density of bees on the dancefloor. There are two solutions to the problem: an active and a passive approach. The latter involves compliant, flexible materials and the former a reactive behavior when a collision is anticipated. Taking the active approach, we have developed a visual feedback system using small embedded cameras. This system furthermore recognizes different body poses of nearby bees and enables the robot to automatically provide food samples to bees approaching frontally or diminish the waggle amplitude when bees are approaching from the sides. The system makes use of two embedded camera modules that are carried by the robot attached to the main metal rod. Each camera observes a hemisphere around the bee replica. Since the cameras are fixed to the central rod the perspective is as well. Hence, irrespective of translational or rotational movements the cameras always observe the same relative space with respect to the replica. The image evaluation is simplified to maximize recognition speed: The comb is back lighted using red or infra red LED lamps. Thus, all objects on the comb cast a shadow towards the cameras. This facilitates the segmentation of the image into obstacle objects, i.e. other bees, and free space by simply thresholding the sum of pixel intensities for a so called "'sensor ROI'" (SROI). Our previous prototype, based on a Atmel ATmega8, allowed measuring occupancy in 7 regions around the robot's body. The new camera prototype can evaluate 700 regions at a higher frame rate. The image processing is distributed among the cameras' CPUs and the robot's CPU. After thresholding, each camera reports a 35x20 binary occupancy map to the main processor. This map is used as a lookup table to determine if certain motions can be executed without running into obstacles. Secondly it is processed further to extract shape information about nearby bees. First the binary image is eroded and then diluted again (a morphological operation called "opening" in image processing) to erase "bridges" connecting blobs. In a second step the connected components are identified leaving out too small or too big objects. Each resulting binary object is then classified according to its position and orientation. Blobs near the abdomen of the robot may be follower bees that are crucial not to overrun or hit in the waggle phase. Blobs facing the robot's head may be followers that try to initiate trophallaxis, the exchange of food. Whenever a potential food exchanger is observed for 2 seconds the robot triggers automatically the trophallaxis pump and stops the dance motion for a maximum of 10 seconds. If a potential follower at the sides is detected the waggle amplitude is reduced by 30 percent to avoid a strong impact.

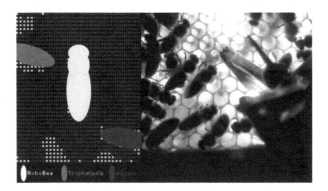

Fig. 4. Visualization of the computer vision system of RoboBee. A robotic dance was conducted and filmed. The output of the sensor cameras was recorded to disk. The left part of the figure shows the sensory data. The sensor stream contains binary information on region occupancy (depicted in yellow) and, optionally, BLOB information for two cases: If a BLOB is "facing" the robot and is located near the head it is considered to be a follower bee ready for trophallaxis. These bees are visualized in the figure as orange ellipses. BLOBs facing the abdominal part of the robot are classified waggle followers and are depicted here as blue ellipses. The image to the right shows the same situation in the hive.

5   Experimental Results

In the summer 2010 we have been conducting the first experiments to test the recruitment ability of the robot. To simplify that task we have pre-trained a group of bees to a feeding place 230 m away from the hive. This group established a memory of that location and might therefore be recruited easier than naive bees. The experimental trials are all arranged in three phases. In training, usually a few days long, the bees fly to an artificial flower offering unscented sugar solution. Every bee that visits the feeder is tagged with a small number plate glued to the thorax. The sugar supply is shut down for a whole day in the pretest phase. Throughout that day the bees learn that the food source is depleted. They keep visiting the feeder, but with decreasing frequency until, at the end of the pre-test, they remain in the hive for most of the time. On the following day, the test, we use the robot to communicate the availability of food at that field spot and count bees at the feeder before and after the robotic dance. This is repeated several times with different stimulus combinations, dance durations and in different areas on the comb surface. Figure 5 depicts the cumulated number of bees visiting that feeding place over the time of day. The robotic dance clearly leads to an increase of arrival rate if and only if exhibiting wing oscillations. This result coincides with previous reports on silent dances ([22]) not beeing able to recruit foragers. We conclude that the robotic dance must at least effect the foraging motivation. Though, this experiment lacks a final proof that the robot communicates the direction and distance to the target. Subsequently, we tackled the question whether the robot can convey directional information and

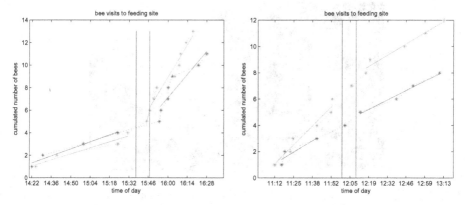

Fig. 5. The cumulated number of individual visits to both feeding sites over time of day. Only bees landing at the feeder were counted. If an individual lands more than once at the feeder within 5 minutes it is counted only once. Each asterisk denotes the time of arrival. The vertical lines delimit the time of the robotic dance to on of the feeders. The solid green and red lines depict the average function for the two intervals, before and after the test. The dotted lines are plotted for convenience to show which events are following each other. The robot danced without scent, trophallaxis and increased body temperature. The left figure shows the effect of the dance with wing vibrations, whereas the figure to the right shows a control with no wing vibrations.

introduced a second feeding source at the same distance to the setup. This second source was located 110° from the first one. Similarly, we trained a new group of bees to the second feeder and repeated the experiment as described above. In the test situation only one location was communicated by the robot and the time of arrival and identity of bees were recorded at both feeders. Intriguingly, on both sites we observe a significant effect: Bees responded to the dance irrespective of the angle it communicated. Bees of group 1 flew to feeder 1 and the same holds for the other group. This observation recently has been reported to be often the case for natural dances too [14]). Under natural conditions bees might not follow the information encoded in the dance and fly to the experienced field spot rather than to the communicated feeder. It seems as the number of waggle runs followed is a crucial predictor for the outcome. Bees following more than 20 waggle runs most likely fly to the new field coordinate. For this row of experiments, individual recruits followed less 5 waggle runs each trial. The above described plastic cover was not yet used and hence, the bees clustered quickly after opening the access window. Using the plastic cover led to drastic improvements to this problem. With the cover the robot is able to continuously dance for many hours. After many hours we could observe following behavior of potential recruits very similar to that of natural dances. This behavior indicates that those bees try to decode the message encoded in the robotic dance. Some bees followed up to 35 waggle runs in a row before exiting the hive. We have tracked their flight using a harmonic radar system ([13]). Bees that followed robotic dances were identified in the hive if they had a number tag. When a particular bee exits the hive, a second person outside catches her, glues a radar transponder to her thorax and releases her immediately after. Although we were not able to track many bees, the few flight we recorded indicate a similar result as reported in [14]. However, the analysis of those flight tracks has not yet finished.

6 Conclusions and Future Work

We have built the first robotic honeybee that uses a reactive vision system and was able to excite sustained following behavior among other foragers. We show that the robot is able to increase foraging motivation. However, the final proof that the robot communicates the direction and distance to a food source could not yet be brought forward. The experimental results are promising, though. Bees could be motivated to fly out and forage at a previously known feeder. However, no bee of group 1 was registered at feeder 2 and vice versa. It can be argued that the missing following behavior in this set of experiments indicates a high level of disturbance and thus might have prevented the bees from decoding the message. We hypothesize that the dance message thus can be modeled as two-staged: First the motivational component, and second, the instructive component. The results of the recruitment experiments indicate that the robot might at least implement the motivational component. However, the experiment exhibits a central flaw: Group 1 was naive for the feeder 2 and vice versa. That might have resulted in an asymmetric threshold to decode the meaning of the dance. To test whether the robot is able to convey angular information one would have to equalize these thresholds by training a single group of bees to both feeders alternatingly such that after a few days this group has experienced reward at both locations equally frequent. However, natural dances are able to recruit naive bees. The more dance circuits are followed by a recruit the more likely she will show up at the communicated place. We could drastically increase the number of robotic dance circuits followed with a plastic cover sheet. The notion that the opening of the hive creates disturbances to the hive with an impact to the readiness to respond to robotic dances was thus confirmed. However, although an interested bee might follow many robotic waggle runs, the behavior is relatively rarely observable at all. Whatever determines the switch in behavior is still unknown to us. It might involve yet unknown signals or the correct display of a sequence of behaviors that in our experiments were played back coincidentally. It also might be related to the individual motivation to forage on new nestsites which varies with the knowledge of existing nest sites. The results so far nonetheless suggest that once the following behavior is excited the follower bee will most likely exit the hive for a long range search flight. The total numbers of flights we were able to record is still too low to make an assertion about the flight direction. We will investigate this further in the upcoming summer period.

Acknowledgments. The authors gratefully acknowledge the contribution of the German Research Foundation and reviewers' comments.

References

1. Bozic, J., Valentincic, T.: Attendants and followers of honeybee waggle dances. J. Apic. Res. (1991)
2. De Marco, R., Menzel, R.: Learning and memory in communication and navigation in insects. Learning and Memory-A Comprehensive Reference 1 (2008)
3. Esch, H., Esch, I., Kerr, W.E.: Sound: An Element Common to Communication of Stingless Bees and to Dances of the Honey Bee. Science 149(3681), 320–321 (1965)
4. Faria, J.J., Dyer, J.R.G., Clément, R.O., Couzin, I.D., Holt, N., Ward, A.J.W., Waters, D., Krause, J.: A novel method for investigating the collective behaviour of fish: introducing Robofish. Behavioral Ecology and Sociobiology, 1–8 (2010)
5. von Frisch, K.: Die Tänze der Bienen, Österr. Zool. Z 1, 1–48 (1946)
6. von Frisch, K.: Tanzsprache und Orientierung der Bienen. Springer, Berlin (1965)
7. Gil, M., De Marco, R.J.: Decoding information in the honeybee dance: revisiting the tactile hypothesis. Animal Behaviour 80(5), 887–894 (2010)
8. Griffin, D.R.: Animal minds. University of Chicago Press (1994)
9. Halloy, J., Sempo, G., Caprari, G., Rivault, C., Asadpour, M., Tache, F., Said, I., Durier, V., Canonge, S., Ame, J.M., Detrain, C., Correll, N., Martinoli, A., Mondada, F., Siegwart, R., Deneubourg, J.L.: Social Integration of Robots into Groups of Cockroaches to Control Self-Organized Choices. Science 318(5853), 1155–1158 (2007)
10. Kirchner, W.H., Towne, W.F.: The sensory basis of the honeybee's dance language. Sci. Am. (1994)
11. Landgraf, T., Rojas, R., Nguyen, H., Kriegel, F., Stettin, K.: Analysis of the waggle dance motion of honeybees for the design of a biomimetic honeybee robot. PloS One 6(8), e21354 (2011)
12. Landgraf, T., Kriegel, F., Nguyen, H., Rojas, R.: Extraction of motion parameters of waggle dancing honeybees for the design of a biomimetic honeybee robot (in prep.)
13. Menzel, R., Greggers, U., Smith, A., Berger, S., Brandt, R., Brunke, S., Bundrock, G., Hulse, S., Plumpe, T., Schaupp, F., Schuttler, E., Stach, S., Stindt, J., Stollhoff, N., Watzl, S.: Honey bees navigate according to a map-like spatial memory. Proceedings of the National Academy of Sciences 102(8), 3040–3045 (2005)
14. Menzel, R., Kirbach, A., Haass, W.-D., Fischer, B., Fuchs, J., Koblofsky, M., Lehmann, K., Reiter, L., Meyer, H., Nguyen, H., Jones, S., Norton, P., Greggers, U.: A common frame of reference for learned and communicated vectors in honeybee navigation. Curr. Biol. 21(8), 645–650 (2011)
15. Michelsen, A.: Signals and flexibility in the dance communication of honeybees. Journal of Comparative Physiology A: Neuroethology, Sensory, Neural, and Behavioral Physiology 189(3), 165–174 (2003)
16. Michelsen, A., Andersen, B.B., Storm, J., Kirchner, W.H., Lindauer, M.: How honeybees perceive communication dances, studied by means of a mechanical model. Behavioral Ecology and Sociobiology 30(3), 143–150 (1992)
17. Rohrseitz, K., Tautz, J.: Honey bee dance communication: waggle run direction coded in antennal contacts? Journal of Comparative Physiology A: Neuroethology, Sensory, Neural, and Behavioral Physiology 184(4), 463–470 (1999)
18. Seeley, T.D.: The Wisdom of the Hive: The Social Physiology of Honey Bee Colonies. Harvard University Press, London (1995)
19. Stabentheiner, A., Hagmller, K.: Sweet food means hot dancing in honeybees. Naturwissenschaften 78(10), 471–473 (1991)

20. Tautz, J.: Honeybee waggle dance: recruitment success depends on the dance floor. Journal of Experimental Biology 199(6), 1375–1381 (1996)
21. Thom, C., Gilley, D.C., Hooper, J., Esch, H.E.: The scent of the waggle dance. PLoS Biology 5(9), e228 (2007)
22. Towne, W.F., Kirchner, W.H.: Hearing in honey bees: detection of air-particle oscillations. Science 244(4905), 686 (1989)
23. Warnke, U.: Effects of electric charges on honeybees. Bee World 57(2), 50–56 (1976)

A Framework for Mobile Robot Navigation Using a Temporal Population Code

André Luvizotto[1], César Rennó-Costa[1], and Paul Verschure[1,2]

[1] The laboratory for Synthetic Perceptive, Emotive and Cognitive Systems - SPECS,
Universitat Pompeu Fabra,
Roc Boronat. 138, 08018 Barcelona, Spain
{andre.luvizotto,cesar.costa}@upf.edu

[2] ICREA - Institució Catalana de Recerca i Estudis Avançats - Barcelona, Spain
paul.verschure@upf.edu
http://specs.upf.edu/

Abstract. Recently, we have proposed that the dense local and sparse long-range connectivity of the visual cortex accounts for the rapid and robust transformation of visual stimulus information into a temporal population code, or TPC. In this paper, we combine the canonical cortical computational principle of the TPC model with two other systems: an attention system and a hippocampus model. We evaluate whether the TPC encoding strategy can be efficiently used to generate a spatial representation of the environment. We benchmark our architecture using stimulus input from a real-world environment. We show that the mean correlation of the TPC representation in two different positions of the environment has a direct relationship with the distance between these locations. Furthermore, we show that this representation can lead to the formation of place cells. Our results suggest that TPC can be efficiently used in a high complexity task such as robot navigation.

Keywords: Temporal Population Code, Navigation, Mobile Robots, Place Cells, Saliency Maps.

1 Introduction

Biological systems have an extraordinary capacity of performing simple and complex tasks in a very fast and reliable way. These capabilities are largely due to the robust processing of sensory information in an invariant manner. In the mammalian brain, the representation of dynamic scenes in the visual world are processed invariant to a number of deformations such as perspective, different light conditions, rotations and scales. This great robustness leads to successful strategies for optimal foraging. For instance, a foraging animal has to explore the environment, search for food, escape from possible predators and produce an internal representation of the world that allows it to successfully navigate through it.

To perform navigation with mobile robots using only visual inputs from a camera is a complex task [1]. Recently, a number of models have been proposed to solve this task [2,3]. Unlike the biological visual system, most of these methods are based on brute force methods of feature matching that are computationally expensive and stimulus dependent.

T.J. Prescott et al. (Eds.): Living Machines 2012, LNAI 7375, pp. 144–155, 2012.

In contrast, we have proposed a model of the visual cortex that can generate reliable invariant representations of visual information. The temporal population code, or TPC, is based on the unique anatomical characteristics of the neo-cortex: dense local connectivity and sparse long-range connectivity [4]. Physiological studies have estimated that only a few percent of synapses that make up cortical circuits originate outside of the local volume [5,6,7]. The TPC proposal has demonstrated the property of densely coupled networks to rapidly encode the geometric organization of the population response, for instance induced by external stimuli, into a robust and high-capacity encoding [8,9].

It has been shown that TPC provides an efficient encoding and can generalize to navigation tasks in virtual environments. In a previous study, TPC model applied to a simulated robot in a virtual arena could account for the formation of place cells [10]. It has also been showed that TPC can generalize to realistic tasks such as handwritten character classification [8]. The key parameters that control this encoding and the invariances that it can capture, are the topology and the transmission delays of the laterally coupled neurons that constitute an area. In this respect TPC emphasizes that the neo-cortex exploits specific network topologies as opposed to the randomly connected networks found in examples of, so-called, reservoir computing [11].

The encoding concept used in the TPC model has been supported by a number of physiological studies of different mammalian cortical areas [12,13,14,15]. The population code has also been observed in the antennal lobe of the moth [16,17]. These studies have provided direct evidence for the notion that the temporal dynamics of the population response in primary sensory areas can serve as a substrate for stimulus encoding. This way the TPC encoding can be applied independent of the properties of the input stimuli. It can thus be seen as a generic or canonical model of cortical encoding.

In this paper, we evaluate whether the TPC applied can be generalized to a navigation task in a natural environment. In particular we assess whether it can be used to extract spatial information from the environment. In order to achieve this generalization we combine the TPC, as a model of the ventral visual system, with a feedforward attention system and a hippocampal like model of episodic memory. The attention system uses early simple visual features to determine which salient regions in the scene the TPC framework will encode. The spatial representation is then generated by a model of the hippocampus that develops place cells. Our model has been developed to mimic the foraging capabilities of a rat. Rodents are capable to optimally explore the environment and its resources using distal visual cues for orientation. The images used in the experiments are acquired in a real-world environment taken by the camera of a mobile robot called the Synthetic Forager (SF) [18].

Our results show that the combination of TPC with a saliency based attention mechanisms can be used to develop robust place cells. These results directly support the hypothesis that the computational principle proposed by TPC leads to a spatial-temporal coding of visual features that can reliably account for position information in a real-world environment. Differently from previous TPC experiments with navigation, here we stress the capabilities of position representation in a real-world environment.

In the next section we present a description of the 3 different systems we use in the experiments. We detailed how the environment for the experiments was setup and how the images for the experiments were generated. In the following section we expose the main results obtained with the experiments and we finish discussing the core findings of the study.

2   Methods

2.1   Architecture Overview

The architecture presented here comprises three areas (Fig. 1): attention system, early visual system and a model of the hippocampus. The attention system exchanges information with the visual system in a bottom-up fashion. It receives bottom-up input of simple visual features and provides salience maps [19]. The saliency maps determine the sub regions that are rich in features such as colors, intensity and orientations that pop-out from the visual field and therefore will be encoded in the visual system.

The visual system also provides input to the hippocampal model in a bottom-up way. The subregions of interest determined by the attention system are translated into temporal representations that feed the hippocampal input stage. The temporal representation of the visual input is generated using the so called Temporal Population Code, or TPC [8,9,20]. The hippocampus model translates the visual representation of the environment in a spatial representation based on the activity of place cells [21,22]. These kinds of granule cells respond only to one or a few positions in the environment. In the following sections we present the details and modeling strategies used for each area.

2.2   Attention System

In our context, attention is basically the process of focusing the sensory resources to a specific point in space. We proposed a deterministic attention system that is driven by the local points of maximum saliency in the input image. To extract these subregions of interest, the attention system uses a saliency-based model proposed and described in detail in [19]. The system makes use of image features extracted by the early primate visual system, combining multi-scale image features into a topographical saliency map (bottom-up). The output saliency map has the same aspect ratio as the input image. To select the center points of the subregions, we calculate the points of local maxima $p_1, p_2, \ldots, p_{21}$ of the saliency map. The minimum accepted distance between peaks in the local maxima search is 20 pixels. This competitive process among the salient points can be interpreted as a top-down pushpull mechanism [23].

A subregion of size 41×41 is determined by a neighborhood of 20 pixels surrounding a local maximum p. Once a region is determined, the points inside of this area are not used in the search for the next p_{k+1}. This constraint avoids processing redundant neighboring regions with overlaps bigger than 50% in area. For the experiments, the first 21 local maxima are used by the attention system. These 21 points are the center of the subregions of interest cropped from the input image and sent to the TPC network. We used a Matlab implementation[1] of the saliency maps [19]. In the simulations we use the default parameters, except for the map width which is set to the same size as the input image: 967x100 (width and height respectively).

---

[1] The scripts can be found at:
http://www.klab.caltech.edu/~harel/share/gbvs.php

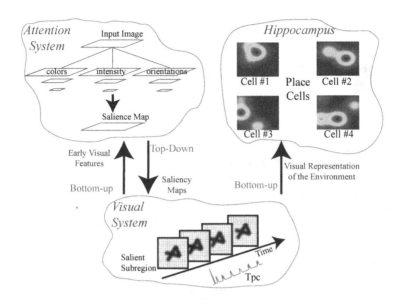

Fig. 1. Architecture scheme. The visual system exchanges information with both the attention system and the hippocampus model to support navigation. The attention system is responsible for determining which parts of the scene are considered for the image representation. The final position representation is given by the formation of place cells that show high rates of firing whenever the robot is in a specific location in the environment, i.e. place cells.

2.3   Visual System

The early visual system consists of two stages: a model of the lateral geniculate nucleus (LGN) and a topographic map of laterally connected spiking neurons with properties found in the primary visual cortex V1 (Fig. 2) [9,8]. In the first stage we calculate the response of the receptive fields of LGN cells to the input stimulus, a gray scale image that covers the visual field. The approximation of the receptive field's characteristics is performed by convolving the input image with a difference of Gaussians operator (DoG) followed by a positive half-wave rectification. The positive rectified DoG operator resembles the properties of *on* LGN center-surround cells [24,25]. The LGN stage is a mathematical abstraction of known properties of this brain area and performs an edge enhancement on the input image. In the simulations we use a kernel ratio of 4:1, with size of 7X7 pixels and variance $\sigma = 1$ (for the smaller Gaussian).

The LGN signal is projected to the V1 spiking model, where the coding concept is illustrated in Fig. 3. The network is an array of NxN model neurons connected to a circular neighborhood with synapses of equal strength and instantaneous excitatory conductance. The transmission delays are related to the Euclidean distance between the positions of the pre- and postsynaptic neurons. The stimulus is continuously presented to the network and the spatially integrated spreading activity of the V1 units, as a sum of their action potentials, results in the so called TPC signal.

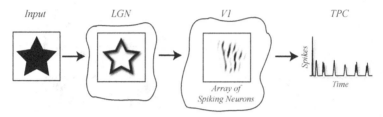

Fig. 2. Visual model overview. The saliency regions are detected and cropped from the input image provided by camera image. The cropped areas, subregions of the visual field, have a fixed resolution of 41x41 pixels. Each subregion is convolved with difference of gaussian (DoG) operator that approximates the properties of the receptive field of LGN cells. The output of the LGN stage is processed by the simulated cortical neural population and their spiking activity is summed over a specific time window rendering the Temporal Population Code.

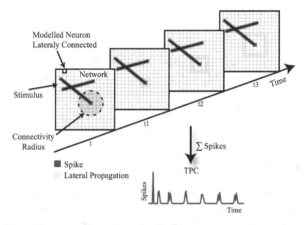

Fig. 3. The TPC encoding paradigm. The stimulus, here represented by a geometric shape, is projected topographically onto a map of interconnected cortical neurons. When a neuron spikes, its action potential is distributed over a neighborhood of a given radius. The lateral transmission delay of these connections is 1 ms/unit. Because of these lateral intra-cortical interactions, the stimulus becomes encoded in the network's spatio-temporal activity trace. The TPC representation is defined by the spatial average the population activity over a certain time window. The invariances that the TPC encoding renders are defined by the local excitatory connections.

In the network, each neuron is approximated using the spiking model proposed by Izhikevich [26]. Relying only on four parameters, our network can reproduce different types of spiking behavior using a system of ordinary differential equations of the form:

$$v' = 0.04v^2 + 5v + 140 - u + I \qquad (1)$$

$$u' = a(bv - u) \qquad (2)$$

with the auxiliary after-spike resetting:

$$\text{if } v \geq 30 \text{ mV, then } \begin{cases} v \leftarrow c \\ u \leftarrow u + d \end{cases} \tag{3}$$

Here, v and u are dimensionless variables and a, d, c and d are dimensionless parameters that determine the spiking or bursting behavior of the neuron unit and $' = \frac{d}{dt}$, where t is the time. The parameter a describes the time scale of the recovery variable u. The parameter b describes the sensitivity of the recovery variable u to the sub-threshold fluctuations of the membrane potential v. The parameter c accounts for the after-spike reset value of v caused by the fast high-threshold K^+, and d the after-spike for the reset of the recovery variable u caused by slow high-threshold Na^+ and K^+ conductances.

The excitatory input I in eq. 1 consists of two components: first a constant driving excitatory input g_i and second the synaptic conductances given by the lateral interaction of the units $g_c(t)$. So

$$I(t) = g_i + g_c(t) \tag{4}$$

For the simulations, we used the parameters suggested in [27] to reproduce regular firing (RS) spiking behavior. All the parameters used in the simulations are summarized in the table 1. The lateral connectivity between V1 units is exclusively excitatory with strength w. A unit $u_a(\mathbf{x})$ connects with u_b if they have a different center position $\mathbf{x}_a \neq \mathbf{x}_b$ and if they are within a region of a certain radius $\|\mathbf{x}_b - \mathbf{x}_a\| < r$.

According to recent physiological studies, intrinsic V1 intra-cortical connections cover portions that represent regions of the visual space up to eight times the size of the receptive fields in V1 neurons [28]. In our model we set the connectivity radius r within this range in 7 units. The lateral synapses are of equal strength w and the transmission delays τ_a are proportional to $\|\mathbf{x}_b - \mathbf{x}_a\|$ with 1 ms/cell.

The temporal population code is generated by summing the network activity in a time window of 64 ms. Thus a TPC is a vector with the spiking count of the network for each time step. In the discrete-time, all the equations are integrated with Euler's method using a temporal resolution of 1 ms.

3   Hippocampus Model

The hippocampus model is an adaptation of a recent study on the rate remapping in the dentate gyrus [21] [29] [22]. In our implementation, the granule cells receive excitatory input from the grid cells of the enthorinal cortex. The place cells that are active for a given position in the environment are then determined according to the interaction of the summed excitation and inhibition using a rule based on the percentage of maximal suprathreshold excitation E%-max winner-take-all (WTA) process. The excitatory input received by the i_{th} place cell from the grid cells is given by:

$$I_i(r) = \sum_{j-1}^{n} W_{ij} G_j(r) \tag{5}$$

where W_{ij} is the synaptic weight of each input. In our implementation, the weights are either random values in the range of $[0, 1]$ (maximum bin size among normalized

Table 1. Parameters used for the simulations

Variable	Description	Value
N	network dimension	41x41 Neurons
a	scale of the recovery	0.02
b	sensitivity of the recovery	0.2
c_{rs}	after-spike reset value of v for RS neurons	-65
c_{bs}	after-spike reset value of v for BS neurons	-55
v	membrane potential	-70
u	membrane recovery	-16
g_i	excitatory input conductance	20
T_i	minimum V1 input threshold	0.4
r	lateral connectivity radius	7 units
w	synapse strength	0.3

histograms) or 0 in case of no connection. In our implementation, we use the TPC histograms as the information provided by the grid cells $G_j(r)$. The activity of the i_{th} place cell is given by:

$$F_i(r) = I_i(r)H(I_i(r) - (1 - k)^{max}_{grid}(r))$$ (6)

The constant k varies in the range of 5% to 15%. It is referred to E%-max and determines which cells will fire [29]. Specifically, the rule states: a cell fires if their feedforward excitation is within E% of the cell receiving maximal excitation.

4 Experimental Environment

For the experiments we produced a dataset of pictures from an indoor square area of $3m^2$. The area was divided in a 5x5 grid of 0.6 m^2 square bins, compromising 25 sampling positions (Fig. 4a). The distance to objects and walls ranged from 30 cm to 5 meters. To emulate the sensory input we approximated the rat view at each position with 360 degree panoramic pictures. Rats use distal visual cues for orientation [30] and such information is widely available for the rat given the spaced positioning of their eyes, resulting in a wide field of view and low binocular vision [31]. The pictures were built using the software Hugin [32] to process 8 pictures with equally distant orientations (45 degrees steps) for each position. The images were acquired using the SF-robot [18]. Each indoor images was segmented in 21 subregions of 41×41 pixels by the attention system as described earlier. The maximum overlap among the subregions is 50 % (Fig. 4b).

4.1 Image Representation Using TPC

For every image, 21 TPC vectors are generated, i.e. one for each subregion selected by the attention system. In total, we acquired 25 panoramic images of the environment, leading to a matrix of 525 vectors of TPCs (25x21). We cluster the TPC matrix into

7 classes using the k-means cluster algorithm. Therefore, a panoramic image in the environment is represented by the cluster distribution of its TPC vectors using a histogram of 7 bins. The histograms are used as a representation of each position in the environment (Fig. 4b). They are the inputs for the hippocampus model.

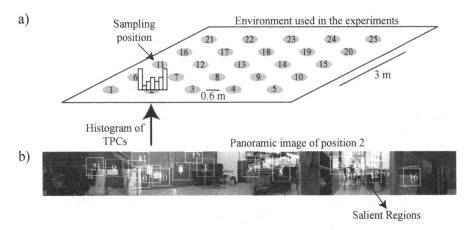

Fig. 4. Experimental environment. a) The indoor environment was divided in a 5x5 grid of 0.6 m^2 square bins, compromising 25 sampling positions. b) For every sampling position, a 360 degrees panorama was generate to simulate the rat's visual input. A TPC responses is calculated for each salient region, 21 in total. The collection of TPC vectors for the environment is clustered into 7 classes. Finally, a single image is represented by the cluster distribution of its TPC vectors using a histogram of 7 bins.

5   Results

In the first experiment, we explored whether the TPC response preserves bounded invariance in the observation of natural scenes, i.e. if the representation is tolerant to certain degrees of visual deformations without loosing specificity. More specifically, we investigate whether the correlation between the TPC histograms of visual observations in two positions can be associated with their distance.

We calculate the mean pairwise correlation among TPC histograms of the 25 sampling positions used for image acquisition in the environment. We sort the correlation values into distance intervals of 0.6 m, the same distance intervals used for sampling the environment.

Our results show that the TPC transformation can successfully account for position information in a real world scenario. The mean correlation between two positions in the environment decays monotonically with the increase of their distance (Fig. 5). As expected, the error in the correlation measure increases with the distance.

In the second experiment, we use the TPC histograms as input for the hippocampus model to acquire place cells. Neural representations of position, as observed in place cells, present a quasi-Gaussian transition between inactive and active locations, which

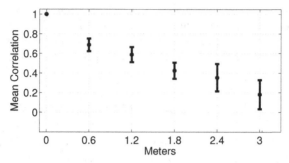

Fig. 5. Pairwise correlation of the TPC histograms in the environment. We calculate the correlation among all the possible combinations of two positions in the environment. We average the correlation values into distance intervals, according to the sampling distances used in the image acquisition. This result suggests that we can produce a linear curve of distance based on the correlation values calculated using the TPC histograms.

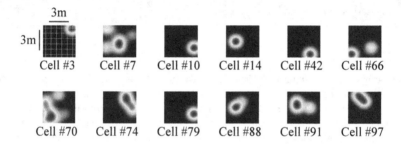

Fig. 6. Place cells acquired by the combination of TPC based visual representations and the E%-max WTA model of the hippocampus

makes bounded invariance a key property for a smooth reconstruction of the position from visual input as observed in nature. Previous work have shown that the conjunction of TPC responses of synthetic figures in a virtual environment do preserve bounded invariance and, with the use of reinforcement learning, place cells could be emerge from these responses [33].

We perform the simulations using 100 cells with random weights. The results show that from 100 cells used, 12 have significant place related activity (Fig. 6). These cells show high activity in specific areas of the environment, also called place fields. Thus the TPC representation of the natural scenes can successfully lead to the formation of place fields.

6   Discussion

We addressed the question whether we could combine the properties of TPC with an attention system and a hippocampus model to reliably provide position information in a navigation scenario. We have shown that in our cortical model, the TPC, the representation of a static visual stimulus is reliably generated based on the temporal dynamics of neuronal populations. The TPC is a relevant hypothesis on the encoding of sensory events given its consistency with cortical anatomy and its consistency with contemporary physiology.

The information carried in this, so called, Temporal Population Code, is efficiently used to pass a complete and dense amount of information. The encoding of visual information is done through a sub-set of regions of high saliency that pops up from the visual field. Combining the TPC encoding strategy with a hippocampus model, we have shown that this approach provides for robust representation of position in a natural environment.

In the specific benchmark evaluated here, the model showed that distance information could be reliably recovered from the TPC. Importantly, we showed that the model can produce reliable position information based on the mean correlation value of TPC based histograms. Furthermore, this method does not use any particular mechanism specific to the stimuli used in the experiments. It is totally generic.

Panoramic images have been proposed in other studies of robot navigation [34]. Similarly, the TPC strategy proposed here also makes use of pixel differences between panoramic images to provide a reliable cue to location in real-world scenes. In both cases differences in the direction of illumination and environmental motion often can seriously degrade the position representation. However, in comparison, the representation provided by TPC is extremely compact. A subregion of 41x41 pixels is transformed in a TPC vector of 64 points. Finally, the visual field, an image of 96.000 pixels, is represented by a histogram of 7 bins.

Also, the processing time needed for classifying an input of 41x41 pixels was about 50 ms using an offline Matlab implementation. Therefore, we conclude that our model could be efficiently used in a real-time task. Hence, our model shows that the brain might use the smooth degradation of natural images represented by TPCs as an estimate of distance.

Acknowledgment. This work was supported by EU FP7 projects GOAL-LEADERS (FP7-ICT-97732) and EFAA (FP7-ICT-270490).

References

1. Bonin-Font, F., Ortiz, A., Oliver, G.: Visual Navigation for Mobile Robots: A Survey. J. Intell. Robotics Syst. 53(3), 263–296 (2008)
2. Jun, S., Kim, Y., Lee, J.: Difference of wavelet SIFT based mobile robot navigation. In: 2009 IEEE International Conference on Control and Automation, pp. 2305–2310. IEEE (December 2009)
3. Koch, O., Walter, M.R., Huang, A.S., Teller, S.: Ground robot navigation using uncalibrated cameras. In: 2010 IEEE International Conference on Robotics and Automation, pp. 2423–2430. IEEE (May 2010)
4. Binzegger, T., Douglas, R.J., Martin, K.A.C.: A quantitative map of the circuit of cat primary visual cortex. The Journal of Neuroscience: the Official Journal of the Society for Neuroscience 24(39), 8441–8453 (2004)

5. Schubert, D., Kötter, R., Staiger, J.F.: Mapping functional connectivity in barrel-related columns reveals layer- and cell type-specific microcircuits. Brain Structure & Function 212(2), 107–119 (2007)
6. Liu, B.H., Wu, G.K., Arbuckle, R., Tao, H.W., Zhang, L.I.: Defining cortical frequency tuning with recurrent excitatory circuitry. Nature Neuroscience 10(12), 1594–1600 (2007)
7. Nauhaus, I., Busse, L., Carandini, M., Ringach, D.L.: Stimulus contrast modulates functional connectivity in visual cortex. Nature Neuroscience 12(1), 70–76 (2009)
8. Wyss, R., Konig, P., Verschure, P.F.M.J.: Invariant representations of visual patterns in a temporal population code. Proceedings of the National Academy of Sciences of the United States of America 100(1), 324–329 (2003)
9. Wyss, R., Verschure, P.F.M.J., König, P.: Properties of a temporal population code. Reviews in the Neurosciences 14(1-2), 21–33 (2003)
10. Wyss, R., König, P., Verschure, P.F.M.J.: A model of the ventral visual system based on temporal stability and local memory. PLoS Biology 4(5), e120 (2006)
11. Lukoševičius, M., Jaeger, H.: Reservoir computing approaches to recurrent neural network training. Computer Science Review 3(3), 127–149 (2009)
12. Samonds, J.M., Bonds, A.B.: From another angle: Differences in cortical coding between fine and coarse discrimination of orientation. Journal of Neurophysiology 91(3), 1193–1202 (2004)
13. Benucci, A., Frazor, R.A., Carandini, M.: Standing waves and traveling waves distinguish two circuits in visual cortex. Neuron 55(1), 103–117 (2007)
14. Gollisch, T., Meister, M.: Rapid Neural Coding in the Retina with Relative Spike Latencies. Science 319(5866), 1108–1111 (2008)
15. MacEvoy, S.P., Tucker, T.R., Fitzpatrick, D.: A precise form of divisive suppression supports population coding in the primary visual cortex. Nature Neuroscience 12(5), 637–645 (2009)
16. Carlsson, M.A., Knusel, P., Verschure, P.F.M.J., Hansson, B.S.: Spatio-temporal Ca2+ dynamics of moth olfactory projection neurones. European Journal of Neuroscience 22(3), 647–657 (2005)
17. Knusel, P., Carlsson, M.A., Hansson, B.S., Pearce, T.C., Verschure, P.F.M.J.: Time and space are complementary encoding dimensions in the moth antennal lobe. Network 18(1), 35–62 (2007)
18. Rennó-Costa, C., Luvizotto, A.L., Marcos, E., Duff, A., Sánchez-Fibla, M., Verschure, P.F.M.J.: Integrating Neuroscience-based Models Towards an Autonomous Biomimetic Synthetic. In: 2011 IEEE International Conference on RObotics and BIOmimetics (IEEE-ROBIO 2011), Phuket Island, Thailand. IEEE (2011)
19. Itti, L., Koch, C., Niebur, E.: A model of saliency-based visual attention for rapid scene analysis. IEEE Transactions on Pattern Analysis and Machine Intelligence 20(11), 1254–1259 (1998)
20. Luvizotto, A., Rennó-Costa, C., Pattacini, U., Verschure, P.F.M.J.: The encoding of complex visual stimuli by a canonical model of the primary visual cortex: temporal population coding for face recognition on the iCub robot. In: IEEE International Conference on Robotics and Biomimetics, Thailand, p. 6 (2011)
21. de Almeida, L., Idiart, M., Lisman, J.E.: The input-output transformation of the hippocampal granule cells: from grid cells to place fields. The Journal of Neuroscience: the Official Journal of the Society for Neuroscience 29(23), 7504–7512 (2009)
22. Rennó-Costa, C., Lisman, J.E., Verschure, P.F.M.J.: The mechanism of rate remapping in the dentate gyrus. Neuron 68(6), 1051–1058 (2010)
23. Mathews, Z., i Badia, S.B., Verschure, P.F.M.J.: PASAR: An integrated model of prediction, anticipation, sensation, attention and response for artificial sensorimotor systems. Information Sciences 186(1), 1–19 (2011)

24. Rodieck, R.W., Stone, J.: Analysis of receptive fields of cat retinal ganglion cells. Journal of Neurophysiology 28(5), 833 (1965)

25. Einevoll, G.T., Plesser, H.E.: Extended difference-of-Gaussians model incorporating cortical feedback for relay cells in the lateral geniculate nucleus of cat. Cognitive Neurodynamics, 1–18 (November 2011)

26. Izhikevich, E.M.: Simple model of spiking neurons. IEEE Transactions on Neural Networks 14(6), 1569–1572 (2003)

27. Izhikevich, E.M.: Which model to use for cortical spiking neurons? IEEE Transactions on Neural Networks 15(5), 1063–1070 (2004)

28. Stettler, D.D., Das, A., Bennett, J., Gilbert, C.D.: Lateral Connectivity and Contextual Interactions in Macaque Primary Visual Cortex. Neuron 36(4), 739–750 (2002)

29. de Almeida, L., Idiart, M., Lisman, J.E.: A second function of gamma frequency oscillations: an E%-max winner-take-all mechanism selects which cells fire. The Journal of Neuroscience: the Official Journal of the Society for Neuroscience 29(23), 7497–7503 (2009)

30. Hebb, D.O.: Studies of the organization of behavior. I. Behavior of the rat in a field orientation. Journal of Comparative Psychology 25, 333–353 (1932)

31. Block, M.: A note on the refraction and image formation of the rat's eye. Vision Research 9(6), 705–711 (1969)

32. D'Angelo, P.: Hugin (2010)

33. Wyss, R., Verschure, P.F.M.J.: Bounded Invariance and the Formation of Place Fields. In: Advances in Neural Information Processing Systems 16. MIT Press (2004)

34. Zeil, J., Hofmann, M.I., Chahl, J.S.: Catchment areas of panoramic snapshots in outdoor scenes. Journal of the Optical Society of America A 20(3), 450 (2003)

Generalization of Integrator Models to Foraging: A Robot Study Using the DAC9 Model

Encarni Marcos, Armin Duff, Martí Sánchez-Fibla, and Paul F.M.J. Verschure

SPECS, Department of Technology, Unversitat Pompeu Fabra,
Roc Boronat 138, 08018, Barcelona, Spain
{encarnacion.marcos,armin.duff,marti.sanchez,paul.verschure}@upf.edu
http://specs.upf.edu

Abstract. Experimental research on decision making has been mainly focused on binary perceptual tasks. The generally accepted models describing the decision process in these tasks are the integrator models. These models suggest that perceptual evidence is accumulated over time until a decision is made. Therefore, the final decision is based solely on recent perceptual information. In behaviorally more relevant tasks such as foraging, it is however probable, that the current choice also depends on previous experience. To understand the implications of considering previous experience in an integrator model we investigate it using a cognitive architecture (DAC9) with a robot performing foraging tasks. Compared to an instantaneous decision making model we show that an integrator model improves performance and robustness to task complexity. Further we show that it compresses the information stored in memory. This result suggests a change in the way actions are retrieved from memory leading to self-generated actions.

Keywords: sequence learning, decision making, integrator models.

1 Introduction

Binary perceptual tasks have been widely used to study the neural mechanisms underlying decision making [1,2]. This kind of task involve a simple decision about a feature of a stimulus that is expressed as a choice between two alternative options. Many models have been proposed to explain this decision making process predicting the relationship between reaction time and accuracy [3,4]. Most of them explain decision making as an accumulation process that takes place over time until a decision bound is reached. These models are known as integrator models and have been generally accepted as an explanation for decision making in perceptual tasks where learning is not required to successfully perform the task. Here, we investigated the interaction between an integrator model and memory in foraging tasks using a well establish cognitive architecture as a framework [5].

One largely used perceptual experimental paradigm is defined by a random-dot motion (RDM) task where humans or monkeys have to select between two

T.J. Prescott et al. (Eds.): Living Machines 2012, LNAI 7375, pp. 156–167, 2012.

possible stimulus categories, such as leftward or rightward motion [6]. Integrator models, such as race models [3] and drift-diffusion models [4], provide a straightforward account of the speed-accuracy trade-off. These models suggest that evidence is accumulated over time until this accumulation reaches a bound, i. e. criterion level, and a decision is made. As the RDM task, many of the experimental paradigms used to study the decision making process are simple perceptual tasks where the correct performance of a trial depends exclusively on the current perceptual information, e. g. color. The proposed integrator models assume that the alternative options are known a priori and therefore learning during the task is not required. However, this would not be the case in more realistic foraging tasks where the information about different targets have to be acquired from the environment and many alternative choices might be available at each decision point. Therefore, a two-fold problem has to be solved during foraging: the appropriate learning of the environment and actions and the appropriate retrieval of information to achieve goal states (targets), i. e. sequences of perception and action need to be learned and retrieved to reach goal positions in an environment.

To study the interaction between decision making and memory, we worked in the framework of the Distributed Adaptive Control (DAC) architecture [7,5]. The decision making in DAC follows the Bayesian principle [8,9]. We extended the architecture with an integrator decision making model (DAC9; see [7,10,11] for details about previous versions of DAC), based on the race model, to investigate its implications during foraging tasks and we further compared it with the decision making in DAC (instantaneous model). We show that the integrator model resulted in a new mechanism of storing and recalling information from memory suggesting that the actions are not stored in memory but self-generated during retrieval of information. In a previous study [12], we assessed the impact of these two decision making models in the learning of event order and interval in a sequence in two foraging tasks. In the current study, we go one step further quantifying (1) the scalability of the two models with task complexity in five different foraging tasks and (2) the implications on the information stored in memory and proposing (3) a new working memory mechanism that accounts for a continuous action space.

2   Materials and Methods

2.1   Cognitive Architecture

The DAC architecture has already proven its suitability to study the problems encountered in biology helping to investigate perception, cognition and behavior in foraging situations in which the access to real neuronal and behavioural data is difficult [5]. DAC is based on the assumption that learning consists of the interaction of three layers of control: reactive, adaptive and contextual, as illustrated in Fig. 1. The reactive layer provides pre-wired responses that allows for a simple interaction with the environment and accomplish simple automatic behaviours. The adaptive layer provides mechanism for the classification of the sensory events (internal representations) and the shaping of responses in simple tasks as in classical conditioning [13]. The internal representations (prototypes; see [14] for details) generated by the adaptive layer are stored in the contextual layer as couplets of sensory-motor states and used to plan future behaviour, as in operant conditioning [15].

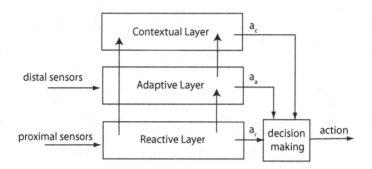

Fig. 1. Schematic representation of the DAC architecture. It is based on the assumption that behavior results from three tightly coupled layers of control: reactive, adaptive and contextual. Abbreviations mean: a_r, reactive layer action; a_a, adaptive layer action, a_c, contextual layer action.

In this study, we mainly focused on the contextual layer of DAC that provides mechanisms for memorizing and recalling information. It consists of two memory structures: the short-term memory (STM) and the long-term memory (LTM), for permanent storage of information. During learning, pairs of prototype-action are stored in the STM as the robot interacts with the environment. When a goal state is reached, i. e. reward or punishment, the content of the STM is copied into the LTM and the STM is reset. The LTM has sequences of pairs of prototype-action that lead the robot to goal states. The prototype-action pairs that form a sequence are called segments. During the recall process, the prototypes stored in LTM are matched against the generated prototypes from ongoing sensory events. The degree of matching of segment l in sequence q determines the input to its, so called, *collector* unit, c_{lq}:

$$c_{lq} = (1 - d(e, e_{lq}))t_{lq} \qquad (1)$$

where $d(e, e_{lq})$ is calculated as the Euclidean distance between stored *prototype* e_{lq} and current *prototype* e and t_{lq} is called *trigger*. The trigger value biases the sensory matching process of the segments and allows chaining through a sequence, i. e. its default value is 1 and it is set to a higher value if the previous segment $l - 1$ is activated.

The activity of the collectors contribute to the action proposed by the contextual layer. We only consider the collectors' activity that satisfy both conditions:

(1) its activity is above a certain threshold (θ^C), (2) its activity is inside a predefined percentage range from the maximum collector's activity, i.e. the collectors compete in an E%-Max Winner Take All (WTA) mechanism [16]. The actual action proposed by the contextual layer (a_c) is calculated as:

$$a_c = \sum_{l,q \in LTM} \pm \frac{c_{lq} H(c_{lq} - \theta^C)}{\delta_{lq}} a_{lq} \qquad (2)$$

where $H(.)$ is a step function that is 0 for values lower than θ^C and is 1 for values higher than θ^C, δ_{lq} is the distance measured in segments between the selected segment l and the last segment in the sequence, i.e. the distance to the goal state and a_{lq} is the action stored in segment l of sequence q. By doing this division the segments closer to the goal state have more impact on the contextual action. The sign is positive if the segment belongs to an appetitive sequence and negative if it belongs to an aversive sequence.

The actions triggered by each of the three different layers are filtered by priority, giving more priority to reactive actions (a_r), then to contextual actions (a_c) and finally to adaptive actions (a_a). The one that takes the control of the motor is stored in STM and afterwards in LTM.

2.2   Integrator Models

Many integrator models have been proposed, but mainly, in all of them, the change in the accumulation of evidence in favour of one alternative $(x_i(t))$ can be described as:

$$\frac{dx_i}{dt} = \mu E_i(t) + \xi \qquad (3)$$

where μ is the growth rate of the accumulation, $E_i(t)$ is the internal estimate of evidence at time t and ξ is a Gaussian noise with mean of zero and variance of σ^2. The proposed models consider the variables $x_i(t)$, $E_i(t)$, μ and ξ in a different manner. We implement a rise-to-threshold model based on the race model. The race model [3] suggests that there are separate variables $x_i(t)$ for each option that accumulate evidence independently until one of them reaches a decision bound and a decision is made.

Our implementation of the race model consisted of a number of independent variables that compete to take the control of the robot. Each variable accumulated evidence in favour of one action, such as right or left. When the value of a variable grew above a criterion level, i. e. decision bound, the action associated with it was performed by the robot. The change in the activity of the variables within a time step dt was defined as:

$$da_i(t) = \begin{cases} dt(\mu_r a_{r_i} + \mu_a a_{a_i} + \mu_c a_{c_i} + \xi) & \text{, if } t - t_{la} > T_{ref} \\ 0 & \text{, if } t - t_{la} \leq T_{ref} \end{cases} \qquad (4)$$

where a_r, a_a, a_c are the actions triggered by the reactive, adaptive and contextual layer respectively, $i \, \epsilon \, \mathbb{N}^N$ and it is the subindex of the N different possible actions, μ_r, μ_a, μ_c are the mean growth rates of the variables units, ξ is a Gaussian noise term with a mean of zero and a variance of σ^2, t_{la} is the time at which the last action was executed and T_{ref} is the refractory period. In our experiments $dt = 1ms$ and $\xi = 0$. When the value of a_i reaches a predefined threshold the associated action is executed. In biology, the refractory period is the amount of time a excitable membrane needs to be ready for a second stimulus once it returns to the resting state. Consistent with this, the T_{ref} term referred to the amount of time necessary to start again the competition between actions after one of them was executed.

2.3   Foraging Tasks

The mobile agent was simulated in C++ and wSim [17] using the 3D Open Graphics Library approximating a Kephera robot [1]. Different previous studies have proven the validity of this simulated robot with respect to a real one [17,18]. The robot has a radio of 5.5 cm and 8 proximity sensors and 8 light sensors. The values captured by both light and proximity sensors decay exponentially. The proximity sensors measure the distance to obstacles while the light sensors measure the intensity of light sources. The robot is equipped with a color camera with a visual angle of 45 deg. of amplitude. The image from the camera is color separated such that there are three channels: red, green and blue, each of them with a resolution of 36x36 pixels. Except otherwise specified the camera is always pointing to the floor with a tilt angle of -60 deg. with respect to the horizontal axis. The robot translates with a speed of $0, 1 \times robotradius$ and it rotates with a speed of 10 deg.

To study the interaction between an integrator model and memory we defined a number of foraging tasks where not only perceptual but also memory information was essential to achieve a performance about chance. The tasks had different rated complexity to assess how the decision making models scaled to it (Figure 2). In all the environments the goal of the task was to go to the light source, i. e. reward. Every trial started from one of the positions shown in Figure 2, randomly selected. The trial ended when the robot hit the light or collided with the wall. A successful trial ended when the light was hit. The environments contained colored patches that served as cues. The light was detected by the light sensors of the robot. However, the light was not strong enough to trigger a reactive action from the side patches. The adaptive layer used reactive layer sub-threshold activity to generate the prototypes and to learn the associations between prototype-action. Once the prototypes were stable the contextual layer started storing sequences of prototype-action that leaded to a goal state. The goal state occurs when the robot reaches the light or collides with the wall. When a collision occurred it was stored as an undesirable state in memory and had a negative influence on the action proposed by the contextual layer.

---

[1] K-Team, Lausanne, Switzerland.

The complexity of the tasks was rated taking into account the number of patches and how ambiguous they were as follows:

$$TC = \frac{n_p}{n_c} n_a \tag{5}$$

where n_p is the number of patches, n_c is the number of different colors and n_a is the number of different turning angle amplitudes needed to be learned. This measure was then useful to compare the robot performance in each of the tasks for the two proposed models. The complexity of the task 1 is 3, the complexity of the task 2 is 4.5 and 5, 7.5 and 11.7 for the tasks 3, 4 and 5, respectively.

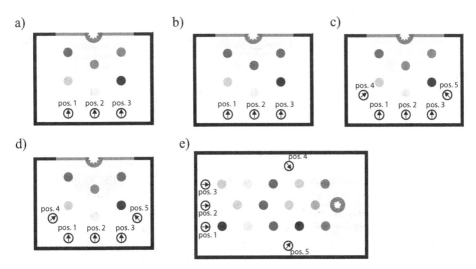

Fig. 2. Foraging tasks ordered by task complexity. (a) Task 1. Unambiguous restricted open arena, $TC = 3$. (b) Task 2. Ambiguous restricted open arena, $TC = 4.5$. (c) Task 3. Ambiguous restricted open arena, $TC = 5$. (d) Task 4. Ambiguous restricted open arena, $TC = 7.5$. (e) Task 5. Ambiguous restricted open arena, $TC = 11.7$.

The first task was an unambiguous restricted open arena foraging task, i.e. no context information was needed because the location of the target was uniquely predicted by the color patches (Fig. 2a). Therefore, this task could be correctly solved by the adaptive layer, but still we tested the performance at the contextual layer level. The rest of the four tasks consisted of ambiguous restricted open arenas, because, in all cases, context information was needed to reach a performance above chance. The actions associated with the patches closest to the light were not unique but depended on the previous context (Fig. 2b, 2c, 2d and 2e) and therefore the problem could only be solved at the contextual layer level. These experiments allowed us to study the detailed performance of each model and its dynamics as well as to evaluate the results in tasks where different kind of actions were required. In all the foraging tasks, when the contextual layer was enabled, the actions from the reactive and the adaptive layers were deactivated to avoid any influence they could have on the results.

To test the system for robustness we added 5% of noise to the motors, following a Gaussian random distribution and we varied the initial position of each trial according to a two dimensional normal distribution with mean 0 and variance $0, 1\times$ robot radius. Moreover, to assess the impact of the camera noise in the information stored in memory, in Task 2 (Fig. 2b and 2e), we added noise to the hue sensed by the camera from 0% to 10% in steps of 1%, following a Gaussian random distribution. For every condition, we ran 10 experiments with 1000 trials each.

To investigate what was the impact on memory of the interaction between the decision making models and memory itself we calculated the degree of compression of information in memory through the entropy of the stored information as follows:

$$E_M = -\sum_{s \epsilon S} p(s) log_2(p(s)) \tag{6}$$

where s is one segment in memory and $p(s)$ is the probability that the segment is selected in a current experiment. This measurement allows us to assess the amount of information needed to encode a visual stimulus in memory.

3   Results

In this study, we investigated the generalization of an integrator model in foraging tasks. We designed a number of foraging tasks with increased complexity to assess the generalization of the integrator model in more realistic tasks. The results in these tasks suggested a new mechanism to store and recall information from memory. We further tested the implications of this new mechanism in a foraging task and we show that it resulted in a more optimal way of learning and exploiting the environment.

3.1   Foraging Tasks

In all the tasks, we recorded the performance of the robot after LTM acquisition. As shown in Fig. 3, as task complexity increased the performance of the robot decreased dramatically in the case of the instantaneous model where it dropped to a mean value of 0.55 for the most complex task. It kept stable in the integrator model, maintaining a mean value of performance above 0.9.

In order to evaluate the impact of the camera noise in the information stored in memory due to the influence of each decision making models, we used Task 2 ($TC = 4.5$) because it was the simplest one that requires the use of the contextual layer. For clarity we also report here the performance of the robot with varying camera noise [12]. As previously reported in [12], the performance of the robot decreased as the camera noise increased in both models (Fig. 4a). The difference between the performance of the two models was significantly different along the different values of camera noise (ks-test, $p < 0.01$). From 0% of camera noise to 6% the instantaneous model was incrementally more affected by the noise than the integrator model. However, from 6% to 10% the noise had an important

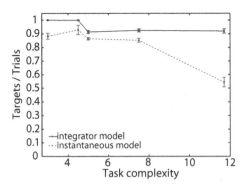

Fig. 3. Robot performance along different task complexity. Ratio targets/trials with instantaneous and integrator models as task complexity increases. Bars represent means ± sem.

impact on the integrator model, resulting in a smaller difference in performance with respect to the instantaneous model. Once the noise of the camera started to critically affect the sequentiality of the actions the performance decayed in both models with a similar slope (Fig. 4a).

To assess the impact of both models at the memory level we calculated the entropy of the stored information, E_M (see Eq. 6). As shown in Fig. 4b, E_M with the integrator model was higher along the different camera noise compared to the instantaneous model (ks-test: $p < 0.001$). Moreover, the dynamics in both cases were opposite: E_M decreased as camera noise increased in the integrator model whereas it increased as camera noise increased in the case of the instantaneous model. Low values of E_M means that segments of memory respond to a small fraction of the stimuli resulting in a higher number of segments in memory. The opposite occurs for high values of E_M. Therefore, the integrator model compressed the memory and less number of segments were necessary to encode same stimulus. As a drawback, explicit representation of time in memory, i. e. the number of steps needed to cross a patch, is lost.

3.2   Self-generated Actions

The compression of information in memory due to the use of the race model changes the way information is stored in memory suggesting a new mechanism to recall it. Instead, of a recall of actions from memory it suggests the recall of goals. Consequently, we hypothesize that the actions are self-generated rather than stored in memory. During the recall period, visual information is retrieved from memory and actions are performed depending on the position of perceptual target with respect to the robot. When the information is selected from memory, we distinguish between two different recall methods: (1) the next prototype in the sequence is retrieved, i.e. sub-goals are progressively achieved; (2) the prototype associated with the final goal is retrieved. In both cases, the retrieved information is stored in working memory and the robot searches for it. To do

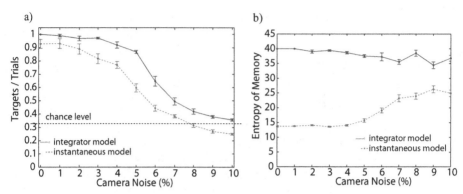

Fig. 4. Performance and entropy along different camera noise. (a) Ratio Targets / Trials distribution for instantaneous and integrator models. (b) Entropy of the memory along different camera noise. In both figures bars represent means ± sem.

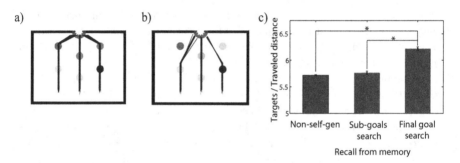

Fig. 5. Robot trajectories and performance. (a) Trajectories generated during sub-goals search. (b) Trajectories generated during final goal search. (c) Ratio Targets / Travelled distance for the integrator model with non-self-generated actions recall, sub-goals search recall and final goal search recall. Bars represent means ± sem.

so, it moves the tilt angle of the camera from -60 deg. to -20 deg. and rotates over its own axis. In this way, the robot can see the visual cues that are far away from its current position. Once the robot sees the sub-goal or final goal it moves again the tilt angle from -20 deg. to -60 deg. and goes straight to the goal, i. e. self-generating the actions. We test this new way of retrieving information from memory in the Task 1 and compare the results to the non-self-generated actions investigated in the section. As shown in Fig. 5a and 5b, the robot follows a different path depending on the mechanism it uses to recall information from memory. We observed that the ratio targets divided into travelled distance was significantly higher in the case of the final goal search mechanism of self-generated actions (Fig. 5c; ks-test: $p < 0.001$), i. e. the robot follows a shorter path to hit the target. On the contrary, there is no significant difference between the sub-goal search and the non-self-generated actions (ks-test: $p > 0.05$). This result shows an optimal way of using the compressed information of the memory when the actions are not stored but self-generated during the retrieval of information from memory. It results in more flexibility in the actions to be taken and allows to account for a continuous action space.

4   Conclusions

We tested the implications of an integrator decision making model in sequence learning tasks with multiple alternatives using a cognitive architecture that we called DAC9, evolving from previous implementations [7]. We compared the results with a Bayesian decision making model which is thought to be optimal for action selection. As a framework we used a robot based architecture which allowed us to understand the behavioural and architectural implications of these alternative models during foraging tasks. We showed that the race model has a more robust task-related performance when perceptual noise is added [12] and when task complexity increases compared to the Bayesian model. Moreover, the race model also implied a compression of information in memory suggesting an alternative way of storing information, i. e. only perceptual information is acquired and the actions are self-generated during recall. The self-generation of actions during the retrieval from memory shows a mechanism able to account for a continuous action space.

In a previous study [12], we reported the differences in the storage of information in memory due to both models and the impact they have in performance. Here, we quantified the difference in the information stored in memory by calculating the memory entropy. We showed that the entropy is higher in the integrator model than in the instantaneous model. In the instantaneous model the actions are continuously recalled and performed. Therefore, in this case, the robot executes a number of actions, generally greater than one, each time it crosses a visual cue. In the case of the integrator model, we proposed a new mechanism to optimally use the information from memory. We implemented a goal oriented mechanism that retrieves visual cues from memory instead of actions. Once the visual cue, i. e. goal, is selected from memory the robot searches for it in the environment. Whenever the robot sees the goal it goes towards it. This new mechanism can be seen as the storage of an abstract object in memory, i. e. a door. If a person wants to leave a room she/he has to first localize the door and then go towards it.

So far the integrator models have been used to explain simple decision making tasks, such as RDM task [6] or the countermanding task [19]. The implementation of the integrator model was based on the race model. Generally, the race model has been mainly used to explain behavior in a countermanding task [19], predicting probability of failure and reaction time. Here, we showed the implication of this decision making model in a more general framework. We observed that it has an important impact on how the memory is constructed and therefore on how the information is used later on.

The main assumptions we made in our proposal of self-generated actions during the recall from memory is that visual cues can always be seen from the current position of the robot. However, in wide open field environments, when this is not the case, our assumption would fail. In those situations, we would rely on head direction accumulator [20] cells. The heading direction information would be stored in memory together with the visual information. During the recall from memory the actions would be also self-generated. Similar to the search of visual prototypes tested in this study, the robot would rotate around its own axis until its current head direction is equal or close enough to the retrieved head direction.

Physiological studies have shown that granular and pyramidal cells in the hippocampus encode information with high sparsity (low entropy), i. e. neurons respond to a small fraction of stimuli [21]. In contrast, cells in the PFC have shown to be selective to particular cues with less sparsity (higher entropy) than the hippocampus and also with distinct temporal profile [22]. We observed that these two mechanisms of encoding memory have some similarity with the implications shown in this study due to the two decision making models, i. e. higher entropy in the integrator model compared to the instantaneous model. One could speculate that there is a distributed control system for sequence learning involving the hippocampus and the PFC connected to an external area which accumulates evidence, as found in the superior colliculus [23], the lateral intraparietal area [24], the frontal eye fields [25] and the PFC itself [26].

Acknowledgements. This work was supported by the EU projects Synthetic Forager-SF (EC-FP7/217148), eSMC (EC-FP7/270212) and Goal-Leaders (EC-FP7/270108).

References

1. Gold, J.I., Shadlen, M.N.: The neural basis of decision making. Annu. Rev. Neurosci. 30, 535–574 (2007)
2. Smith, P.L., Ratcliff, R.: Psychology and neurobiology of simple decisions. Trends Neurosci. 27, 161–168 (2004)
3. Logan, G.D., Cowan, W.B.: On the ability to inhibit thought and action: A theory of an act of control. Psychol. Rev. 91, 295–327 (1984)
4. Ratcliff, R., Rouder, J.N.: Modelling response times for two-choice decisions. Psychol. Sci. 9, 347–356 (1998)
5. Verschure, P.F.M.J., Voegtlin, T., Douglas, R.J.: Environmentally mediated synergy between perception and behavior robots. Nature 425, 620–624 (2003)
6. Shadlen, M.N., Newsome, W.T.: Neural basis of a perceptual decision in the parietal cortex (area lip) of the rhesus monkey. J. Neurophysiol. 86, 1916–1936 (2001)
7. Verschure, P.F.M.J., Althaus, P.: A real-world rational agent: unifying old and new ai. Cognitive Science 27, 561–590 (2003)
8. Bayes, T.: An essay towards solving a problem in the doctrine of chances. Transactions of the Royal Society 53, 370–418 (1763)
9. Verschure, P.F.M.J.: Distributed adaptive control: A theory of the mind, brain, body nexus. In: BICA (in press, 2012)

10. Duff, A., Fibla, M.S., Verschure, P.F.M.J.: A biologically based model for the integration of sensory-motor contingencies in rules and plans: A prefrontal cortex based extension of the distributed adaptive control architecture. Brain Res. Bull. (2010)
11. Mathews, Z., i Badia, S.B., Verschure, P.F.: Pasar: An integrated model of prediction, anticipation, sensation, attention and response for artificial sensorimotor systems. Information Sciences 186(1), 1–19 (2012)
12. Marcos, E., Duff, A., Sánchez-Fibla, M., Verschure, P.F.M.J.: The neuronal substrate underlying order and interval representations in sequential tasks: a biologically based robot study. In: IEEE World Congress on Computational Intelligence (2010)
13. Pavlov, I.P.: Conditioned reflexes: an investigation of the physiological activity of the cerebral cortex. Oxford University Press (1927)
14. Duff, A., Verschure, P.F.M.J.: Unifying perceptual and behavioral learning with a correlative subspace learning rule. Neurocomputing 73(10-12), 1818–1830 (2010)
15. Thorndike, E.: Animal intelligence. Macmillan, New York (1911)
16. de Almeida, L., Idiart, M., Lisman, J.E.: A Second Function of Gamma Frequency Oscillations: An E%-Max Winner-Take-All Mechanism Selects Which Cells Fire. J. Neurosci. 29(23), 7497–7503 (2009)
17. Wyss, R.: Sensory and motor coding in the organization of behavior. PhD thesis, ETHZ (2003)
18. Wyss, R., König, P., Verschure, P.: A model of the ventral visual system based on temporal stability and local memory. PLoS Biol. 4 (2006)
19. Hanes, D.P., Schall, J.D.: Neural control of voluntary movement initiation. Science 274, 427–430 (1996)
20. Mathews, Z., Lechón, M., Calvo, J.M.B., Dhir, A., Duff, A., Badia, S.B.: Verschure, P.F.M.J.: Insect-like mapless navigation based on head direction cells and contextual learning using chemo-visual sensors. In: Proceedings of the 2009 IEEE/RSJ International Conference on Intelligent Robots and Systems, IROS 2009, pp. 2243–2250. IEEE Press (2009)
21. Jung, M.W., McNaughton, B.L.: Spatial selectivity of unit activity in the hippocampal granular layer. Hippocampus 3, 165–182 (1993)
22. Asaad, W.F., Rainer, G., Miller, E.K.: Neural activity in the primate prefrontal cortex during associative learning. Neuron 21, 1399–1407 (1998)
23. Ratcliff, R., Hasegawa, Y.T., Hasegawa, R.P., Smith, P.L., Segraves, M.A.: Dual diffusion model for single-cell recording data from the superior colliculus in a brightness-discrimination task. J. Neurophysiol. 97, 1756–1774 (2007)
24. Roitman, J.D., Shadlen, M.N.: Response of neurons in the lateral intraparietal area during combined visual discrimination reaction time task. J. Neurosci. 22, 9475–9489 (2002)
25. Gold, J.I., Shadlen, M.N.: The influence of behavioral context on the representation of a perceptual decision in developing oculomotor commands. J. Neurosci. 23, 632–651 (2003)
26. Kim, J.N., Shadlen, M.N.: Neural correlates of a decision in the dorsolateral prefrontal cortex of the macaque. Nat. Neurosci. 2, 176–185 (1999)

The Emergence of Action Sequences from Spatial Attention: Insight from Rodent-Like Robots

Ben Mitchinson[1], Martin J. Pearson[2], Anthony G. Pipe[2], and Tony J. Prescott[1]

[1] ATLAS Research Group, The University Of Sheffield, UK
[2] Bristol Robotics Laboratory, Bristol, UK

Abstract. Animal behaviour is rich, varied, and smoothly integrated. One plausible model of its generation is that behavioural sub-systems compete to command effectors. In small terrestrial mammals, many behaviours are underpinned by foveation, since important effectors (teeth, tongue) are co-located with foveal sensors (microvibrissae, lips, nose), suggesting a central role for foveal selection and foveation in generating behaviour. This, along with research on primate visual attention, inspires an alternative hypothesis, that integrated behaviour can be understood as sequences of foveations with selection being amongst foveation targets based on their salience. Here, we investigate control architectures for a biomimetic robot equipped with a rodent-like vibrissal tactile sensing system, explicitly comparing a salience map model for action guidance with an earlier model implementing behaviour selection. Both architectures generate life-like action sequences, but in the salience map version higher-level behaviours are an emergent consequence of following a shifting focus of attention.

Keywords: brain-based robotics, action selection, tactile sensing, behavioural integration, saliency map.

1 Introduction

The problem of behavioural integration, or behavioural coherence, is central to the task of building life-like systems [1,2]. Animals display rich and varied behaviours serving many parallel goals; for instance, maintenance of homeostatic equilibrium, gathering of food, meeting social or sexual goals. These behaviours require integration over space and time such that the animal controls its effector systems in a co-ordinated way and generates sequences of actions that do not conflict with one another. How animals achieve behavioural integration is, in general, an unsolved problem in anything other than some of the simplest invertebrates. What is clear from the perspective of behaviour is that the problem is under-constrained since similar sequences of overt behaviour can be generated by quite different underlying control architectures. This implies that to understand the solution to the integration problem in any given organism is

T.J. Prescott et al. (Eds.): Living Machines 2012, LNAI 7375, pp. 168–179, 2012.

Fig. 1. (Left) A laboratory rat, showing the prominent macrovibrissae arrayed around the snout. (Right) Shrewbot, the robot used in the reported experiments, pictured alongside the cardboard cylinder that is also described. Visible are: eighteen large macrovibrissae surrounding the small microvibrissae (six fitted, here); the head on which they are mounted at the end of a 3 d.o.f. neck; the Robotino mobile platform.

going to require investigation of mechanism in addition to observations of behaviour. In this regard, physical models—such as robots—can prove useful as a means of embodying hypotheses concerning alternative control architectures whose behavioural consequences can then be measured observationally [3]. Research with robots has long demonstrated (see [4], for one example) forms of emergent behaviour—the appearance of integrated behavioural sequences that are not explicitly programmed—demonstrating the value of this embodied testing for suggesting and testing candidate mechanisms.

The biological literature provides for a range of different hypotheses concerning the mechanisms that can give rise to behavioural integration; here, we highlight two, and explore and discuss their behaviour in a robot. The neuroethology literature suggests a decomposition of control into behavioural sub-systems that then compete to control the animal (see [5] for a review). This approach has been enthusiastically adopted by researchers in behaviour-based robotics as a means of generating integrated patterns of behaviour in autonomous robots that can be robust to sensory noise, or even to damage to the controller. An alternative hypothesis emerges from the literature on spatial attention, particularly regarding visual attention in primates, including humans [6]. This approach suggests that actions, such as eye movements and reaches towards targets, are generated by first computing a 'salience map' that integrates information about the relevance (salience) to the animal of particular locations in space into a single topographic representation. Some maximisation algorithm (such as winner-takes-all [7]) is then used to select the most salient position in space towards which action is then directed. It is usual in this literature to distinguish between the computation of the salience map, the selection of the target within the map, and orienting actions that move the animal, or its effector systems, towards the target. In the mammalian brain these different functions may be supported by distinct (though overlapping) neural mechanisms [5,8]. Of course, the approaches of behavioural competition and salience map competition are not mutually exclusive and it is possible to imagine various hierarchical schemes, whereby, for instance, a behaviour is selected first and then a point in space to which the behaviour will be directed. Alternatively, the target location might be selected and then the action to be directed at it. Finally, parallel, interacting sub-systems may simultaneously converge on both a target and suitable action [9].

One line of our own recent research in this area has centred around behavioural observations of 'rodent-like' animals [10] and our development of a lineage of robots modelled on these animals and their behaviour [11]. By 'rodent-like' we indicate animals that share some essential features with rodents. Key amongst these features is a heavy reliance on a facial vibrissal sensory system, consisting of large motile *macrovibrissae* and smaller immotile *microvibrissae*, which forms the primary sensory system of many such animals [12]. This feature is prominent in rodents such as rats and mice, but is present in the same broad form in animals from other orders including shrews [13] and opossums [10]. For such animals, it is useful to identify the region around the mouth as a 'multi-sensory fovea' [14]. This region has been identified as a 'tactile fovea' owing to the presence of the fine microvibrissae [12]; other local organs playing a sensory role include the teeth, lips, tongue and nostrils. A behavioural feature that is consistently observed in discrimination experiments, then, is bringing this multi-sensory fovea to bear on the discriminandum [12]. Although small mammals can grasp and manipulate objects with their forelimbs, a key effector system for foraging and object investigation is the mouth itself. Thus, 'foveation' (bringing this region to an object) constitutes also the majority movement of some actions that might be elicited towards it once it has been identified, and the overall pattern of movement can then be approximated as movements solely of the foveal region [14]. We label this ethological consequence of the morphology by the approximation that *foveation is action*. This feature, along with the gross body pattern of a stable mobile base and an agile orientable head, completes the set of features which we denote by the term 'rodent-like'.

One goal of this line of research is to generate a physical model of the exploratory and orienting behaviour of these animals as measured in our laboratory [15,16,10] using high-speed video recordings. At the same time, we seek to implement and test hypotheses about how the mammalian brain generates this behaviour. Accordingly, these robots are biomimetic both in their morphology and in some aspects of their control architecture. In the current study we investigate the hypothesis that a salience map model can be used to generate action sequences for a biomimetic robot snout mounted on a mobile robot platform, and compare this with an earlier control model based on behaviour selection. Below, we describe the two models of behavioural integration in the context of the more general control architecture used by our robots. We go on to describe two experiments using the robot 'Shrewbot' (Figure 1). Both control systems generate life-like sequences which alternate between exploration and orienting behaviour, but in the salience map version these higher-level behaviours are an emergent consequence of actions determined by following a shifting focus of spatial attention (determined by a salience map) rather than being explicit control primitives. In the mammalian brain sensimotor loops involving the cortex, the superior colliculus, the cerebellum and the basal ganglia (as well as other centres) may interact to implement a control system similar to the hypothesised salience map model.

2   Models

Figure 2 summarises the multi-level loop architecture used in our biomimetic robots, which approximates the structure of the neural substrate [17]. There is no general agreement on the function of many neural centres, or even whether description in such terms is possible. Since the robot must 'function' in some sense if we are to experiment with it, our break-down of the control system into modules is by function, with the particular break-down chosen driven by the anatomy. This places us in a strong position to hypothesise relationships between structure and function in the neural system, and these hypotheses are a major outcome of our robot work [3]. Here, we focus on the component 'Selection Mechanism' which is responsible for the majority of movements of the robot's body (neck and wheels). Below this system, low-level reflex loops effect rapid responses to current conditions (for instance, whisker protraction is inhibited by contact with the environment [18]). Above this system, we are beginning to add more cognitive components that modulate selection, either by gleaning detailed information about what has been contacted from the sensory signals (in the

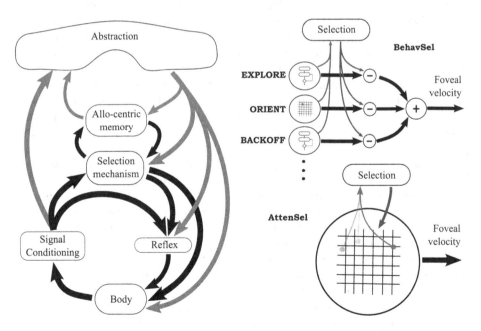

Fig. 2. (Left) Multi-level loop architecture common to the discussed models. (Right) Detail of the component labeled 'Selection Mechanism' in the left-hand panel.

component labelled 'Abstraction' [19]), or by retaining a memory of the robot's past spatial experience (in the component labelled 'Allo-centric memory' [20]). The details of the wider system are covered elsewhere [11].

We identify a 'tactile fovea' [12], a short distance in front of the microvibrissae (henceforth, just 'fovea'). The selection mechanism drives movements of this fovea, its output being the instantaneous foveal velocity. Our focus is on the key component of behaviour in rodent-like mammals, highlighted above, of bringing the fovea to a target. For instance, when faced with a task of discriminating between multiple objects, rat behaviour can be described as foveation to each discriminandum in sequence [12]. Beyond that it supports this foveation, the movement of the remaining nodes of the animal/robot is unconstrained, so these nodes (neck joints, body) are slaved to the fovea in our models—that is, our robots are 'led by the nose'. This is, of course, a simplification of biological behaviour, though we have been surprised by how life-like (and practical) the resulting behaviour is. Our robots do not have manipulators, so that foveation is the only behaviour that they express: the remainder of this section compares two approaches we have taken to generating the foveal velocity vector.

The first approach [14,21,18] is a model of selection between individual distinct behaviours—that is, it is biomimetic at the ethological level. It consists of a list of pre-defined behaviours which are arbitrated by a model of Basal Ganglia (BG), after a previous model of robot foraging [22], see Figure 2. The BG chooses one behaviour to take control of the motor plant (i.e. provide the foveal velocity) at any one time. The individual behaviours themselves are responsible for 'bidding' for this privilege, with an intensity that reflects the degree to which they are appropriate given the immediate conditions. Note that this model has minimal memory—its response, modulation from higher systems aside, is a function only of its recent inputs (and endogenous noise). Thus, this model fits neatly into the paradigm of 'behaviour-based robotics'; we denote it as **BehavSel**.

Whilst, in principle, a list of any number of behaviours could be included, in practice we have only ever implemented four (EXPLORE, ORIENT, BACKOFF, GOTO). This was based on need—we found that we were able to simulate rat behaviour to a sufficiently good approximation for our experiments with only these (in fact, the heuristic behaviour GOTO was included as an experimental convenience, rather than to correspond to any particular animal behaviour, and is not described here). Each behaviour in **BehavSel** is a 'fixed action pattern' in the sense that, once initiated, it will usually complete [22]. A higher priority behaviour can interrupt an ongoing, lower priority, behaviour. For instance, detection of an obstacle causes EXPLORE to be interrupted by ORIENT. Furthermore, some of the behaviours are parametrised such that immediate sensory information can change their details (owing to this, the term 'modal action pattern' may be preferred by some ethologists [22]). For instance, ORIENT directs the fovea to a recently-detected item in the environment.

In an early form of this model, the selection mechanism was the only source of foveal control. Since arbitration takes time [22], and processing resources on the robot are limited, this system did not prove sufficiently fast to protect the robot from damage when it foveated inaccurately. For this reason, we added a low-level reflex such that strong contact on the microvibrissae leads the neck to immediately and rapidly concertina, retracting the snout and protecting it from damage in the period of a few tens of milliseconds before the selection mechanism is able to respond to the same signals. This approach has been very satisfactory, both because we no longer damage microvibrissae and because it does not interfere at all in our investigations of the broader behaviour of the system (that is, we can investigate selection without thinking about the influence of this reflex).

The long-term behaviour of **BehavSel** can be summarised, as follows. Without stimulus, EXPLORE has the highest salience, and the robot proceeds forward sweeping its fovea from side to side. On contacting a stimulus, the salience of ORIENT is raised, and the robot foveates the point of contact. Meanwhile, the salience of BACKOFF is raised somewhat, so that when the ORIENT action completes, BACKOFF is chosen and the robot moves away from the contacted obstacle, before falling back into EXPLORE. With this arrangement, the robot can be safely left to freely explore a simple environment, so long as the geometry of the environment is fairly smooth (the robot has no reversing cameras).

One question raised by the **BehavSel** model, from a biomimetic standpoint is: where are the EXPLORE and BACKOFF behaviours implemented in the brain? Behaviours, after all, are things we observe from outside the animal, and are not necessarily explicitly represented in any algorithm. In the field of bio-inspired robotics, being unable to answer this question is not problematic. In biomimetic robotics, however, communicability between biological data (with the exception of ethological data) and the design of ànd results from models suffers if they do not share the same encoding and algorithms. As discussed, all of the behaviours in **BehavSel** are expressed through foveation. (Visual) foveation in primates is well studied and is mediated by the Superior Colliculus (SC) [6]. In rats, stimulation of the SC can evoke not only eye movements [23], but also orienting-like movements of the snout, circling, and even locomotion, amongst other behaviours [24]. Inspired by these facts, we have developed a second model of foveal velocity vector generation that mirrors the features of the SC—that is, a topographic saliency map driven by sensory input and modulated by information from mid- and upper-brain, with a simple motor output transform that drives foveation to the most salient region of local space. Selection, then, is between foveation targets in local space, rather than between behaviours (see Figure 2). In the case of our robots, salience is excited by whisker contact and endogenous noise and suppressed by a top-down 'inhibition-of-return' (IOR) signal from an allo-centric memory component which lowers the salience of regions that have recently been foveated (after a related model of IOR in the primate visual system [7]). This model is denoted **AttenSel**.

The long-term behaviour of **AttenSel** can be summarised, as follows. Initially, salience is driven only by endogenous noise, and the robot foveates stochastically. The noise is spatially-biased, such that foveations in a forward direction are more likely, and the robot tends to proceed forward. On contacting a stimulus, the salience in the corresponding region of the map is raised, and the robot foveates the point of contact. That location then becomes less salient, owing to the IOR signal. The foveation targets selected following this, then, drive the robot away from that location. Some time after, the robot has returned to the stochastic foveation pattern observed initially. That is, the observed behaviour of the robot using **AttenSel** is quite similar to that observed using **BehavSel**—or, to put it another way, quite similar behaviours emerge from the latter model to those that are explicitly designed into the former model.

3   Results

In Experiment 1, we placed Shrewbot nearby and facing a cardboard obstacle (150mm diameter), with each of the two models in control, in turn. In each case, the robot proceeds forwards, exploring, until it contacts the obstacle with its macrovibrissae. It then orients to the obstacle. Finally, it backs away and moves off exploring in another direction. The results of each experiment are

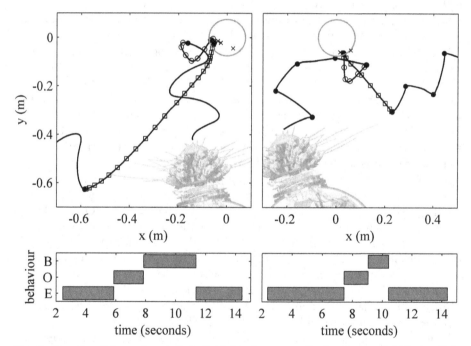

Fig. 3. Results of Experiment 1. Left/right shows results for **BehavSel/AttenSel**. (Upper) Projected onto the horizontal plane; time course of the fovea location (solid line), moments of re-selection (dots), and the ranges of **EXPLORE** (no markers), **ORIENT** (circles), and **BACKOFF** (squares) behaviours. The obstacle boundary is marked in grey, contacts as crosses. (Lower) Ethograms of behaviour against time (E, O, B, for **EXPLORE, ORIENT, BACKOFF**).

shown in Figure 3. In the case of **BehavSel**, the ethogram is recovered from recorded signals; in the case of **AttenSel**, the ethogram was generated by an observer, charged with reviewing the video *post hoc* and judging which of the three behaviours explicit in **BehavSel** was being exhibited over time.

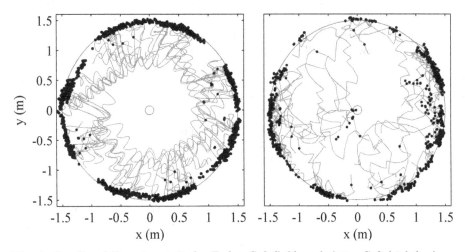

Fig. 4. Results of Experiment 2, for **BehavSel** (left) and **AttenSel** (right). Arena boundary and cardboard tube at the centre are marked as grey circles. The foveal location over time, projected onto the horizontal plane, is shown as a grey trace. Contact locations over the course of the whole experiment are marked as black dots. Some odometry drift is apparent (groups of dots not aligned with obstructions), as is some noise in the contact signal (isolated dots far away from obstructions).

In Experiment 2, we placed Shrewbot inside a circular arena (1500mm radius, bounded by 500mm high smooth vertical walls) at the centre of which was the cardboard tube from Experiment 1. The robot was then allowed to behave freely for ten minutes. The mobile platform suffers from significant odometry drift—to mitigate this for the sake of this experiment, we halted the robot every thirty seconds and re-measured its position and orientation using an overhead camera. To avoid calibration problems that arise if the whiskers are in contact with something when the robot is restarted, the robot was occasionally moved away from a wall before being restarted, which is why the foveal location traces in the figure are discontinous. The general result of this experiment is that both models perform a high-level behaviour that might be described as 'exploration' of the arena. Some differences in the details of the exploratory behaviour expressed by the two models, visible in the figure, are not relevant to our theme and so we do not discuss them, here.

4 Discussion

The **BehavSel** and **AttenSel** models can be distinguished by what is represented explicitly (or, what is 'encoded') and what is represented implicitly (or, what 'emerges'). In **BehavSel**, behaviours are represented explicitly (and selected between) whilst 'spatial attention', as measured by the location that is foveated, emerges from the interaction of those behaviours with the physical environment. Conversely, in **AttenSel**, locations of spatial attention are represented explicitly (and selected between) whilst behaviours, as measured by a human observer, emerge from interaction with the environment. In the same way as short-term behaviours emerge from **AttenSel** (Experiment 1), long-term behaviours emerge in both models (Experiment 2). Modelling at multiple levels of abstraction is undoubtedly useful; indeed, parts of the wider control system outlined above but not described are represented using a variety of encodings. Here, we discuss the two approaches to modelling that we have explored in the context of biological research, robotic performance, and the controller design process.

Biological Research. The correspondence in representation between biomimetic models and the biology mean that observations about structure and function in the robot control system can potentially be useful in helping us to understand the biological control system. Conversely, hypotheses about structure and function that arise from biological research can be tested very directly in the robot. The structure of both of the models presented here is biomimetic at a high level, consisting of nested sensorimotor loops (see Figure 2). The most prominent feature of the SC, common to most accounts in most species, is its domination by topographically-organised encodings of sensory and motor space [6] (though it can, at the same time, be understood to be involved in the mediation of behaviour [24]). In reflecting its topographical organisation, the **AttenSel** component is biomimetic at a low level. This type of correspondence between model and system can only favour communicability between experiments in the two contexts, allowing us to approach questions such as whether behaviours are encoded in the SC, or whether they emerge from the temporal patterns of activation in general purpose motor maps.

Since behaviours are not specified in advance in the **AttenSel** model, the implicit behavioural space is large and continuous. In contrast, the **BehavSel** model behavioural space—parameterisation aside—is populated by a limited number of singular points. As a result, no time window of observation of the **AttenSel** model can be said to have precisely isolated 'behaviour X'. The human observer can and will pick out particular behaviours, certainly—we describe behaviour using behavioural terms—and we have chosen to describe the behaviour of **AttenSel** in the language of **BehavSel** in Experiment 1, above. But, the behaviour of the model at any one time is chosen from that continuous implicit behavioural space. Another way of stating this is that the things we might describe as discrete behaviours—ORIENT, EXPLORE, BACKOFF—bleed into one another, so that the specific behaviour of the robot in any time window is an overlapping integration of these. A certain class of animal behavioural experiments can be summarised, methodologically, as long periods of watching the animals freely behave punctuated by brief recording opportunities where clean examples of behaviours of interest, defined in advance, are expressed [15,16]. This methodology reflects the fact that stereotyped descriptions of the behaviour of the animal under such conditions are approximations, at best. In this sense, **AttenSel** is a more complete model of foveation behaviour than **BehavSel**.

Robot Performance. The **AttenSel** model retains a desirable aspect of the behavioural-based robotics paradigm. That is, the immediate information about current conditions remains largely in the outside world, rather than being internalised, and the long-term behaviour represents overlapping contributions from—implicitly represented—simple behaviours. However, its encoding permits, very naturally, interaction between behaviours and the modulation of behaviour by systems that deal, explicitly, in something other than behaviour. For instance, a BACKOFF-like behaviour emerges from the interaction between endogenous noise in the saliency map and the spatial memory that the area right in front of the robot has previously been visited. A higher-level planning system that was explicitly concerned with reaching a particular location (recalling the GOTO behaviour mentioned above) could evoke locomotion from the model by modulating the salience map appropriately. This communicability between components of the system is analogous to that between the model and the biological data—all of the listed influences encode space.

The continuum of possible behaviours in the **AttenSel** model represents behavioural flexibility. Speaking in terms more suited to the **BehavSel** model substrate, the use of a little bit of BACKOFF in a particular ORIENT behaviour, which in a particular situation might be advantageous, would be the result of an automatic mixing of different influences on the robots attention. To achieve the same thing in the **BehavSel** model, the model would require a specification in advance of how behaviours should mix under different conditions, an exponentially-increasing specification as the list of behaviours increases. In **AttenSel**, a specification is required, conversely, for how different influences on attention should mix—whilst this is not a trivial specification to derive either, it may be easier to derive than that for **BehavSel**, owing to the common encoding used by those influences. Thus, and in analogy to the richness of behaviour observed in animals, the **AttenSel** model might favour flexibility to the particular conditions faced by the robot in the environment, a key requirement for autonomous systems.

The Design Process. One significant characteristic of an **AttenSel**-like control model is that it may be difficult to get the robot to do 'the right thing'. Behaviours cannot be directly designed; rather, they emerge from the interaction between the environment and signal encodings and transforms that are specified by the designer. Thus, we cannot 'program the robot to do X'; rather, we can program how the robot generates and processes encodings, in a condition-dependent way. The design loop is closed by critiquing behaviour and modifying the way the robot computes those encodings, accordingly. Being divorced from behaviour in this way might not suit the goals of industrial robotics, for example, where 'the right thing' is tightly specified in advance. However, one of the major challenges of contemporary robotics is persuading robots to behave sensibly under conditions that are unknown *a priori*, where 'the right thing' may be, accordingly, unknown. One approach to this challenge is to use learning, applicable—in general—regardless of the choice of encoding, in the hope of side-stepping the design process to some degree. But, designing learning algorithms that are able to derive appropriate behaviour from scratch is difficult even when the task is very constrained (see [25] for several practical examples). When the task is much more general, this problem is exacerbated by the sizes of the sensory and motor spaces that must be related—learning often doesn't 'scale well' [25]. As Brooks notes: 'Most animals have significant behavioral expertise built in without having to explicitly learn it all from scratch.' [4]. The **AttenSel** model is an example of borrowing some of that expertise to help us to choose the structure of a complex control system. In this report we have shown that, given a structure derived from biology, a design process is sufficient—even without any learning—to produce behaviour as sensible as that which we designed explicitly in an earlier model.

Acknowledgments. The authors would like to thank Jason Welsby (Bristol Robotics Laboratory) for his work towards the construction of Shrewbot as well as his help in performing the experiments. This work was supported by the FP7 grant BIOTACT (ICT-215910).

References

1. Prescott, T.: Forced moves or good tricks in design space? landmarks in the evolution of neural mechanisms for action selection. Adaptive Behavior 15(1), 9–31 (2007)
2. Brooks, R.: Coherent behavior from many adaptive processes. From Animals to Animats 3, 22–29 (1994)
3. Mitchinson, B., Pearson, M.J., Pipe, A.G., Prescott, T.J.: Biomimetic robots as scientific models: A view from the whisker tip. In: Krichmar, J.L., Wagatsuma, H. (eds.) Neuromorphic and Brain-Based Robots, pp. 23–57. Cambridge University Press (2010)
4. Brooks, R.A.: A robot that walks; emergent behaviors from a carefully evolved network. Neural Computation 1, 253–262 (1989)
5. Redgrave, P., Prescott, T.J., Gurney, K.: The basal ganglia: a vertebrate solution to the selection problem? Neuroscience 89(4), 1009–1023 (1999)
6. Gandhi, N.J., Katnani, H.A.: Motor functions of the superior colliculus. Annu. Rev. Neurosci. 34, 205–231 (2011)
7. Itti, L., Koch, C.: Computational modelling of visual attention. Nature Reviews Neuroscience 2, 194–203 (2001)
8. Posner, M.I., Peterson, S.E.: The attention system of the human brain. Annual Review of Neuroscience 13, 25–42 (1990)
9. Cisek, P., Kalaska, J.: Neural mechanisms for interacting with a world full of action choices. Annual Review of Neuroscience 33, 269–298 (2010)

10. Mitchinson, B., Grant, R.A., Arkley, K., Rankov, V., Perkon, I., Prescott, T.J.: Active vibrissal sensing in rodents and marsupials. Phil. Trans. R. Soc. B 366, 3037–3048 (2011)

11. Pearson, M.J., Mitchinson, B., Sullivan, J.C., Pipe, A.G., Prescott, T.J.: Biomimetic vibrissal sensing for robots. Phil. Trans. R. Soc. B 366, 3085–3096 (2011)

12. Brecht, M., Preilowski, B., Merzenich, M.M.: Functional architecture of the mystacial vibrissae. Behavioural Brain Research 84, 81–97 (1997)

13. Anjum, F., Turni, H., Mulder, P.G., van der Burg, J., Brecht, M.: Tactile guidance of prey capture in etruscan shrews. Proc. Natl. Acad. Sci. USA 103(44), 16544–16549 (2006)

14. Mitchinson, B., Pearson, M., Melhuish, C., Prescott, T.J.: A Model of Sensorimotor Coordination in the Rat Whisker System. In: Nolfi, S., Baldassarre, G., Calabretta, R., Hallam, J.C.T., Marocco, D., Meyer, J.-A., Miglino, O., Parisi, D. (eds.) SAB 2006. LNCS (LNAI), vol. 4095, pp. 77–88. Springer, Heidelberg (2006)

15. Mitchinson, B., Martin, C.J., Grant, R.A., Prescott, T.J.: Feedback control in active sensing: rat exploratory whisking is modulated by environmental contact. Royal Society Proceedings B 274(1613), 1035–1041 (2007)

16. Grant, R.A., Mitchinson, B., Fox, C.W., Prescott, T.J.: Active touch sensing in the rat: Anticipatory and regulatory control of whisker movements during surface exploration. J. Neurophys. 101, 862–874 (2009)

17. Kleinfeld, D., Berg, R.W., O'Connor, S.M.: Anatomical loops and their electrical dynamics in relation to whisking by rat. Somatosens Mot. Res. 16(2), 69–88 (1999)

18. Pearson, M.J., Mitchinson, B., Welsby, J., Pipe, T., Prescott, T.J.: SCRATCHbot: Active Tactile Sensing in a Whiskered Mobile Robot. In: Doncieux, S., Girard, B., Guillot, A., Hallam, J., Meyer, J.-A., Mouret, J.-B. (eds.) SAB 2010. LNCS, vol. 6226, pp. 93–103. Springer, Heidelberg (2010)

19. Sullivan, J., Mitchinson, B., Pearson, M., Evans, M., Lepora, N., Fox, C., Melhuish, C., Prescott, T.J.: Tactile discrimination using active whisker sensors. IEEE Sensors Journal 12(2), 350–362 (2011)

20. Fox, C.W., Evans, M.H., Lepora, N.F., Pearson, M., Ham, A., Prescott, T.J.: CrunchBot: A Mobile Whiskered Robot Platform. In: Groß, R., Alboul, L., Melhuish, C., Witkowski, M., Prescott, T.J., Penders, J. (eds.) TAROS 2011. LNCS, vol. 6856, pp. 102–113. Springer, Heidelberg (2011)

21. Pearson, M.J., Pipe, A.G., Melhuish, C., Mitchinson, B., Prescott, T.J.: Whiskerbot: A robotic active touch system modeled on the rat whisker sensory system. Adaptive Behaviour 15(3), 223–240 (2007)

22. Prescott, T.J., Gonzalez, F.M.M., Gurney, K., Humphries, M.D., Redgrave, P.: A robot model of the basal ganglia: Behavior and intrinsic processing. Neural Netw. 19(1), 31–61 (2006)

23. McHaffie, J.G., Stein, B.E.: Eye movements evoked by electrical stimulation in the superior colliculus of rats and hamsters. Brain Research 247, 243–253 (1982)

24. Sahibzada, N., Dean, P., Redgrave, P.: Movements resembling orientation or avoidance elicited by electrical stimulation of the superior colliculus in rats. J. Neurosci. 6(3), 723–733 (1986)

25. Kaelbling, L.P., Littman, M.L., Moore, A.W.: Reinforcement learning: A survey. J. Artificial Intelligence Research 4, 237–285 (1996)

How Past Experience, Imitation and Practice Can Be Combined to Swiftly Learn to Use Novel "Tools": Insights from Skill Learning Experiments with Baby Humanoids

Vishwanathan Mohan and Pietro Morasso

Robotics, Brain and Cognitive Sciences Department,
Istituto Italiano di Tecnologia, Via Morego 30,Genova, Italy
{vishwanathan.mohan,pierto.morasso}@iit.it

Abstract. From using forks to eat to maneuvering high-tech gadgets of modern times, humans are adept in swiftly learning to use a wide range of tools in their daily lives. The essence of 'tool use' lies in our gradual progression from learning to act 'on' objects to learning to act 'with' objects in ways to counteract limitations of 'perceptions, actions and movements' imposed by our bodies. At the same time, to learn both "cumulatively" and "swiftly" a cognitive agent (human or humanoid) must be able to efficiently integrate multiple streams of information that aid to the learning process itself. Most important among them are social interaction (for example, imitating a teacher's demonstration), physical interaction (or practice) and "recycling" of previously learnt knowledge (experience) in new contexts. This article presents the skill learning architecture being developed for the humanoid iCub that dynamically integrates multiple streams of learning, multiple task specific constraints and incorporates novel principles that we believe are crucial for constructing a growing motor vocabulary in acting/learning robots. A central feature further is our departure from the well known notion of 'trajectory formation' and introduction of the idea of 'shape' in the domain of movement. The idea is to learn in an abstract fashion, hence allowing both "task independent" knowledge reuse and task specific "compositionality" to coexist. The scenario of how iCub learns to bimanually coordinate a new tool (a toy crane) to pick up otherwise unreachable objects in its workspace (recycling its past experience of learning to draw) is used to both illustrate central ideas and ask further questions.

Keywords: Skill learning, imitation, shape, Passive motion paradigm, tool use, iCub.

1   Introduction: On Learning to Use Tools and Learning 'Green'

The essence of tool use lies in our gradual progression from acting 'on objects' to acting 'with objects' in ways to counteract limitations of 'perceptions, actions and movements' imposed by our bodies. Emerging research from animal cognition suggests that tool use is a sophisticated behavior highly evolved in primates [1-2]. Even though many animals use simple tools to extend their physical capabilities (usually

T.J. Prescott et al. (Eds.): Living Machines 2012, LNAI 7375, pp. 180–191, 2012.

extension of reach and/or amplification of force), humans are unique in having established a culture that revolves around actively 'manufacturing, learning and using' tools. In this context, understanding tool use behavior could be a window for not just uncovering the mechanisms underlying human intelligence/dexterity but also for revealing the evolutionary path that created these mechanisms from the 'raw material' of the non-human primate brain [3]. Even though research on tool use behavior has been wide spread in animal and infant cognition [1], electrophysiological studies on nonhuman primates [4] and neuropsychological studies on brain damaged patients with motor deficits [5-6], studies on autonomous tool use in cognitive robots are still rare. Industrial robots perform tasks like cutting, welding, screwing, and painting, but do that on the basis of carefully scripted programs that only work in controlled conditions. With cognitive robots being envisaged as future companions/assistants in our homes, offices, eldercare facilities, hospitals and public environments in the near future, it is hard to imagine how effective they will be in these unstructured environments if the capability to learn to use external objects as tools is absent. To this effect, these problems have not been well addressed by the embodied robotics community. Few valiant attempts to incorporate tool use in cognitive robots have been made [7-8] but a general computational foundation is still elusive. Even more so are the dual problems of a) how use of tools can be learnt by a combination of self exploration/physical interaction and observation of a teacher using the tool; b) the problem of skill transfer or how motor knowledge acquired while learning one skill is "recycled" to speed up learning some other skill (so that the robot doesn't start from zero every time it learns something new). In general there is a need for architectures that integrate multiple streams of information that arises from social and physical interactions, learn green (recycle experience actively) and learn cumulatively.

In this article, we attempt to address these problems and present a general framework for skill learning in cognitive robots. The proposed architecture extends the Passive Motion Paradigm framework [9, see10 for a recent review] that aims to create a unified computational machinery for 'generation, simulation, learning and reasoning' about Action. In general, learning to use a tool requires an understanding of several aspects, for example, the physical characteristics of a tool, its relationship to the body, its effects on the world and finally using all this information to generate goal directed motor commands (for a highly redundant body like iCub). Going beyond use of stick, rake like tools that have been attempted previously [8, 11], we present a novel and more complex example of iCub learning to bimanually coordinate a "toy crane" in order to pick up otherwise unreachable objects with the magnetized tool tip. The further goal is to demonstrate how teacher's inputs, practice and past experience (of learning to draw [12]) can be meaningfully exploited and integrated to both aid the learning process and perform goal directed "body+tool" actions. The sections that follow gradually build up the discussion starting from the building blocks (section 2), learning to coordinate a new tool by a combination of physical and social interactions (section 3) and using the learnt knowledge to coordinate the tool in a goal directed fashion (section 4) under different conditions (normal, pathological etc).The final section presents few concluding remarks.

2   Context: Functional Coupling of the Body with Tools, Primitive Building Blocks of the iCub Skill Learning Architecture

In this section we describe the primitive subsystems needed to close the perception-action loop that facilitates robust socio-physical interactions during the learning process to use a 'novel' object as a tool. Specifically we enlist two modules: 1) Forward/Inverse model for coordinating the movements of the iCub upper body; 2) Multiple visual perception systems to process visual information related to the task.

(1) Forward/Inverse Model for iCub Upper body coordination: The Passive Motion Paradigm : Passive Motion Paradigm (PMP) is a computational framework for action generation inspired by the equilibrium point hypothesis [13] and theory of impedance control [14]. The PMP model basically solves the 'degrees of freedom problem' [16] i.e. coordinating a highly redundant body in a task specific fashion. Hence it is a core building block for tool use and skill learning (where additional degrees of freedom are coupled to the body). The flexibility and robustness of the PMP system while coordinating movements of iCub has been demonstrated through a range of manipulation tasks like reaching, bimanual coordination [9, 10]. For reasons of space, we refer the interested reader to a recent review on PMP rationale [10] for detailed description of the computational framework. Figure 1 shows the PMP network used to coordinate the iCub upper body. In the context of the work presented in this paper, this subsystem is used for three purposes: 1) bimanually reach and grasp the tool (toy crane); 2) generate motor commands for upper body movements during the learning phase (practice) with the new tool; 3) generate goal directed "body+tool" actions. Points 2 and 3 listed above will be described in detail in the sections that follow.

(2) Visual Perception systems: To close the basic perception-action loop, the PMP based action system has to be coupled to vision that at present is the dominant source of sensory input in iCub. In this work, three different visual modules are deployed to process visual information. The first is a simple color segmentation module that is used to recognize objects based on their color information (for example, the magnetized distal tip of the toy crane is recognized by the red color). The second subsystem is a motion detector module that captures trajectories of moving objects in the visual workspace of the robot [12]. This subsystem has already been employed successfully in a range of tasks that require extracting end effector trajectories of a teacher, most recently to teach iCub to draw after observing the teacher's demonstration. The third visual module is a 3D reconstruction system that maps salient points (centroids of color blobs, motion trajectories etc) computed in the image planes of the two cameras $(U_{left}, V_{left}, U_{right}, V_{right})$ into corresponding points in the iCub's egocentric space (x, y, z). Further algorithmic details of the basic visual modules can be found in [12]. To summarize, the three visual modules provide object recognition, motion extraction and 3D reconstruction/localization of salient points in the visual scene to iCub's egocentric space necessary to initiate action.

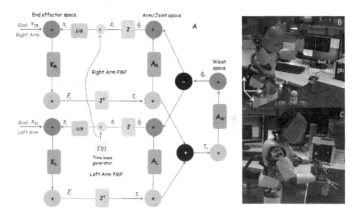

Fig. 1. PMP network for iCub upper body coordination. As seen, the network is grouped into multiple motor spaces (end effector, arm joint, waist), each motor space consisting of a displacement node (blue) and a force node (pink). Vertical connections (purple) denote impedances (K: Stiffness, A: Admittance) in the respective motor spaces and horizontal connections denote the geometric relation between the two motor spaces represented by the Jacobian (Green). Observe that the network is fully connected and all computations are local. The goal induces a force field that causes incremental elastic configurations in the network analogous to the coordination of a marionette with attached strings. Multiple constraints can be coupled to the network at any node, to bias the final solution. The network is a multi-referential system of action representation, which integrates a Forward/Inverse internal Model.

3 Learning to Control a Toy Crane: Insights into iCub Skill Learning Architecture

The basic action and perceptual subsystems described in the previous section have been used to create a more complex skill learning architecture for iCub. In this section we will describe the principles and novel features in the architecture that enables iCub to combine multiple learning streams, multiple task constraints while learning a new skill. We focus on the task of learning to bimanually coordinate a new tool i.e. a toy crane in order to pick up otherwise unreachable objects with its magnetized tip. Figure 3 explains the scenario. *A central goal is to combine information coming from teachers' demonstration, past experience (or knowledge previously gained from drawing [12]) and information acquired by practicing with the new tool, to "quickly" learn to use this new toy in a goal directed fashion.* In general, while learning to use any new tool iCub has to learn atleast three things:

1) To perform appropriate 'spatiotemporal' trajectories using the toy as demonstrated by the teacher (for example, performing synchronized quasi circular trajectories with both hands while turning the toy crane);

2) While performing such coordinated movements with the tool, learn the geometric relationship between the movements of the body effectors and the corresponding consequence on the tool effector (magnetized tip) encoded by the tool Jacobians J_T;

3) The third issue is of course related to using this learnt knowledge to generate 'goal directed' body+tool movements (given a goal to pick up an otherwise 'unreachable' environmental object using the new tool).

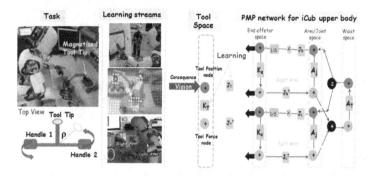

Fig. 2. Left panel illustrates the scenario of learning to bimanually coordinate the toy crane. Middle panel show the various learning streams involved: Past experience, teachers demonstration and practice. The right panel shows an extended PMP network that in addition to upper body network (figure 1) includes a tool space (that is represented exactly the same way as other motor spaces with a position and force node). The network is not yet connected as the connecting links at the interface have still to be learnt as a result of movements produced by the upper body and consequence observed through vision.

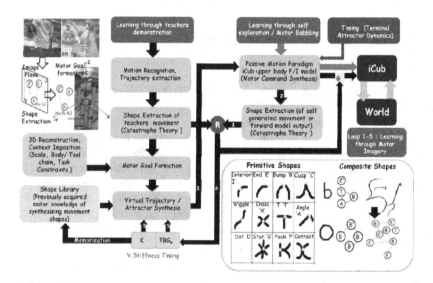

Fig. 3. Motor Skill learning architecture of iCub: Building blocks and Information flows. Top left box pictorially shows the demonstration-shape extraction-3D reconstruction loop. Bottom right box shows the 12 primitive shapes derived using catastrophe theory and more complex trajectories represented as composition of the primitives.

With the help of figure 3, we outline central features in the skill learning architecture that allows iCub to integrate multiple learning streams and learn to use the new toy, by playing with it, observing the teacher doing so.

Learning through Imitation, Exploration and Motor Imagery: Multiple streams of learning i.e. learning through teacher's demonstration (information flow in black arrow), learning through physical interaction (blue arrow), use of past experience (through information in the shape library) and learning through motor imagery (loop 1-5) are integrated into the architecture. The imitation loop initiates with the teachers demonstration and ends with iCub reproducing the observed action. The motor imagery loop is a sub part of the imitation loop, the only difference being that the motor commands synthesized by the PMP are not transmitted to the actuators. Instead, the forward model output is used to close the learning loop.

From Trajectory to Shape, towards 'Context Independent' Motor Knowledge: Most skilled actions involve synthesis of spatio-temporal trajectories of varying complexity. *A central feature in our architecture is the departure from the well known concept of 'trajectory formation' and introduction of the notion of 'Shape' in the domain of movement.* So when the teacher demonstrates the use of a tool, while the motion recognition system extracts the trajectories of the teachers end effector motion, the shape extraction system extracts the shape of the trajectory demonstrated by the teacher. The same system also extracts the shape of iCub's self generated end effector trajectories (forward model output of PMP). Shape of teacher's movements and iCub's own movement can directly be compared to evaluate how closely iCub imitates the teacher. The shape extraction system is based on Catastrophe theory [18] and it can be shown that a set of 12 primitive shape features or critical points (figure 4, bottom right panel) are sufficient to describe the shape of any trajectory in general [19]. Interested reader is referred to a recent article [12] for detailed formal analysis. The transition from learning at the level of movement trajectories to movement shapes offers many advantages that we detail below. In general, a trajectory may be thought as a sequence of points in space, from a starting position to an ending position. 'Shape' is a more abstract description of a trajectory, which captures only the critical events in it. By extracting the 'shape' of a trajectory, it is possible to liberate the trajectory from task specific details like scale, location, coordinate frames and body effectors that underlie its creation and make it 'context independent'. For example, the critical events in a trajectory like 'U' is the presence of a minima (or Bump 'B') in between two end points ('E'). Thus, the shape is represented as a graph 'E-B-E' (see figure 4). If the 'U' was drawn on a paper or if someone runs a 'U' in a playground, even though the trajectories are of different locations, scales, executed by different end effectors (legs, hands, a pen etc) *the shape representation is 'invariant'* (there is always a minima in between two end points). More complex shapes can be described as 'combinations' of the basic primitives, like a circular trajectory is a composition of 4 bumps. In short, using the shape extraction system it is possible to move from the visual observation of the end effector trajectory of the teacher to its more abstract 'shape' representation. Inversely, instead of learning to generate movement trajectories, if a humanoid robot learns to generate movement shapes, it basically has the graphical grammar to compose a wide range of spatiotemporal trajectories based on context. Consider for example, actions like using a screw driver, steering wheel, uncorking, paddling a bicycle, writing a 'U', 'C' 'O' all result in circular trajectories that result in similar shape representations. If the knowledge to synthesize a particular shape already exists then this can be exploited to generate any 'context specific' movement trajectory that results in the same shape representation, without any further learning (because shape information is invariant). *Hence transition from trajectory to shape allows us to conduct motor learning at an abstract level, store knowledge in a context independent fashion and thus speed up learning by exploiting the power of knowledge 'reuse' and 'compositionality'.*

From Context Independent Motor Knowledge to Context Dependent Motor Actions: As the teacher demonstrates the use of the toy crane, the shape extraction system basically extracts the shape of the teacher's end effector trajectories. The extracted 'shape' representation may be thought of as an 'abstract' visual goal created by iCub after perceiving the teacher's demonstration. To begin practicing with the new tool itself, iCub has to transform this visual goal into a motor goal and generate the appropriate motor commands for its upper body to reproduce the observed movement. This involves a sequence of next three stages of figure 4: 1) Motor goal creation 2) Virtual trajectory (or attractor) synthesis and 3) Motor command synthesis by PMP.

Motor Goal Formation: The first stage of motor goal formation is related to transforming the shape representation computed visually *i.e* with respect to image planes of the two cameras into corresponding representation in the iCub's egocentric space (x,y,z) through a process of 3D reconstruction (see figure 4, top left box). Note that 'shape' is conserved by this transformation for example, a minima still remains a minima. Additional task specific constraints like the end effectors/ body chain involved in generating the action are added to the goal description.

Virtual Trajectory Synthesis: The motor goal basically consists of a discrete set of shape critical points (their spatial location in iCub's ego centric space and type), that describe in abstract terms the 'shape' of the spatiotemporal trajectory that iCub must now generate (with the task relevant body chain). While the visual shape extraction system transforms a continuous trajectory of the teacher into a discrete set of shape critical points, the virtual trajectory generation system (VTGS) does the inverse operation. It transforms the discrete set of shape critical points into a continuous set of equilibrium points (that act as moving point attractor to the PMP system). *In this sense, the virtual trajectory is like a skilled puppeteer who is pulling the task relevant effector (in the PMP network) in a specific fashion.* In a recent work [12], we have shown that by modulating just two parameters i.e virtual stiffness 'K' and timing τ (figure 4, green boxes), virtual trajectories for the 12 primitive shapes derived in [19] (figure 4) can be learnt by iCub very quickly. A central idea here is to exploit 'compositionality' in the shape domain i.e since complex trajectories can be perceptually 'decomposed' into combinations of the primitive shape features, inversely motor actions needed to create them be 'composed' using combinations of the corresponding 'learnt' primitive actions. *During the synthesis of more complex trajectories, composition and recycling of previous knowledge takes the front stage (considering that the correct parameters to generate the primitive shapes already exist in memory).*

Using Past Motor 'Experience' to Generate Virtual Trajectories on the Fly: When iCub learnt to draw trajectories like 'U', 'C' etc , it acquired the correct parameters (K and γ) to synthesize virtual trajectories for shapes that result in 'Bump' critical points (or quasi circular trajectories). *When the teacher demonstrates iCub to bimanually steer the toy crane by performing quasi circular trajectories with his two arms, the correct parameters to generate such movement shapes already exist in the shape memory since they have been acquired during the past experience of learning to draw trajectories that resulted in similar shape features.* Using the previously learnt parameters of K and γ from the shape library, iCub is able to instantaneously generate virtual trajectories (or attractors) to generate such movements. Generated virtual trajectories now couple with the PMP network for the upper body to generate the motor commands to perform the action and begin practicing with the toy crane.

From Virtual Trajectory to Motor Commands: Learning to Use the New Tool
The PMP system transforms every point in the virtual trajectory into motor commands in the intrinsic space (upper body chain), hence enabling iCub to mimic the teacher's action of bimanually steering the toy crane. *Of course, this is just the starting point. iCub has to now learn the consequence and utility of the action in this new context.* Practicing with the toy crane now generates sequences of sensorimotor data:

1) the instantaneous position of the two hands $Q \in (x_R, y_R, z_R, x_L, y_L, z_L)$ coming from proprioception (and cross-validated by forward model output of PMP i.e position node in end effector space)
2) the resulting consequence i.e the location of the tool effector X $:(x,y,z)_{Tool}$, perceived through vision and reconstructed to Cartesian space (using the same technique to reconstruct teachers movement). As iCub acquires this sensorimotor data by practicing with the tool, a neural network can be trained to learn the mapping X= f(Q). We used a standard multilayer feedforward network with one hidden layer, where $Q=\{q_i\}$ is the input array (end effector position), $X=\{x_k\}$ is the output array (tool position), and $Z=\{z_j\}$ is the output of the hidden units, W the weight matrices.

$$X = f(Q) \Rightarrow \begin{cases} h_j = \sum_i \omega_{ij} q_i \\ z_j = g(h_j) \\ x_K = \sum_j w_{jk} z_j = \sum_j w_{jk} \left(g \left(\sum_{ij} \omega_{ij} q_i \right) \right) \end{cases}$$

(5)

The trick here is that once the neural network is trained on the sequences of sensorimotor data generated by the robot, the tool Jacobian J_T can be extracted from the learnt weight matrix by applying chain rule in the following way:

$$J_T = \frac{\partial x_k}{\partial q_i} = \sum_j \frac{\partial x_k}{\partial z_j} \frac{\partial z_j}{\partial h_j} \frac{\partial h_j}{\partial q_i} = \sum_j w_{jk} g'(h_j) \omega_{ij}$$

(6)

Once the tool Jacobians are learnt by iCub, the Tool+Body PMP network (of figure 3) is complete and fully connected to allow goal directed maneuvering of the toy crane (see figure 5). Note that, the tool admittance 'A$_T$' is a property of the tool itself and can be approximately estimated as the ratio of the total force exerted by iCub with its two hands and the corresponding displacement of the tool (approximated as an identity matrix under normal conditions).

4   Connectivity between "Body" and "Tool": Normal Case, Pathological Case and What If the Toy Crane Was an Animate Human Instead..?

An interesting point to observe in figure 4 is that PMP framework does not make any special distinction between the 'body' and 'tool'. The tool space is represented exactly in the same manner as any other motor space with a force node and position node, linked vertically by impedance and horizontally by tool Jacobian J_T that is learnt based on the tool and retrieved from memory (if already learnt). During goal directed coordination the body and the tool act as one cohesive unit. The goal now acts on the 'tool effector' which is the most distal part of the PMP chain. The pull of the goal acting on the tool tip is incrementally circulated to the proximal spaces (end effector, arm joints, waist etc) according to the information flow in figure 4. As the magnetized tip is being pulled towards the goal target, iCub's end effectors are simultaneously being pulled towards the required positions so as to allow the tool tip to reach the goal. These positions are the goals for the end effector space. As a consequence, the joints are concurrently pulled so as to allow the end effectors to reach the position that allows the tool tip to reach the goal. These are the goals for the intrinsic space. If motor commands derived through this incremental internal simulation of action are transmitted to the robot, it will reproduce the motion, hence allowing iCub to perform goal directed movements using the 'body + toy crane' network. This kind of goal-centered functional organization of action is reminiscent of the results of Iriki and colleagues [20], who showed that, with practice, a rake becomes a part of the acting monkey body schema and recent work of *Umiltà [4].*Figure 5 presents 3 sets of results related to goal directed coordination of the toy crane using the new body+tool PMP network (fig 5). The rows reflect different situations: *normal condition (row 1), pathological condition (row 2) and a hypothetical situation that simulates the resulting behaviour if the toy crane is replaced by a human instead whose hands are coupled to the hands of iCub (row 3).*

In the normal condition (panels 1-4) both the tool is compliant (A_T=0.01) and both arms equally functional to generate force (K_e=0.01 for both arms). The observed behavior is characterized by the following patterns: the green tool angle faithfully tracks the planned red attractor or goal (panel 1); the tool tip is successfully steered to the goal (panel 2); the components of the   forces transmitted by the two hands to the tool are approximately bell-shaped and terminate with null values (panel 3); a similar evolution characterizes the torque applied to the tool (panel D) as well as the tool rotation speed. **The second row (panels 5-8) illustrates a "pathological case": the right arm is** *functionally compromised,* in the sense that its force generating capabilities is reduced. In the reported experiment the elastic coefficient Ke is reduced 10 times (from 0.01 to 0.001 N/m) while keeping the same value for the other arm. Panels 5-8 illustrate the resulting behavior: in spite of the strongly different patterns of force delivered by the two arms (panel 7) the tool tip behaves in a consistent way, although with some error (panels 1,2) and the tool torque is still approximately bell-shaped (panel 8). Summing up, the PMP-based synergy formation mechanism of fig. 1 has self-adapting properties that allow the robot to exhibit acceptable performance for large variations of the system's parameters. Remarkably, *no learning is needed to* *accomplish this; it is in fact the property of the attractor dynamics of the 'elastic' PMP system to take into account unaccounted situations and yet do the best in achieving the goal.*

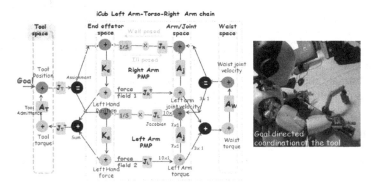

Fig. 4. Fully connected 'Body + Toy crane' PMP network, goal now acting on the tool tip

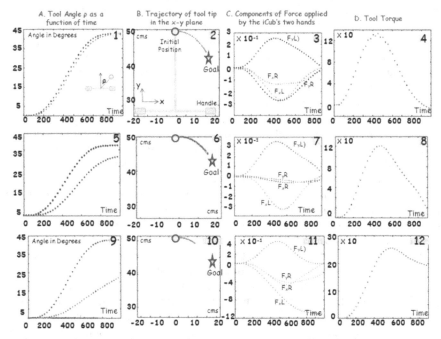

Fig. 5. Row 1.Tool use under Normal Conditions (Tool is compliant $A_T=0.1$, Both arms are equally active to generate force $K_e=0.01$); Row 2: Tool use when right arm functionality is compromised (Tool is compliant $A_T=0.1$, but $K_{e, Right Arm}=0.001$, $K_{e, Left Arm}=0.01$) or the force generating capability of the right arm is reduced 10 times; Row 3. What if: 'toy crane' is replaced by a 'human' who chooses not to 'Comply'?

Finally, let us consider the scenario where iCub interacts not with an inanimate tool (the toy crane) but an animate organism (whether a human or another robot). In either case, the organism has a choice: *comply* or *not comply* with the action planned by iCub. In the framework of the present experimental scenario this is equivalent to setting the value of A_T: high value (0.1) for the *comply choice* and low value (0.01) for the *not comply choice*. In the former case we obtain a behavior similar to the nominal situation (row 1 of fig. 2). In the latter case we obtain the behavior illustrated in row 3. *The two organisms (robot-human) do not cooperate and although the PMP-based controller is attempting to do its best to persuade the partner*, given the condition, the goal is not achieved and, at the end of the action, there is still energy which could not be channelized to the task achievement. This result we believe is a stepping stone to further investigate the connectivity between bodies and tools (that may be other animate bodies as well). Future work will be directed in this regards.

5   Concluding Remarks

Learning to coordinate and use external objects as tools to realize otherwise unrealizable goals is a sophisticated adaptive behavior observed in several species   and highly evolved in humans who are unique in having established a culture that revolves around actively manufacturing and using tools. Computationally, the effective utilization of a tool requires an understanding of several aspects, for example, the physical characteristics of a tool, its relationship to the body, its effects on the world and finally how to coordinate it in the context of a goal. To go beyond 'machine' learning and pave way for 'humanlike' learning systems, we believe it is crucial to exploit and integrate multiple learning channels i.e. social interactions, physical interactions and past experience (memory) to constrain, speed up and shape the overall learning process. A novel skill learning architecture based on this idea was presented in this article with an example of iCub learning to use a toy crane. While the teacher's demonstration constrains the space of exploration with the new toy, the past experience provides the correct parameters (stiffness and timing) to imitate the demonstrated action with the tool. Interacting and practicing with the tool allows to learn the causal relationship between the body and the tool being coordinated (i.e. the tool Jacobian JT and represented in a sub symbolic fashion). The PMP network is the natural site where all this motor knowledge related to stiffness, timing, Jacobians and admittance is integrated in the context of a goal. The attractor dynamics of PMP relaxation allows to systematically go down to the directly controlled elements of the body (actuators in the robot) to compute the necessary motor commands, at the same time characterized with self adaptive and compensatory properties to deal with multiple constraints at runtime as seen in the results. In this sense, the skill learning architecture is also a preliminary attempt to unify task independent knowledge acquisition/reuse (through the shape perception/synthesis hypothesis) and task specific compositionality (both at the level of shapes and at the level of force fields i.e. PMP) in a single computational framework. This article is just a small step and further efforts in this direction will pave way for autonomous robots to acquire a rich, growing 'action' vocabulary, both learn from us and meaningfully assist us in our various needs in the environments we inhabit and create.

Acknowledgments. The research presented in this article is supported by the EU FP7 projects EFAA (www.efaa.upf.edu, Grant No: FP7-270490) and DARWIN (www.darwin-project.eu, Grant No: FP7-270138).

References

1. Visalberghi, E., Tomasello, M.: Primate causal understanding in the physical and in the social domains. Behavioral Processes 42, 189–203 (1997)
2. Weir, A.A.S., Chappell, J., Kacelnik, A.: Shaping of hooks in New Caledonian Crows. Science 297, 981–983 (2002)
3. Iriki, A., Sakura, O.: Neuroscience of primate intellectual evolution: natural selection and passive and intentional niche construction. Philos. Trans. R Soc. Lond. B Biol. Sci. 363, 2229–2241 (2008)
4. Umiltà, M.A., Escola, L., Intskirveli, I., Grammont, F., Rochat, M., Caruana, F., Jezzini, A., Gallese, V., Rizzolatti, G.: When pliers become fingers in the monkey motor system. Proc. Natl. Acad. Sci. USA 105(6), 2209–2213 (2008)
5. Johnson-Frey, S.H.: The neural bases of complex human tool use. Trends in Cognitive Sciences 8, 71–78 (2004)
6. Johnson-Frey, S.H., Grafton, S.T.: From "Acting On" to " Acting With": the functional anatomy of action representation. In: Prablanc, C., Pelisson, D., Rossetti, Y. (eds.) Space Coding and Action Production. Elsevier, New York (2003)
7. Stoytchev, A.: Robot Tool Behavior: A Developmental Approach to Autonomous Tool Use. Ph.D. Dissertation, College of Computing, Georgia Institute of Technology (August 2007)
8. Stoytchev, A.: Learning the Affordances of Tools Using a Behavior-Grounded Approach. In: Rome, E., Hertzberg, J., Dorffner, G. (eds.) Towards Affordance-Based Robot Control. LNCS (LNAI), vol. 4760, pp. 140–158. Springer, Heidelberg (2008)
9. Mohan, V., Morasso, P., Metta, G., Sandini, G.: A biomimetic, force-field based computational model for motion planning and bimanual coordination in humanoid robots. Autonomous Robots 27(3), 291–301 (2009)
10. Mohan, V., Morasso, P.: Passive motion paradigm: an alternative to optimal control. Front. Neurorobot. 5, 4 (2011), doi:10.3389/fnbot.2011.00004
11. Mohan, V., Morasso, P., Metta, G., Kasderidis, S.: Actions & Imagined Actions in Cognitive robots. In: Cutsuridis, V., Hussain, A., Taylor, J.G. (eds.) Perception-Reason-Action Cycle: Models, Algorithms and Systems. Springer Series in Cognitive and Neural Systems, vol. 1, ch. 17, pp. 539–572 (2011)
12. Mohan, V., Morasso, P., Zenzeri, J., Metta, G., Chakravarthy, V.S., Sandini, G.: Teaching a humanoid robot to draw 'Shapes'. Autonomous Robots 31(1), 21–53 (2011)
13. Asatryan, D.G., Feldman, A.G.: Functional tuning of the nervous system with control of movements or maintenance of a steady posture. Biophysics 10, 925–935 (1965)
14. Hogan, N.: Modularity and Causality in Physical System Modeling. ASME Journal of Dynamic Systems Measurement and Control 109, 384–391 (1987)
15. Zak, M.: Terminal attractors for addressable memory in neural networks. Phys. Lett. A 133, 218–222 (1988)
16. Bernstein, N.: The coordination and regulation of movements. Pergamon Press, Oxford (1967)
17. Shapiro, R.: Direct linear transformation method for three-dimensional cinematography. Res. Quart. 49, 197–205 (1978)
18. Thom, R.: Structural Stability and Morphogenesis. Addison-Wesley, MA (1975)
19. Chakravarthy, V.S., Kompella, B.: The shape of handwritten characters. Pattern Recognition Letters (2003)
20. Maravita, A., Iriki, A.: Tools for the body (schema). Trends in Cognitive Science 8, 79–86 (2004)

Towards Contextual Action Recognition
and Target Localization with Active Allocation
of Attention

Dimitri Ognibene, Eris Chinellato, Miguel Sarabia, and Yiannis Demiris

Imperial College London, South Kensington Campus, London SW7 2AZ, UK
{d.ognibene,e.chinellato,miguel.sarabia,y.demiris}@imperial.ac.uk

Abstract. Exploratory gaze movements are fundamental for gathering the most relevant information regarding the partner during social interactions. We have designed and implemented a system for dynamic attention allocation which is able to actively control gaze movements during a visual action recognition task. During the observation of a partner's reaching movement, the robot is able to contextually estimate the goal position of the partner hand and the location in space of the candidate targets, while moving its gaze around with the purpose of optimizing the gathering of information relevant for the task. Experimental results on a simulated environment show that active gaze control provides a relevant advantage with respect to typical passive observation, both in term of estimation precision and of time required for action recognition.

Keywords: active vision, social interaction, humanoid robots, attentive systems, information gain.

1 Introduction

The introduction of active vision [2,1] was a fundamental step towards overcoming the limits of the classical vision paradigm as formulated by Marr [11]. Nevertheless, the perception of dynamic events still poses fundamental problems, such as the timely detection of the relevant elements, and the recognition of the discriminant dynamics. An archetypal and behaviorally relevant example of event perception is the recognition of an action executed by another agent. In order to deliver a behavioural advantage, and allow for timely action selection, the target and the end effector of an action should be predicted in advance, notwithstanding the limited perceptual and computational resources of the observer, and its knowledge of the environment, which is never optimal, due to occlusions and inner visual complexity.

We present here an attention system which integrates top-down with bottom-up attentional mechanisms. Starting from the simulation theory of mind point of view for action perception, we manage attention allocation in an active way, according to the predicted plausibility of candidate actions and possible targets. For a given action, the information that the attention system extracts during

T.J. Prescott et al. (Eds.): Living Machines 2012, LNAI 7375, pp. 192–203, 2012.

action observation is the state of the variables that the corresponding inverse model would control if it was executing that same action. For example, the inverse model for executing an arm movement will request the state of the arm when used in perception mode. This novel approach provides a principled way for supplying top-down signals to the attention system, which is to be integrated with bottom-up signals such as saliency maps or movement detectors. The influence of different attention biases can be modulated according to the task, the perceived interaction stage, what we know regarding the partner, and so forth.

We consider top-down attention as a competition of resources between multiple inverse models that seek to confirm their hypotheses about what the demonstrator's action/intention is. The saliency of a request for resources from each inverse mode can be linked to the quality of the predictions it offers. The computational and sensorimotor resources of the robot are distributed to the different inverse models as a function of the quality of the predictions they offer about forthcoming states of the interaction. In the meanwhile, a continuous estimation of environmental affordances allows for a dynamical update of which inverse models are applicable to the current state of the interaction.

In this way, the system is able to provide a prediction of the position of the observed agent effector, and thus an interpretation of his action, and also an estimation of the location of objects in the environment which constitute potential targets for the action being executed. Saccadic movements are performed according to a certain confidence level attributed to each of the competing models, and to the saliency of a feature (either hand or object).

Differently from previous approaches, which required the knowledge of the features of the different targets present in the environment to detect them [5], or the knowledge of their positions, in this work we propose a model that can overcome these limits, allowing for simultaneous exploration of the environment and recognition of the actions, exploiting both source of information to achieve faster action recognition. Also, according to a foveal model of vision, we consider that visual information gets more reliable and less noisy moving from the periphery to the center of the visual field.

2   The Problem

Perception in active vision is constituted by a sequence of visual shots interleaved by saccadic movements[1,2], aimed at purposefully exploring the environment, in order to extract the information relevant for pursuing the current goals. Given this strategy, the quality of the obtained information is due in great part to efficient and intelligent gaze control. A fundamental issue on this regard is the implicit indetermination of attending to something we cannot precisely locate yet. This requires a concurrent evolution of both the knowledge regarding the environment and the quality of the attention strategy. In our case, for achieving a shared, dynamical attention allocation during a social interaction, decisions on where to look are strictly linked to the movements of the partner. This adds further complexity to the task, which now has to account for a changing visual scenario. Human behavioural studies, on tasks like face recognition [7] and visuo-motor control [15,10], have shown that humans are able to adapt their visual exploration to the specific requirements of the task at hand.

Robotic studies on adaptive active vision so far have focused on the previously mentioned topics in isolation. In [17] an artificial fovea is controlled by an adaptive neural controller. Without a teacher, this learns trajectories causing the fovea to find targets in simple visual scenes and to track moving targets. The model in [9] solve active sensing problems under uncertainty. A reinforcement learning algorithm allows it to develop active sensing strategies to decide which uncertainties to reduce. However, in this study the model of the task is known a-priori and motor control is hardwired. Other works (e.g. [3,19]) employ evolutionary learning techniques for developing adaptive active vision systems. These approaches are robust to the *perceptual aliasing problem*, however they do not allow on-line adaptation to changing environments. In [12] a neural architecture for eye arm coordination is proposed which learns autonomously task-specific attentional policies, exploiting a strong link between attention and execution of actions. The authors also proposed that a bottom-up attention system can be exploited to bootstrap learning, and hypothesised on the basis of neural simulations that the limited size of fovea can play an important role in the efficiency of learning [14].

In this work, we deal with the above issues by letting an integrated attention system assume gaze control while observing a partner performing a reaching action toward one of a small set of target objects. Neither the goal of the action, nor the exact location of the potential targets are known beforehand, so that hand trajectory and target position have to be estimated contextually while trying to understand what is the action goal, i.e. where the partner is moving its hand towards. An example of a possible experimental setup is provided in Fig. 1, where the humanoid robot iCub is observing a human partner starting a reaching movement towards one of three potential target objects placed in the common working space. The robot has to decide where to observe (estimated hand or object position) in order to 1) estimate the objects exact positions and 2) understand where the partner's hand is reaching at.

3   Action Recognition with Dynamic Allocation of Attention

In this work, we build on some of the concepts introduced with the HAMMER model for action perception and imitation based on the direct matching hypothesis [4]. The system described here is based on the integration of the latest HAMMER framework implementation [16] with a gaze controller which directs attention in order to maximise discrimination performance, while maintaining robustness to noise, and a contextual estimation of both end effector location and position of all potential targets.

A number of different models, at least one for each of the possible targets of the action, concur for both attention allocation and for the final discrimination of

Fig. 1. Example of experimental setup, with iCub looking at target objects and arm movement (its own pointing movement is not relevant here)

the action goal. Following HAMMER guidelines [5], the discrimination between the available action hypotheses is based on the computation of a confidence value that measures the overall Euclidean distance between the predicted action trajectories and the observed motion trajectories. To compute such prediction, HAMMER uses a combination of forward and inverse model pairs which are the same models that can be used for action control.

Formally, an inverse model is a function that, given a certain goal g, maps the current state S to the action A the agent has to execute to achieve the goal: $i_g : S \to A$. A forward model is a function that maps the current state and the action being executed to the next expected state $f_g : S \; times A \to S$.

In this work, the candidate actions among which the observer has to chose are different reaching movements toward different targets in space, g. A reaching model m_g, composed of a pair of inverse and forward models (i_g, f_g) is required for each different g. Each reaching model works directly in the space of the end-effector, and the action space is coincident with the state space $(A \equiv S)$, because the used inverse model computes the next desired end-effector position $\mathbf{p}^{t+1}$, and the forward model returns the same value, too.

The following is the equation, akin to a PID controller, employed by a model m_g to compute the next position $\mathbf{p}^{t+1}$, when the target is at position $\mathbf{p}_g$:

$$\mathbf{p}^{t+1} = \mathbf{p}^t + \tau\{\dot{\mathbf{p}}^t + \tau[K(\mathbf{p}_g - \mathbf{p}^t) - D\dot{\mathbf{p}}^t]\}. \tag{1}$$

This equation leads to a motion with a smooth linear trajectory that brings asymptotically toward the target. The confidence function for each model and time step c_t^g is updated employing the difference between the predicted end-effector position $\tilde{\mathbf{p}}^{t+1}$ and the perceived one $\mathbf{p}^{t+1}$:

$$c_{t+1}^g = \frac{1}{1.0 + \|\tilde{\mathbf{p}}^{t+1} - \mathbf{p}^{t+1}\|} + c_t^g. \tag{2}$$

For increased plausibility, we assume that the observations of the end-effector and of the affordances are affected by noise that is dependent on the sensors configuration, i.e. gaze position. Thus, if the observer gaze position is $\mathbf{pos}_o$, the actual observation of an object at position $\mathbf{p}$ is distributed according to:

$$N(\mathbf{p}, 0.15\|\mathbf{p} - \mathbf{pos}_o\|^2). \tag{3}$$

This noise model is an approximation of human foveal vision, where the most of the visual receptors are located in the central area (fovea) of the retina and their density, and thus the visual resolution, decreases departing from the fovea.

In order to manage the noisy input, each model uses Kalman filters for the estimation of the end-effector and affordance positions, and an active vision system is integrated that exploits the estimation of the uncertainty produced by the Kalman filters. The main assumption is that the observer can use features which allow to discriminate between the different affordances and the effector. This approach is similar to that of [8] and [18] but, instead of being limited to track or to find objects in a dynamic environment, it allows for the active recognition of a dynamic event.

In this work, independent Kalman filters are used for each action element. An action element has position $\mathbf{p}$ and produces an observation z. The associated Kalman filter produces an estimated probability distribution $b_{\mathbf{p}}$ and a corrected probability $\hat{b}_{\mathbf{p}}$, which uses the observation received at the current time step.

In order to produce these estimations the Kalman filter uses a process model and noise model. The process model has the form $\mathbf{p(t + 1)} = \mathbf{Ap}(t) + \mathbf{b}(t)$. For both the effector and the targets, matrix $\mathbf{A}$ is the identity matrix $\mathbf{I}$. We assume that the targets are still, thus their process noise $\mathbf{b}(t)$ is zero. The effector process noise is $0.01\ \mathbf{I}$ to model small changes in the trajectories.

In this work, we decoupled the prediction and the correction phases of the Kalman filters for the end-effector in each model. The mean position of the end-effector is updated in accordance with the action model in eq. 1, after the correction of estimated position of the target and the prediction (not corrected) of the end-effector using the process model as described above. The variance estimated by the process model is not modified. After this update the Kalman correction phase takes place also for the end-effector.

In our task, we need to take into account the implicit imprecision of the sensory information, together with the lack of exact knowledge regarding hand and targets position. As a consequence, the typical Kalman formulation have to be adapted, so that the observation models of the Kalman filters are able to account for the change of the sensory configuration and the uncertainty regarding the real environment. While observation noise increases with the distance between observation point and real object position, the latter is not known, and only a prior estimation $b_{\mathbf{p}} = N(\bar{\mathbf{p}}, \Sigma_p)$ is available before saccade execution. Thus, the resulting observation model depends on current gaze position and belief state:

$$P(\mathbf{z}|\mathbf{pos}_o, b_{\mathbf{p}}) = \int p(\mathbf{z}|\mathbf{p}, \mathbf{pos}_o)b_{\mathbf{p}}d\mathbf{p}. \tag{4}$$

The implemented observation model is expressed by the following normal distribution, considering that Kalman filters assume Gaussian distributions and linear dynamics:

$$P(\mathbf{z}|\mathbf{pos}_o, b_{\mathbf{p}}) \approx \mathbf{N}(\bar{\mathbf{p}}, 0.15(\|\bar{\mathbf{p}} - \mathbf{pos}_o\|^2 \mathbf{I} + 0.9\Sigma_p')). \tag{5}$$

The attention system uses the probability distribution estimated by the Kalman filters in all the models and the confidence value of each action hypothesis. The attention system minimises uncertainty on the most probable action. The effects of reducing uncertainty between the different action hypotheses is not directly taken into account by the current system for computational reasons. Gaze point selection currently considers only instantaneous saccades even if the system is allowed to select a new saccade target only after the previous attentive action has been completed.

Each element of each model, e.g. target and effector , is considered independently by the attention system instead of integrating the different probability distributions associated to elements shared by different models. e.g. the different expected positions of the end-effector for the different models. In this implementation each action hypothesis has two elements, effector and affordance, and the attention system selects targets from a set of $2 * n_a$ elements, where n_a is the set of action hypotheses.

The selected target, with estimated position $\bar{\mathbf{p}}$ related to hypothesis g, is the one which maximizes the following objective function:

$$\log(|\Sigma_p|)(1 + c^g). \tag{6}$$

This objective function accounts at the same time for both the reduction of uncertainty, which can be measured using entropy (i.e., in the case of a Gaussian distribution, by the logarithm of the determinant of the covariance matrix), and the relevance for the most probable action hypothesis.

There are several approximations in this objective function: a) it does not consider the residual entropy of the target assuming that after the saccade the object will be perfectly centered; b) it does not consider the information gain on the other targets. At the same time, this formulation allows to select only positions corresponding to estimated targets, while other positions may allow to increase the overall information gain.

4   Experimental Evaluation

The proposed model has been implemented on the iCub Simulator where the simulated robot head was controlled by the attentional system. In the simulated environment (Fig. 2) three target objects were created (small coloured boxes). The system receives noisy observations, represented as small spheres of the same colour as the actual objects. The simulated sensor noise is proportional to the square of the distance between the real position (boxes) and the robot gaze point (red cylinder), following Eq. 3. The end-effector of the other agent is displayed

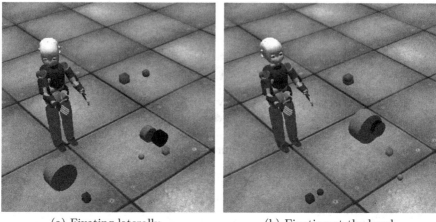

(a) Fixating laterally   (b) Fixating at the hand

Fig. 2. Experimental setup with example of different gaze points. The big red cylinder represents the gaze point of the robot. The boxes represent the real position of the observed action target while the small spheres represent the related observations that the robot senses. The dark small cylinder represents the end effector that is executing the action and the brighter one the related observation.

by a black cylinder, while its observation is a gray cylinder. The head and eye are controlled using the fixation point and the *iKinGazeCtrl* iCub module[1]. A new fixation point is sent only when the previous movement has ended.

The end-effector of the other agent is moving, according to equation 1 with $K = 3$ $D = 0.5$ and $\tau = 0.04$, towards one of the target object, randomly selected. The system is provided with three models, with the same parameters for the three different objects, and with prior $\mathbf{N}(o, I)$, where o is an observation sampled according to the noise model, constituting the initial gaze point. Examples of hand trajectories devised by the concurrent models are depicted in Fig. 3. It can be observed how the hand terminates in each of the three target objects corresponding to the three candidate models. The top trajectory is the one which is actually performed in this case. In order to characterise the behaviour of the system in different working conditions and with different setup and parameters, a number of analyses can be performed on the system performance. The results presented in this section take into account different aspects of the system behaviour, and are currently being employed to improve the reliability and generalization skills of our action recognition module.

First of all, we need to verify whether exploratory gazing movements performed according to the attention control provide an actual advantage with respect to a typical passive perception paradigm. We have performed experiments with moving gaze, performed as described above, and with steady gaze, in which the robot was fixating the same random location of its working environment for the whole test. In both moving and steady gaze experiments we made sure that

---

[1] http://eris.liralab.it/iCub/main/dox/html/group_iKinGazeCtrl.html

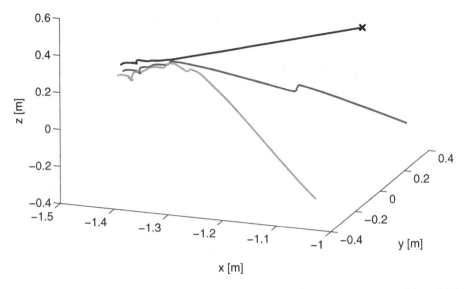

Fig. 3. Estimated hand trajectories according to the three competing models, which converge asymptotically to the three targets. Sudden changes in the two lower estimated trajectories are due to gazing actions that change the related observation models.

all relevant stimuli were always visible to the robot. We executed twenty trials with each of the paradigms, and averaged their outcomes, obtaining the results summarized in Table 1. **Max confidence error** is the overall mean distance between the real effector position and the estimation provided by the most credited model (i.e. the one with highest confidence) at each time step. **Winner error** is the same mean distance computed at each time step for the winning model (this error can only be computed *a posteriori*, when it is known what model has prevailed). There is a clear difference between the methods in this regard, as the moving gaze paradigm error is about half of the steady gaze error. It is also interesting to observe that, in both paradigms, the two different errors assume similar values, indicating that the **Max confidence error** represents a good on-line approximation of the actual performance of the winner model.

Indexes **Choice TS** in Table 1 represent potential decision moments, according to two different thresholds. More exactly, they show the time step at which the confidence of the dominant model, computed according to Eq. 2, is 20% and 10% higher than the others, respectively. The performance of the attention-based protocol is again clearly better than the passive protocol. The 20% threshold is achieved more than 5 time steps earlier on average by the former (corresponding to a 10% improvement). Even more significantly, the 10% threshold is attained about 8 time steps earlier on average, which is like saying that the moving gaze system decides 24% more quickly than the steady gaze system.

A typical evolution of the confidence level for each of the three competing models can be observed in Fig. 4 for both Moving gaze and steady gaze protocols. It is again possible to observe how the active paradigm is able to differentiate

Table 1. Comparison of performance in trials with moving and fixed gaze

	Moving gaze	Steady gaze
Max conf. error	0.049	0.118
Winner error	0.048	0.099
Choice TS (20%)	44.6	49.9
Choice TS (10%)	24.4	32.1

the goal action around time step 20 (Fig. 4(a)), whilst the passive paradigm seems to fail completely in the task (only at the very end of the trial a small, still non-significant prevalence of one of the models can be spotted, Fig. 4(b)).

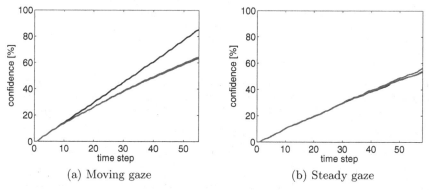

(a) Moving gaze   (b) Steady gaze

Fig. 4. Evolution in time of confidence with (a) and without (b) attentional gaze control. With gaze control the right action is clearly identified before time-step 20. Without gaze control the agent is not able to recognize the performed action.

To better understand how the active exploration of the environment through the execution of saccadic movements is performed, Fig. 5 shows an example of the evolution of gaze direction during an experiment, computed by the attention model as described in the previous section. Three different phases are highlighted. During the first half of the trial (dots in Fig. 5), gaze moves rather erratically all around the task space, but after this bootstrapping phase more regular behaviours can be observed. Time steps from half to three quarters of action execution show gaze points approximately distributed along the dominant hand trajectory, suggesting that the system has now understood where the action is going on (plus symbols in Fig. 5). Finally, in the last quarter of the trial, one of the model seems to be clearly dominant over the others. Estimations of both hand trajectory and location of target object are reasonably accurate, and the dominant model makes the system move forth and back between these two locations, which are now definitely considered the most interesting, to further improve their estimation (circles in Fig. 5).

The last results we present concerns the actual prediction capabilities of the system in terms of estimation of hand trajectory and object position. Fig. 6 shows

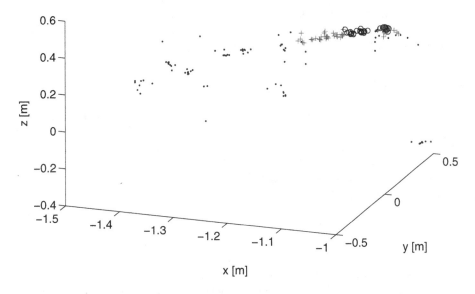

Fig. 5. Evolution of gaze point during different stages of action observation (dots: first half; plus: third quarter; circles: fourth quarter)

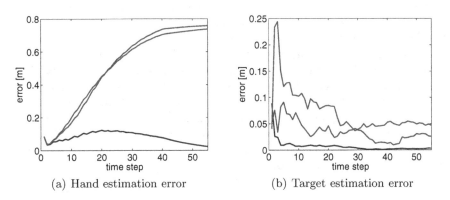

(a) Hand estimation error   (b) Target estimation error

Fig. 6. Evolution of estimation error on hand position (a) and on target position (b) according to the three competing models. End-effector position is estimated correctly only by one of the models (a), while target location is estimated with good approximation by each model, reaching a 50mm error in the worst case.

the error observed in the approximation obtained by each model in its estimation of the effector trajectory and target location. It can be observed that only one of the models achieves a correct estimation of the actual hand position (i.e. the model that correctly recognize the action), whilst the others wrongly convergence towards the wrong targets (Fig. 6(a)). Nevertheless, Fig. 6(b) shows that even the targets associated to the losing models are detected with a good approximation, and the first of the phases depicted in Fig.. 5 is critical in this regard, as it allows to achieve a good representation of all stimuli in the environment while gradually shifting the focus toward the supposedly most interesting one.

5   Conclusions

The results reported in this work show that the proposed approach is viable for the problem of action recognition in unknown environments. A relevant contribution given by this paper is related to the importance of using the simulation approach when perceiving actions with limited perception. Differently from the teleological and associative approaches (see [13] for the distinction), the simulation approach describes mechanisms that implement action recognition by producing dynamic internal representations that can be used also to direct attention. This is particularly evident in this model, which actually employs the simulated position to drive attention, whilst in previous models prediction was used to modulate bottom-up attention (see e.g. [6]). Experimental results have confirmed the advantages of the attention-based action recognition system, both in terms of precision and, maybe more importantly, in terms of decision time. This latter aspect is indeed critical when an agent has to interpret or recognise a partner's action. For a robot, being able to understand what a human partner is doing some tenth of a second earlier can be of fundamental importance in order to achieve a meaningful interaction, avoiding the inconvenient obligation for the human to wait for the robot to interpret his movements.

We are now working on the recognition of real human actions, and the comparison with human performance in the same task. We aim to substitute the current action models with more realistic ones that account for more peculiarities of human movements, that can thus allow for higher accuracy and faster recognition, i.e. using the pre-shaping of hand for gasping. Then, we plan to test the robustness of the system with respect to incorrect models, and to the presence of a high number of action hypotheses, differentiated also for the different parameters chosen, e.g. different execution speeds. Another interesting improvement would be the use non myopic target selection, for a better estimation of the relative importance of targets and effectors in visual perception of actions.

Acknowledgments. This research has received funding from the European Union Seventh Framework Programme FP7/2007-2013 Challenge 2 Cognitive Systems, Interaction, Robotics under grant agreement No [270490]- [EFAA].

References

1. Bajcsy, R.: Active perception. Proceedings of the IEEE 76(8), 966–1005 (1988)
2. Ballard, D.H.: Animate vision. AI 48, 57–86 (1991)
3. de Croon, G.C.H.E., Postma, E.O., van den Herik, H.J.: Adaptive gaze control for object detection. Cognitive Computation 3(1), 264–278 (2011)
4. Demiris, Y., Khadhouri, B.: Hierarchical attentive multiple models for execution and recognition of actions. Robotics and Autonomous Systems 54, 361–369 (2006)

5. Demiris, Y., Khadhouri, B.: Content-based control of goal-directed attention during human action perception. Journal of Interaction Studies 9(2), 353–376 (2008)
6. Demiris, Y., Simmons, G.: Perceiving the unusual: temporal properties of hierarchical motor representations for action perception. Neural Networks 19(3), 272–284 (2006)
7. Heisz, J.J., Shore, D.I.: More efficient scanning for familiar faces. J. Vis. 8(1), 1–10 (2008)
8. Kastella, K.: Discrimination gain to optimize detection and classification. IEEE Transactions on Systems, Man and Cybernetics, Part A: Systems and Humans 27(1), 112–116 (1997)
9. Kwok, C., Fox, D.: Reinforcement learning for sensing strategies. In: IEEE/RSJ International Conference on Intelligent Robots and Systems, IROS 2004 (2004)
10. Land, M.F.: Eye movements and the control of actions in everyday life. Prog. Retin. Eye Res. 25(3), 296–324 (2006)
11. Marr, D.: Vision: A Computational Investigation into the Human Representation and Processing of Visual Information. W. H. Freeman, New York (1982)
12. Ognibene, D., Balkenius, C., Baldassarre, G.: Integrating epistemic action (active vision) and pragmatic action (reaching): A neural architecture for camera-arm robots. In: Proceedings of the Tenth International Conference on the Simulation of Adaptive Behavior (2008)
13. Ognibene, D., Wu, Y., Lee, K., Demiris, Y.: Hierarchies for embodied action perception. Under review (2012)
14. Ognibene, D., Pezzulo, G., Baldassarre, G.: How can bottom-up information shape learning of top-down attention control skills? In: Proceedings of 9th International Conference on Development and Learning (2010)
15. Sailer, U., Flanagan, J.R., Johansson, R.S.: Eye-hand coordination during learning of a novel visuomotor task. J. Neurosci. 25(39), 8833–8842 (2005)
16. Sarabia, M., Ros, R., Demiris, Y.: Towards an open-source social middleware for humanoid robots. In: Proc. 11th IEEE-RAS Int Humanoid Robots (Humanoids) Conf., pp. 670–675 (2011)
17. Schmidhuber, J., Huber, R.: Learning to generate artificial fovea trajectories for target detection. Int. J. Neural Syst. 2(1-2), 135–141 (1991)
18. Sommerlade, E., Reid, I.: Information theoretic active scene exploration. In: Proc. IEEE Computer Vision and Pattern Recognition (CVPR) (May 2008)
19. Suzuki, M., Floreano, D.: Enactive robot vision. Adapt. Behav. 16(2-3), 122–128 (2008)

Robot Localization Implemented with Enzymatic Numerical P Systems

Ana Brândușa Pavel, Cristian Ioan Vasile, and Ioan Dumitrache

Department of Automatic Control and Systems Engineering,
Politehnica University of Bucharest,
Splaiul Independenței, Nr. 313, sector 6, 060042, Bucharest, Romania
{apavel,cvasile,idumitrache}@ics.pub.ro

Abstract. Membrane computing is an interdisciplinary research field focused on new computational models, also known as P systems, inspired by the compartmental model of the cell and the membrane transport mechanisms. Numerical P systems are a type of P systems introduced by Gh. Păun in 2006 for possible applications in economics. Recently, an extension of numerical P systems, enzymatic numerical P systems, has been defined in the context of robot control. This paper presents a new approach to modeling and implementing autonomous mobile robot behaviors and proposes a new odometry module implemented with enzymatic numerical P systems for robot localization. The advantages of modeling robot behaviors with enzymatic membrane controllers and the experimental results obtained on real and simulated robots are also discussed.

Keywords: robot, membrane controller, odometry, localization, enzymatic numerical P systems.

1 Introduction

Membrane computing is a new computational paradigm inspired by the transport mechanisms of the cell's membranes. The cell is delimited by a bio-membrane and contains other compartments with specific functions (nucleus, mitochondria, etc.) [5]. The tree-like structure of the cell's membranes, which bound the compartments, and the trans-membrane transport mechanisms are the fundamental features of the membrane systems. The computational model, introduced by Gh. Păun, operates on symbols (symbolical P systems)[9,11] or variables (numerical P systems) [10]. Each membrane of the P system contains either a list of symbols or variables and a set of rules used for computation and inter-membrane communication. P systems are said to be naturally parallel and distributed systems because of their structure and their computation mechanism.

Most of the research effort has been focused on symbolical P systems. Numerical P systems (NPS) have been introduced in 2006 for possible applications in economics [10]. An extension of the NPS model, *enzymatic numerical P systems* (ENPS), has been proposed by the authors [7] in order to enhance the modeling

T.J. Prescott et al. (Eds.): Living Machines 2012, LNAI 7375, pp. 204–215, 2012.

power of NPS. In this paper, the ENPS model is used to implement an obstacle avoidance and odometric localization modules for autonomous mobile robots. The membrane system model for obstacle avoidance was proposed in [8] as an application for ENPS. Both modules have been tested on real and simulated robots and the experimental results are presented.

This paper presents a new way of modeling robot behavior using a biomimetic computational paradigm inspired by the cell's structure. To the authors' knowledge, this approach is new to the robotics field.

2   Formal Definition of the ENPS Model

2.1   NPS Model

A membrane system has a tree-like structure (figure 1) and computation takes place in parallel in all its nodes (membranes). Each membrane of a numerical P system has variables, which store information, and a set of rules (programs), which are responsible for the computation and the transfer of information between the nodes. The rules' action is inspired from chemical reactions which take place within the cell. A rule consumes the variables that are involved in it and produces a quantity that is distributed to other variables. The variables which receive new values from the rule must be contained within the current, the parent or a child membrane. This process is inspired by the membrane transport mechanisms of the cell [1] and it is responsible for the communication between the artificial membranes.

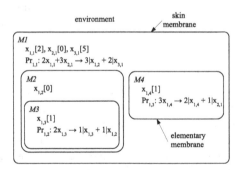

Fig. 1. Numerical P system with 4 membranes, M_1, M_2, M_3, M_4. Membrane M_1 is the skin membrane of the system and M_3 and M_4 are elementary membranes because they don't have children. Each membrane has a number of variables labeled with x_{ij} and one or no rule (program Pr_{ij}). The initial values of the variables are represented between brackets.

The ENPS model is based on the NPS model which is formally defined as follows:

$$\Pi = (m, H, \mu, (Var_1, Pr_1, Var_1(0)), \dots, (Var_m, Pr_m, Var_m(0))) \quad (1)$$

where:

- m is the number of membranes used in the system, degree of Π; $m \geq 1$;
- H is an alphabet that contains m symbols (the labels of the membranes);
- μ is a membrane structure;
- Var_i is the set of variables from compartment i, and the initial values for these variables are $Var_i(0)$;
- Pr_i is the set of programs (rules) from compartment i. Programs process variables and have two components, a production function and a repartition protocol.

The j-th program has the following form:

$$Pr_{j,i} = (F_{j,i}(x_{1,i}, ..., x_{k_i,i}), c_{j,1}|v_1 + ... + c_{j,n_i}|v_{n_i})$$

where:

- $F_{j,i}(x_{1,i}, ..., x_{k_i,i})$ is the production function;
- k_i represents the number of variables in membrane i;
- $c_{j,1}|v_1 + ... + c_{j,n_i}|v_{n_i}$ is the repartition protocol;
- n_i represents the number of variables contained in membrane i, plus the the number of variables contained in the parent membrane of i, plus the number of variables contained in the children membranes of i.

The coefficients $c_{j,1}, \ldots, c_{j,n_i}$ are natural numbers (they may be also 0, case in which it is omitted to write "$+0|x$") [10] which specify the proportion of the current production distributed to each variable $v_1, ..., v_{n_i}$. Let us consider the sum of these coefficients: $C_{j,i} = \sum_{n=1}^{n_i} c_{j,n}$. A program $Pr_{j,i}$ is executed as follows. At any time t, the function $F_{j,i}(x_{1,i}, ..., x_{k_i,i})$ is computed. The value $q = \frac{F_{j,i}(x_{1,i}, ..., x_{k_i,i})}{C_{j,i}}$ represents the "unitary portion" to be distributed to variables $v_1, \ldots, v_{n_i}$, according to coefficients $c_{j,1}, \ldots, c_{j,n_i}$ in order to obtain the values of these variables at time $t + 1$. Specifically, variable v_s which belongs to the repartition protocol of program j, will receive: $q * c_{j,i}, 1 \leq s \leq n_i$.

If a variable belongs to membrane i, it can appear in the repartition protocol of the parent membrane of i and also in the repartition protocol of the child membranes of i. After applying all the rules, if a variable receives such "contributions" from several neighboring compartments, then they are added in order to produce the next value of the variable. A production function which belongs to membrane i may depend only on some of the variables from membrane i. Those variables which appear in the production function become 0 after the execution of the program.

Deterministic NPS have only one rule per membrane ($card(Pr_i) = 1$) or must have a selection mechanism that can decide which rule to apply. The NPS model with multiple rules per membrane is a non-deterministic system. However, by their structure, NPS are well suited for applications which involve numerical variables and require a deterministic behavior, such as control systems for mobile robots. Thus a selection mechanism for the active rules is defined in the ENPS model [7].

2.2   ENPS Model

ENPS is defined as a NPS with special enzyme-like variables which control the execution of the rules. Thus, an ENPS is defined as follows:

$$\Pi = (m, H, \mu, (Var_1, E_1, Pr_1, Var_1(0)), \ldots, (Var_m, Pr_m, E_m, Var_m(0))) \quad (2)$$

where:

- E_i is a set of enzyme variables from compartment i, $E_i \subset Var_i$
- Pr_i is the set of programs from compartment i. Programs have one of the two following forms:
    1. non-enzymatic form, which is exactly like the one from the standard NPS: $Pr_{j,i} = (F_{j,i}(x_{1,i}, \ldots, x_{k_i,i}), c_{j,1}|v_1 + \ldots + c_{j,n_i}|v_{n_i})$
    2. enzymatic form: $Pr_{j,i} = (F_{j,i}(x_{1,i}, \ldots, x_{k_i,i}), e_{t,i}, c_{j,1}|v_1 + \ldots + c_{j,n_i}|v_{n_i})$, where $e_{t,i} \in E_i$

The enzymatic mechanisms of the ENPS model is used for the selection of the valid rules. The enzyme-like variables are inspired by biological enzymes, which are molecules that control most of the biochemical process in living cells. By catalyzing reactions, enzymes synchronize the steps of a biological process.

To understand how ENPS work, let us consider one membrane $M1$ with the following variables: $x_{11}[3]$, $x_{21}[2]$, $e_{11}[4]$ and one production function: $2 * x_{11} + x_{21}(e_{11} \rightarrow)$, where one may notice a specific variable attached, e_{11}, which is the enzyme (figure 2).

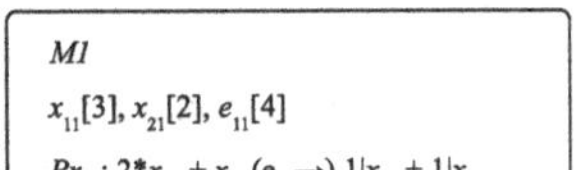

Fig. 2. Membrane with an enzyme variable

In this case the condition is $e_{11} > min(x_{11}/2, x_{21})$, but because $min(x_{11}, x_{21}) > min(x_{11}/2, x_{21})$, the simplified and more general condition $e_{11} > min(x_{11}, x_{21})$ ensures that the amount of enzyme is more than enough and that the reaction can take place. Thus, a rule is active if the associated enzyme variable has a greater value than the minimum of the variables involved in the production function. Because the values of the variables can be sometimes negative, in some applications it is more convenient to test if the value of the enzyme is greater than the absolute value of one of the variables contained in the production function. For example, the rule $Pr_{1,1}$ in membrane $M1$ (figure 2) is active if $e_{11} > min(|x_{11}|, |x_{21}|)$. This last condition is used in this paper. There can be more than one active rule in a membrane or none. In a computational step, all active rules in all membranes are executed in parallel. The universality of the NPS and ENPS computational models is proven in [10] and [14].

3   ENPS Controllers

Membrane controllers can be used to control autonomous mobile robots and to generate various desired behaviors and cognitive abilities, like obstacle avoidance, localization, moving to a given position, wall following, following another robot, etc. The major advantage of using NPS as a modeling tool is that P systems are naturally parallel and distributed systems. Membranes of a NPS can be distributed over a grid or over a network of microcontrollers in a robot. The computation done in each membrane region (the execution of a membranes rules) can also be done in parallel [3].

Another important property is that membranes can only communicate with their parents and their children membranes. Thus, communication in distributed P systems can be implemented efficiently.

Furthermore, membrane controllers can be integrated easily in the control program of the robotic system. Membrane systems can be added or changed to embed the desired functionality without changing the code of the control program. The membrane system definition can be stored in separate files (for example in xml format) and loaded as needed. A simulator for ENPS systems is necessary to execute the membrane systems. The simulator can, however, be optimized for the platform it runs on, taking advantage of the underlying hardware, microcontroller/procesor/DPS, memory architecture and communication technology (TCP/IP, I2C, TWI, Bluetooth, etc.). For instance, a parallelized GPU-based simulator for ENPS was recently proposed in [4].

A robot controller which uses two modules, obstacle avoidance and localization, is presented in Pseudocode 3. In each loop of the controller, the sensors are read (infrared and motor encoders), then the position of the robot is updated based on the previous position and on the values of the encoders. The proximity sensors are used to compute the motors' speeds such that obstacles are avoided. Finally, the motors' speeds are sent to the robot.

```
while(True) {
 read_sensors()
 position = odometry(position, encoders)
 motors_speeds = avoid(proximity_sensors)
 set(motors_speeds)
}
```

This paper proposes a new ENPS module for odometric localization which was implemented and tested on simulated e-puck and KheperaIII robots and on real e-puck robots. Experimental results obtained so far for an autonomous robot with both obstacle avoidance behavior and localization ability will further be presented. The ENPS model for obstacle avoidance is detailed in [8].

4   Odometric Localization for an Autonomous Mobile Robot

An autonomous mobile robot should be able to know its position at any time. Localization is one of the most important and difficult problems in autonomous mobile robotics. Different localization systems used to determine the position of an autonomous robot at any time are presented in [12,13]. Localization can be implemented in many ways, using different devices like: the encoders of the motors, accelerometers, beacons, GPS [13].

An ENPS model which implements odometric localization for an autonomous mobile robot, using the information received from the motors' encoders, is proposed. The ENPS for odometric localization has been designed as a membrane system with 5 membranes (figure 3). The enzyme-like variables have an essential role in the control of the program flow, synchronization and parallel computation. As the robot must know its position at any time, the module receives in the begining of each cycle of the controller the following input information: the initial position of the robot (x_i, y_i, θ_i), where x_i, y_i, represent the coordinates in reference coordinate system and θ_i is the angle made by the direction of the robot and the x axis counterclockwise (the orientation of the robot). Also, the distances traveled by each wheel are input values: dL for the left wheel and dR for right wheel. The output of the module is the updated position of the robot: (x_f, y_f, θ_f) in the same coordinate system. The trajectory of the robot on a short distance can be approximated with a circular arc. Each wheel travels a distance which is given by its encoder. The encoder returns the number of steps which can be converted into a distance. Thus, the following position update formulas are obtained for a differential wheeled robot [13]:

$$\Delta\theta = \frac{d_R - d_L}{wheelDist}$$

$$\Delta s = \frac{d_R + d_L}{2}$$

$$\Delta x = \Delta s \cdot \cos(\theta + \frac{\Delta\theta}{2})$$

$$\Delta y = \Delta s \cdot \sin(\theta + \frac{\Delta\theta}{2})$$

The updated position of the robot, (x_f, y_f, θ_f), is computed in the following way:

$$x_f = x_i + \Delta x \qquad y_f = y_i + \Delta y \qquad \theta_f = \theta_i + \Delta\theta$$

The position of the robot is measured in the point situated in the middle of the segment which joins the two wheels. The distance between the two wheels of the robot is a fixed parameter that depends on the type of robot and is passed to the membrane system via the b variable in membrane *Odometry* (figure 3).

The membranes *Cosine* and *Sine* approximate the values of sine and cosine using their power series:

$$\sin(x) = \sum_{n=0}^{\infty} (-1)^n \cdot \frac{x^{2n+1}}{(2n+1)!} \qquad \cos(x) = \sum_{n=0}^{\infty} (-1)^n \cdot \frac{x^{2n}}{(2n)!}$$

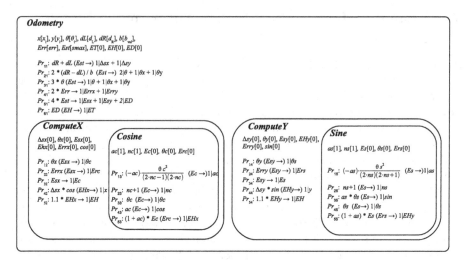

Fig. 3. Enzymatic membrane system for odometric localization

The general term of the power series is computed recursively, in rule $Pr_{1,3}$ for *Cosine* and in rule $Pr_{1,5}$ for *Sine*, and added to the final value, variable *cos* from membrane *ComputeX* and *sin* from *ComputeY*. After a number of steps, both *Cosine* and *Sine* finish their computations. The two membranes are executed in a number of steps which depends on the input value, θc for *Cosine* and θs for *Sine*, and the error, *Erc*, respectively *Ers*. When the general term of the series becomes less than the given error parameter, the computation stops.

In *ComputeX*, the value of Δx is computed by the production function $Pr_{4,2}$ which is activated by *Ehx*. The final value of *x* is computed by summing Δx, received from membrane *ComputeX*, with the initial value, x_i. Similarly, the value of *y* is computed. Then the values of *EHx* and *EHy* are transferred to *EH* ($EH \leftarrow 1.1 \cdot EHx$; $EH \leftarrow 1.1 \cdot EHy$). The factor 1.1 has been chosen in such a way that, when both membranes *ComputeX* and *ComputeY* finish their computations, the enzyme *EH* from the membrane *Odometry* have a greater value than the variable *ED* from the same membrane. *EH* enzyme from the skin membrane sums the values of *EHx* from *ComputeX* and *EHy* from *ComputeY*. If both membranes finished their computations, *EH* has a value close to $2.2 \cdot smax$, while *ED* is $2 \cdot smax$. It is important to note that in general membranes *ComputeX* and *ComputeY* do not finish at the same time. When *EH* becomes greater than *ED*, the rule $Pr_{6,1}$ from the membrane *Odometry* is executed, therefore varible *ET* receives a positive value which generates the termination of the program which returns the current position of the robot, stored in variables *x*, *y* and θ.

The two membranes, *ComputeX* and *ComputeY*, are synchronized in the skin membrane, *Odometry*, by the enzymatic mechanism. If only one of the two membranes finished the computation, then $ED = 2 \cdot smax > EH = 1.1 \cdot smax$, so the rule $Pr_{6,1}$ is not active yet. It is activated only when both membranes have finished their computations and $ED = 2 \cdot smax < EH = 2.2 \cdot smax$.

The odometric localization implemented with ENPS is a parallel computation structure. The enzymatic mechanism enhances the computation power of the membrane system and also allows the control of the program flow and synchronization between the parallel computations.

5   Advantages of ENPS Model

Both NPS and ENPS models can be used for modeling autonomous mobile robot behaviors. The numerical nature, the distributed and parallel structure and the computing power, make membrane controllers suitable candidates for robotic control systems.

ENPS controllers have a less complex structure than NPS controllers. Designing a classical NPS controller is difficult and requires a lot of tricky design mechanisms. If the order of the generated values is changed, the system would not work. By using enzyme-like variables, the model of the controller is clearly simplified, easier to implement and more efficient than the one modeled with classical NPS. Enzyme variables control the program flow. Therefore, they can be used for conditional trans-membrane transport, as stop conditions and synchronization mechanism. In ENPS, if the result is generated, the computational process stops due to the stop conditions implemented by the enzymatic mechanism, while in NPS all the membranes have to finish all their computations in a given number of steps. The design of an ENPS controller requires less effort and the performance of the controller is increased by reducing the computational procedure. Enzyme variables can also filter the noise from the sensors. In the ENPS controller for obstacle avoidance [8], the values lower (greater, after rescaling) than a given number are ignored because they are not considered to indicate a real obstacle detection.

The membrane representation is an advantage for both NPS and ENPS because membrane structures are very efficient for designing and modeling robotic behaviors in a parallel and distributed manner. The controllers are designed independent of how the membrane system is distributed and executed in parallel. Only the simulator for ENPS structures is responsible for the parallel and distributed execution of the membranes. The enzymatic mechanism provides stop conditions, rules selection conditions, and synchronization between the computations performed in different membranes.

6   Experiments and Results

A Java simulator, SimP, which computes ENPS and NPS structures has been implemented in order to test the membrane controller models on robots [6]. Other simulators which can be used to execute ENPS models are SNUPS, which is free and has a graphical user interface [2], and the parallelized GPU-based ENPS simulator proposed in [4]. ENPS is an extension of NPS, thus the simulators can execute NPS structures as well. The membrane controllers are stored in xml files which are parsed by SimP simulator. Different behaviors can be stored in xml format which is a uniform representation. XML representation does not depend on the implementation of the simulator or of the robotic system which integrates the controllers. Thus, implementation changes in the membrane simulator or in the robotic system do not interfere with the membranes. If the membrane simulator is optimized, the performance of all membrane controllers increases.

A framework has been developed in order to test the ENPS modules on real and simulated robots. The framework transfers the information received from the sensors (infrared, motors' encoders) to the membrane controllers. The membrane structures are then simulated with SimP and return the output information (motors' speeds, robot position and orientation) to Webots robotics simulator. Using Webots to route the control program on real robots, the behaviors have been tested on real robots as well.

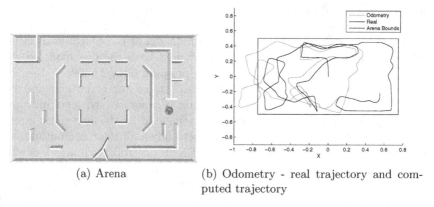

(a) Arena

(b) Odometry - real trajectory and computed trajectory

Fig. 4. Execution time of the controllers' cycle in simulated experiments

The ENPS modules have been successfully tested on both simulated KheperaIII and e-puck robots and on real e-puck robots. Multiple tests have been performed with one robot and more robots in an arena with obstacles. An example of a simulated experiment is shown in figure 4(a). In all experiments, the robots avoided all the obstacles in the arena, updating their position.

In figure 5, the maximum value read by the sensors ($S_{max} = max(S_1, \ldots, S_8)$) and the speeds of the two wheels are displayed for each cycle (a cycle corresponds to one loop of the controller presented in Pseudocode 3). The peaks in the graph of maximum sensors' value (top plot of figure 5) represent detected objects and generate the corresponding spikes in the two graphs of the motors' speeds because the robot has to turn in order to avoid the obstacle. Other interesting features in the maximum sensors' graph are regions of non-zero, near constant values which correspond to the case when the robot passes through a narrow corridor. One such example can be noticed around the 10000 cycle. In this case, the speeds of the motors do not present spikes and are close to the value of the cruise speed which is 200. Therefore, the controller is able to stabilize the trajectory of the robot while passing through tunnels. The mean absolute values of the two speeds are very close to the cruising speed as shown table 1.

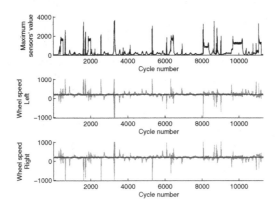

Fig. 5. Maximum sensors' value and motors' speeds in each cycle

Figure 4(b) illustrates the real trajectory of the robot obtained from Webots simulator and the one computed using the odometry module in an experiment done in the arena shown in figure 4(a). In order to prove that the odometry localization module implemented with ENPS returns good estimates of the position, the values computed by the membrane module have been compared to the values returned by the Webots simulator. The positioning error is low at the beginning of the experiment, but increases in time due to numerical and approximation errors and wheel slippage. As shown in figure 4(b), the two trajectories (real and computed) of the robot are similar, but the computed one is shifted to the left. The robot has traveled a total time of 362 seconds and a distance of 8.9849 meters as shown in table 1. The positioning error at the end of the experiment is 0.4943 meters.

The execution time of both modules (avoid and odometry) is very low compared to the cycle time of the robot controller. The cycle time is defined as the duration of a loop (Pseudocode 3) which is about 32 ms in Webots simulator for KheperaIII and e-puck robots and about 180 ms for real e-puck robots. The cycle time in the real experiments is greater because it includes the bluetooth communication. In figure 6 the execution time of both membrane modules for each robot controller cycle is represented. For obstacle avoidance the mean execution time is 0.1884 ms and for the odometric localization, the mean execution time is 0.4719 ms (table 1). The designed membrane controllers can be used with other differential wheeled robots as well, by changing the weights of the sensors for obstacle avoidance and the distance between the wheels for the odometric localization module. The weights used in the avoid procedure depend on the type of sensors and on their placement around the robot. Thus, membrane controllers can be successfully used in autonomous mobile robot applications.

Table 1. Summary of results from experiments with simulated

	Execution time	mean	0.1884 ms
		stddev	0.1441 ms
		max	5.5870 ms
		min	0.1460 ms
Avoid (9 membranes)	S_{max}	mean	316.7189
		stddev	485.7635
		max	3647.0
	Speed left	mean	213.9461
		stddev	144.3022
	Speed right	mean	206.8451
		stddev	144.3022
Odometry (5 membranes)	Execution time	mean	0.4719 ms
		stddev	0.2351 ms
		max	7.7400 ms
		min	0.2830 ms
Total distance traveled			8.9849 m
Total time traveled			362 s
Final positioning error			0.4943 m

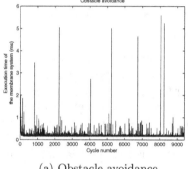

(a) Obstacle avoidance

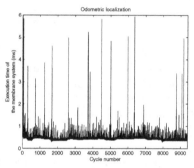

(b) Odometric localization

Fig. 6. Execution time of the controllers' cycle in simulated experiments

7   Conclusions and Future Work

Two ENPS modules for obstacle avoidance and odometric localization have been designed and tested on real and simulated robots. By using a suitable simulator for ENPS, the membrane systems are distributed over computational nodes (like microcontrollers in a robot) in a transparent way. The programmer does not have to worry about communication and synchronization issues or how to distribute the system.

Future work include further testing of the proposed ENPS controllers integrated in the local agent of a cognitive multi-robot architecture. Other robot behaviors will also be modeled with ENPS and tested on autonomous robots. It is a research goal of the authors to develop different components of a robotic system: localization, navigation, planning, etc as membrane systems which will run on a virtual machine distributed over networked microcontrollers. The virtual machine will run the membranes (the code) in a distributed and parallel way, transparently from the user in a similar way as Java and Python. It is a current effort to also develop standard libraries that provide basic and often used functionalities, such as a math library (for sine, cosine, etc.). In this regard, the long term objective is to create a controller programming environment that is completely distributed and parallel.

References

1. Alberts, B., Johnson, A., Lewis, J., Raff, M., Roberts, K., Walter, P.: Molecular Biology of the Cell, 4th edn. Garland Science, NY (2002)
2. Arsene, O., Buiu, C., Popescu, N.: Snups – a simulator for numerical membrane computing. Intern. J. of Innovative Computing, Information and Control 7(6), 3509–3522 (2011)
3. Buiu, C., Vasile, C.I., Arsene, O.: Development of membrane controllers for mobile robots. Information Sciences 187, 33–51 (2012)
4. Garcìa-Quismondo, M., Pèrez-Jimènez, M.J.: Implementing enps by means of gpus for ai applictions. In: Proceedings of Beyond AI: Interdisciplinary Aspects of Artificial Intelligence (BAI 2011), Pilsen, Czech Republic, pp. 27–33 (December 2011)
5. Lodish, H., Berk, A., Kaiser, C.A., Krieger, M., Scott, M.P., Bretscher, A., Ploegh, H., Matsudaira, P.: Molecular Cell Biology (Lodish, Molecular Cell Biology), 6th edn. W.H. Freeman (June 2007)
6. Pavel, A.B.: Membrane controllers for cognitive robots. Master thesis, Department of Automatic Control and Systems Engineering, Politehnica University of Bucharest (February 2011)
7. Pavel, A.B., Arsene, O., Buiu, C.: Enzymatic Numerical P Systems - A New Class of Membrane Computing Systems. In: The IEEE 5th International Conference on Bio-Inspired Computing: Theory and Applications, Liverpool, UK (2010)
8. Pavel, A.B., Buiu, C.: Using enzymatic numerical P systems for modeling mobile robot controllers. Natural Computing (in press), doi: 101007/s11047-011-9286-5
9. Păun, G.: Membrane Computing. An Introduction. Springer, Berlin (2002)
10. Păun, G., Păun, R.: Membrane computing and economics: Numerical p systems. Fundamenta Informaticae 73, 213–227 (2006)
11. Păun, G., Rozenberg, G., Salomaa, A.: The Oxford Handbook of Membrane Computing. Oxford University Press, Inc., New York (2010)
12. Siciliano, B., Khatib, O. (eds.): Springer Handbook of Robotics. Springer (2008)
13. Siegwart, R., Nourbakhsh, I.R.: Introduction to Autonomous Mobile Robots. Bradford Company, Scituate (2004)
14. Vasile, C.I., Pavel, A.B., Dumitrache, I., Păun, G.: On the Power of Enzymatic Numerical P Systems (submitted)

How Can Embodiment Simplify the Problem of View-Based Navigation?

Andrew Philippides[1,*], Bart Baddeley[1], Philip Husbands[1], and Paul Graham[2]

[1] Centre for Computational Neuroscience and Robotics, Department of Informatics,
University of Sussex, Brighton, UK
{andrewop,B.T.Baddeley,philh}@sussex.ac.uk
[2] Centre for Computational Neuroscience and Robotics, School of Life Sciences,
University of Sussex, Brighton, UK
paulgr@sussex.ac.uk

Abstract. This paper is a review of our recent work in which we study insect navigation as a situated and embodied system. This approach has led directly to a novel biomimetic model of route navigation in desert ants. The model is attractive due to the parsimonious algorithm and robust performance. We therefore believe it is an excellent candidate for robotic implementation.

Keywords: Insect navigation, visual homing, route navigation, view-based navigation.

1   Introduction

Many species of ants are expert visual navigators, notably desert ants where the extreme heat means that pheromone trails are not used. Despite their low-resolution vision [1], these ants use information provided by the visual panorama to find their nest or food [2-3] and to guide habitual routes between the two [4-8]. This plainly impressive navigation ability provides proof that robust spatial behaviour can be produced with limited neural resources [9-10]. As such, desert ants have become an important model system for understanding the minimal cognitive requirements for navigation [10]. This is a goal shared by those studying animal cognition using a bottom-up approach to the understanding of natural intelligence [11] as well as biomimetic engineers seeking to emulate the parsimony and elegance of natural solutions [12]. In this spirit, we develop models of desert ant navigation both to gain a fuller understanding of the biological system and also in the expectation that a more detailed understanding will lead to more robust and efficient biomimetic algorithms.

The first generation of biomimetic algorithms for insect-like navigation were inspired by the fact that an insect's use of vision for navigation is often a retinotopic matching process (Ants: [2]; Bees: [13]; Hoverflies: [14]; Waterstriders: [15]; Review: [16]) where remembered views are compared with the currently experienced visual scene in order to set a direction or drive the search for a goal. Insect-inspired robotic models of visual navigation are dominated by snapshot-type models where a

T.J. Prescott et al. (Eds.): Living Machines 2012, LNAI 7375, pp. 216–227, 2012.
© Springer-Verlag Berlin Heidelberg 2012

single view of the world, as memorized from the goal location, is compared to the current view in order to drive a search for the goal ([13]; for review see [17]). Although these models can be shown to work in a variety of environments, there are two limitations that suggest both that they are not satisfactory models of desert ant navigation and that they will not form the basis of efficient long distance artificial navigation systems. (1) Snapshot approaches allow for navigation in the immediate vicinity of the goal. Trying to achieve robust route navigation over longer distances using a chaining technique is a deceptively difficult process [18] and for robust operation requires algorithms to know which snapshot is currently 'active' – essentially some form of place recognition – an additional feature not needed in the original model [19] and not suggested by behaviour [9,10]. (2) Many snapshot type models require snapshots to be aligned to the same orientation before matching. This is behaviourally possible for slow-flying insects but much more complicated for walking ants and insects flying at speeds characteristic of bees' foraging journeys.

Given these problems with the major class of models of visual navigation, we need to look further at the problem to see if there is a different style of model that can be used. A full understanding of any behaviour (and the mechanisms that generated it) requires a full understanding of the natural environments, as perceived through biological sensors, within which animals behave. As well as an understanding of how an animal's embodiment influences its perspective on the world, the possible movements it can make, and the interactions between the two.

In this paper we describe how our considerations of embodiment and situatedness are leading to a fuller understanding of the biology of desert ant navigation, which in turn has led to new biomimetic navigation algorithms. We first consider the information that is available to desert ants navigating in their natural habitats and show that information useful for route guidance could be simply learnt and used. We then use ant behaviour as inspiration for a route following algorithm that captures many properties of ant routes. Finally, by considering a further innate ant behaviour – learning walks – we show that our algorithm can exhibit both place homing and route navigation. We conclude that, in this domain, as in many others, successful bio-inspiration comes when we consider the embodied animal in its natural environment.

2 Understanding the Information in Natural Scenes

In order to gain experimental control of the visual scene experienced by an ant, traditional navigation experiments would use artificial, geometrically simple, landmarks for experiments. Thus, we do not have a general understanding of the mechanisms by which ants use information from natural visual panoramas. An important beginning to answering these questions is to quantify the visual information available to desert ants as they move through the world. Following Zeil and colleagues [20-21] we captured sets of grey-scale images using a panoramic imaging device within the natural environment of navigating desert ants (Fig 1. A,B). By measuring the image difference between a reference image and images from surrounding points, we can build an image difference function (IDF) that shows how images change with distance from a goal view [22]. The image difference between two images X and Y is defined as:

$$IDF(X,Y) = \frac{1}{P}\sqrt{\sum_i \sum_j (X(i,j) - Y(i,j))^2} \qquad (1)$$

where $X(i,j)$ is the pixel in the i'th row and j'th column of image X. Notice that this value is dependent on the alignment of image X and Y. The majority of image- models assume images are aligned to a common frame of reference and we therefore roughly aligned the camera to a reference point (within 5°) before taking the images. Before comparison, images were minimally processed to mitigate the influence of varying light levels or persistent light gradients from sun position, as they may have biased our recorded catchment areas. Firstly, contrast was normalized using histogram equalization ('histeq' function in Matlab) of grey-scale images resulting in integer-valued pixels in the range 0-255. Second the ground is identified as regions with a blue value under a high threshold (set by hand) connected to the bottom of the image, and all 'sky' pixels set to 250.

Fig. 1C shows that image differences increase smoothly with increasing distance from the reference image over a few metres. The presence of a smoothly increasing IDF is significant as it shows that the information needed for view-based homing is available when images are aligned, albeit approximately, to an external reference. For instance, an agent sensitive to the image difference could return home by gradient descent-style algorithms (eg [20]) if equipped with a celestial compass. What's more, these natural scenes contain information for homing over behaviourally significant distances without the additional processing cost of segregating objects from background or matching features between images. We also found that that there is no loss of information (as measured by the size of the catchment area) when the analysis is performed using skylines rather than full scenes. This is important as recent experiments show that the skyline profile is sufficient for homing in ants [23], is easily identifiable using a UV-filter and is potentially less dependent on varying light levels.

The general significance of these results is ensured by the parsimony of the analysis. Evaluating differences between current and reference images using an intentionally simplistic measure (the RMS pixel difference) means that usable information is available without the need for complex visual processing. Notably there is no need to segment views into discrete objects or features. Moreover, the result implies that other models of visual homing, which preserve retinotopic information, will be successful. Inspection of the panoramic scenes along these routes shows why this is. Fig. 1A shows the smooth and gradual change of the scene in an open environment which underpins a smooth gradient in the IDF. Fig. 1A also shows images from a cluttered route and the changing skyline as a function of distance (Fig. 1E). Although the skyline changes rapidly we still see sequences where features in the skyline persist and move slowly within the visual scene. These transiently stable features are enough to underpin gradients in the IDF [22].

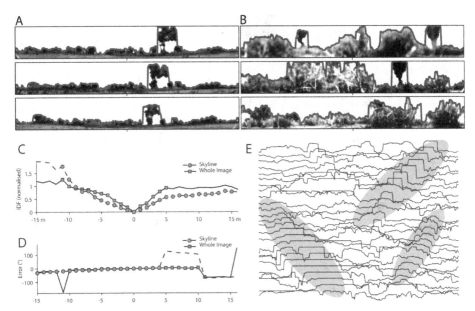

Fig. 1. Panoramic scenes from the perspective of a desert ant. (A) and (B): The panoramic view from an ants'-perspective of three locations (3 m apart) along open and cluttered routes respectively. Images show grey-scale intensities between [0 and 255] and a field of view of 360° by 50° (10° below and 40° above the horizon). Thick line on each image is the skyline defined as the highest 'ground' pixel in each column. (C) Image Difference Function (IDF) for a single reference image at the midpoint of the 30m open route in (A). IDF's are calculated for whole image (solid line, □) and skyline only (dashed line, ○). For comparison, IDFs are normalized to the 80th percentile. Symbols denote catchment area of the reference image, defined as the region within which the IDF increases monotonically to either side of the goal. (D) Orientation errors when all images along the 30m open route in (A) are rotated to find the best match with a reference image at the midpoint. Whole images (□) and skylines (○).Symbols mark points are within the rotational catchment area of the reference image, defined as the continuous set of points around the goal within which absolute orientation error < 45°. (E) 360° wide panoramic skylines extracted from images taken evry metre (top to bottom) along the 25m cluttered route in (A). Although skylines change rapidly, we see sequences where skyline features persist and move slowly within the visual scene (shaded areas). Adapted, with permission, from [22].

3   Using Views as a Visual Compass

Analysis of the catchment area of natural images demonstrates that the information required for visual navigation is present in low resolution panoramic scenes when they are aligned to an external reference. However, it does not reveal what mechanism or model of navigation is used by insects. To get insights into this, we must turn to the behaviour of the intact animal in its natural habitat. One striking observation is of a scanning behaviour in Australian desert ants. The ants can be seen to perform a saccadic scanning where they rotate on the spot, intermittently pausing to view the panorama (Wysrach and Graham, *Pers. Obs.*). Scanning is linked to visual familiarity and ants are seen to scan more in locations where the panorama is unfamiliar.

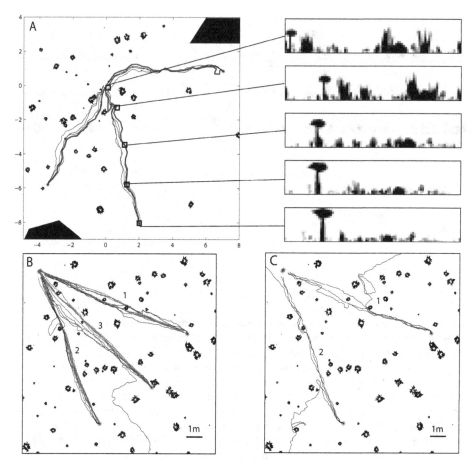

Fig. 2. Route navigation with a holistic memory. The simulated world is viewed from above and is comprised mainly of small tussocks and a few larger more distant objects (trees and bushes). In all panels, red lines are training paths, black lines recapitulations. A: Successful return paths for three different routes. The panels to the right show example views from points (indicated by squares) along the training route. B, C: Learning multiple routes. B) Route recapitulation performance (black lines) for each of three routes (red lines) that are learned with the same network. Testing of each of the routes is performed immediately following training on that route and prior to any subsequent learning. The order in which the routes were learnt is indicated by the numbers next to the training routes. C) Performance on the first two routes following learning of all routes, indicating that the route knowledge gained during the first two phases of learning is retained. Having learnt all 3 routes the network must encode 30m of route information. Adapted, with permission, from [26].

What information could be gained by an ant during a scanning manoeuvre? One intriguing possibility comes from is that the scan could be used to implement a form of *visual compass*. The idea of a visual compass stems from the fact that the alignment of a camera when a reference image was taken can be robustly recovered at nearby locations by comparing the reference image to rotated versions of the current image.

The orientation at which the current image best matches the reference image will be close to the orientation of the reference image [20]. Orientation errors and best matches are calculated by rotating the current image through $360°$ in $1°$ steps and calculating the IDF between goal and rotated current image at each step. The best matching heading is the rotation with the lowest IDF. Analysis of images from the natural habitat of desert ants shows that a single view can be used as an accurate visual compass over a large region (Fig. 1D and [22]). Further, the headings can also be recalled if only skyline profiles are stored. This means that an agent can recover a heading by rotating until it finds the best match between the current view and a snapshot stored when the agent was near the current location and facing in the heading direction, without the need for images to be pre-aligned to an external compass.

The implications for route homing are stark. For an agent with a visual system which is fixed relative to its body axis, the direction of movement is the facing direction. Thus, the correct heading for a portion of a learnt route can be specified by a snapshot stored when the agent was previously moving in the correct direction along the route. A mechanism such as path integration [9-10] which allows traversal of a route in roughly 'correct' directions could scaffold learning of the 'correct' visual information required for a visual compass implemented with a scanning mechanism.

4   Route Navigation with a Holistic Memory

An attractive property of a mechanism that uses stored views to recall an orientation, is that information from comparisons with multiple views can be sensibly polled if they recall similar directions, as they might well be if stored when traversing a route. This is because, unlike the classic snapshot model where views are used to navigate to a discrete position in space, the heading being recovered is not pointing at the waypoint, and thus not necessarily dependent on the agent's position. Thus, views stored at different locations could be simultaneously compared to the current view with heading set by either a voting process or an average, perhaps weighted by similarity, of the outputs across the comparisons. However, the problem of how to select the most appropriate views to define a route remains. This issue is important when chaining together multiple waypoints that define discrete points to be homed to as limiting the number of snapshots and spreading them out reduces the problems of conflicting heading information from different views [18-19].

When using images as visual compasses to recall headings that keep one on the route, the same problem does not arise. Indeed, instead of defining routes in terms of discrete waypoints, all views experienced during training can be used to make a holistic route memory [24-26]. An ant's embodiment means that it is constrained to face in the direction it is moving. Thus, for an ant whose first traversal of a route is governed by path integration (PI) or similar, it is moving, and therefore facing, in the overall route direction most of the time during learning. These familiar viewing directions implicitly define the movement directions required to stay on the route and it is therefore sufficient to learn all views as they are experienced.

Using views as visual compasses means that the problem of navigation is re-framed in terms of a rotational search for the views associated with a route. By visual-ly scanning the environment and moving in the direction that is most similar to the views encountered during learning, an ant should be able to reliably retrace her route. Note that this process associates the current view not with a particular place but in-stead with a particular action, that is, "what should I do?" not "where am I?". In addi-tion, it means that geocentric compass information is not necessary during either learning or recall to learn a holistic route representation.

This leads to the following algorithm (see [26] for details). The agent first traverses the route using a combination of noisy PI (true heading + $N(0,5°)$) and obstacle avoid-ance during which the views used to learn the route are experienced. For the routes shown here, training views were stored after every 4cm 'step' of PI and obstacle avoidance. Subsequently, the agent navigates by visually scanning the world and identifying the direction which is deemed most familiar. Gaussian noise (s.d. 15°) is added to his direction and 10cm step is made, and the scanning routine repeats. To determine view familiarity we train an artificial neural network to perform familiarity discrimination using the training views via the InfoMax algorithm [27].

Fig. 2 shows that by storing the views along a training path and using these to drive a subsequent recapitulation of the route, robust behaviour is achievable. We used our algorithm to learn three routes through an environment containing both small and large objects randomly distributed across the environment. Three subsequent naviga-tion paths were attempted for each route and, despite the noise added to the move-ments during the recapitulation, paths are idiosyncratic though inexact (Fig. 2A). Within a corridor centred on the original route both a good match and a sensible head-ing are recovered that will, in general, drive the simulated ant towards the goal.

The model is also able to learn multiple routes to a single goal. It has been shown that Melophorus bagoti are able to learn and maintain more than one route memory when forced to learn distinct return paths to their nest from a series of different feed-ers [28]. In experiments performed by Sommer et al. [28], seven training runs along a first route were followed by a control run to test whether the ant had learnt the route. This training schedule was repeated for a further two routes that each led back to the same location - the nest. Finally, the ants were tested on the first two routes to see if they had retained the original route memories. Here we attempt to replicate this expe-riment using our route learning algorithm using three 10m routes.

To do this we train a network using the first route. The network is then tested be-fore we continue to train the network using views from the second learning route. The network is then again tested before the final training session using views from the third route, before being tested on all three routes. The performance can be seen in Fig. 2B and C. The network is able to learn and navigate multiple routes without for-getting the earlier ones. It is interesting to note that when the third route is recapitu-lated following learning, the paths tend to get drawn back onto the previously learnt route 2, representing a possible confabulation of these two memories within the net-work. The individual route memories are not held separately and the return paths for route 3 are drawn back to route 2 as, at that point in the world, views from routes 2 and 3 are similar. This property of routes can be seen in the original paper [28].

5   Place Search with a Visual Compass

The visual compass idea is a neat match with route guidance but is not intuitively suited to homing to a specific location, like an inconspicuous nest entrance. However, a recent behavioural observation suggested that a visual compass method could also be used to home to a place. Müller and Wehner observed the behaviour of the ant *Ocymyrmex* which lives in a featureless desert [29]. Its home runs often result in a prolonged search for inconspicuous nest entrances and thus any information provided by local landmarks is readily learnt. Müller and Wehner prompted a bout of learning by introducing a prominent landmark to the nest surroundings of an *Ocymyrmex* nest. Upon noticing the change ants perform a neatly choreographed "learning walk" before departing on their foraging run. Ants loop around the nest entrance and, at a series of points, stop and rotate to accurately face the nest. The brief periods where ants fixate the nest are an ideal opportunity to store snapshots. We hypothesised that a collection of views, acquired on one of these learning walks could be used as visual compasses and collectively guide a search for the nest. A simple simulation using natural images shows that four reference snapshots, stored so that they are centred on the nest, can be used to provide nestward headings from a large area (Fig. 3 and [30]).

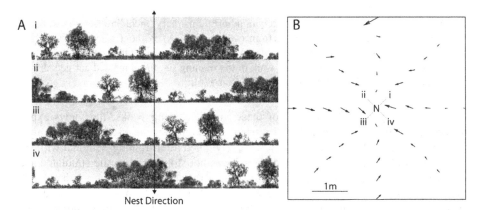

Fig. 3. Visual compasses can be used to navigate to a goal location. We simulate snapshot homing using panoramic images collected from ground level around a desert ant nest. (A) Four images (i-iv) are designated as snapshots, each of which is aligned to face the nest. (B) From the other locations, we derive headings by using each snapshot as a visual compass. To recover a nestwards orientation, we rotate images and find the orientation at which the rotated image best matches a snapshot. This process is performed independently for all four snapshots and the heading is a weighted average (shown in vector plot) showing that an agent navigating with this strategy could return to the nest from any location. Adapted, with permission, from [30].

6   Place Search and Route Navigation with a Holistic Memory

As place homing can be achieved with a visual compass mechanism, the next task is to integrate it with the navigation algorithm described in Section 4. The problem with this algorithm described in Section 4 is that if the simulated ant overshoots the goal it will, in general, carry on heading in the same direction and move further and further away from the goal location (Fig. 4A). This is because there are no training views that point back towards the nest once it has been passed. This problem can be mitigated by including an exploratory learning walk during the training phase [29, 31-32]. These initial paths take the form of a series of loops centred on the nest as can be seen in Fig. 4B which shows the learning walk of a Melophorus bagoti worker taken from a paper by Muser et al. [31]. Essentially, this process means that in the region around the nest there will always be some views which are oriented back towards the nest.

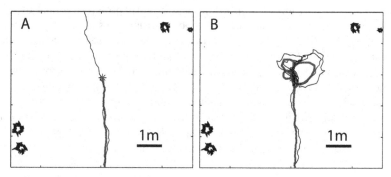

Fig. 4. Route navigation and place search with a holistic memory. Including learning walks in the training route prevents return paths from overshooting. A) Without a learning walk the simulated ant continues in the direction it was heading as it approached the nest. B) By including views experienced during a learning walk in the training set, the simulated ant gets drawn into the nest location. Adapted, with permission, from [26].

To explore the possible effects of these initial learning walks, the views experienced along them were added to the set of inbound views used for route learning [26]. Fig. 4 shows the end section of a route navigated after training with and without a learning walk. In these tests the simulation was not stopped when the simulated ant reached the nest location, analogous to blocking the nest entrance in a behavioural experiment. With the addition of a learning walk (Fig. 4B), as the simulated ant passes the nest, rather than the best match being from the training route and oriented upwards (as in Fig. 4A), the best match comes from the learning walk. The simulated ant is drawn into the loop of the learning walk it first encounters, leading to looped paths (Fig. 4B). Close to the nest, the density of points from the learning walk increases and there are multiple views from nearby locations oriented in a variety of directions. The best match at subsequent points will then likely be from different learning walk loops and so the ant stops following a single loop and enters a search-type path around the nest. Thus, our algorithm demonstrates both route following and place search with a single memory represented by the weights of a neural network.

7   Conclusions and Future Work

In our recent research, we have studied navigation as an embodied situated system which has led to a novel model of route navigation inspired by the behavior of the Australian desert ant. By utilising the interaction of sensori-motor constraints and observed innate behaviours we show that it is possible to produce robust behaviour using a learnt holistic representation of a route. Furthermore, we show that the model captures the known properties of route navigation in desert ants. These include the ability to learn a route after a single training run and the ability to learn multiple idiosyncratic routes to a single goal. Importantly, navigation is independent of odometric or compass information, and the algorithm does not specify when or what to learn, nor separate the routes into sequences of waypoints, so providing proof of concept that route navigation can be achieved without these elements.

The models presented here are inspired by the behaviour and environment of desert ants and so use the appearance of objects against the sky, a salient feature in these environments. However, we believe the model could also be applied in situations where there is overhanging canopy, such as would be experienced by Wood ants, or for robotic navigation either indoors or through urban environments where the skyline is dominated by buildings. In both situations, it is possible that using the whole image as a snapshot, as in Fig. 1, would enable successful navigation. However, in preliminary work in the habitat of wood ants, the pattern of light through the canopy appears to also provide enough visual information for homing. Over what distances this information persists, it's stability to environmental disturbance and what other features exist in wooded, indoor and urban environments is the subject of on-going work. Similarly, we have sampled images from a camera at bee flight-height in open environments and demonstrated that the information for visual homing, or for recovering heading, exists over large distances (~50m). The next stage of this project is to gather data from the paths of radar-tracked bees to couple visual input to behaviour.

A second issue which needs to be addressed for outdoor robotic navigation is how to overcome tilt in the camera. Several solutions are immediately suggested. First, images could be re-aligned by using knowledge of the robot's pose from gyroscopic or inertial sensors. Second, images could be realigned by matching image features. Alternatively, if using skyline height, it might be sufficient to accurately extract the level of the horizon. Once solutions to this issue has been explored, we will be able to thoroughly test a robot navigating in complex outdoor environments to assess the algorithm's performance in the face of environmental noise and in challenging environments. While our algorithm is tolerant to relatively large amounts of simulated sensor and motor noise, the true test will come with a real robot.

A final area of on-going work is the form of visual input and the motor program used to follow and learn routes. We have explored grey-scale, binary and skyline height input representations. Essentially, the richer the visual information, the less chance there is of visual aliasing. However, there is a trade-off with the amount of information that needs to be stored and how easily this information is degraded by noise. Again, this trade-off needs to be explored in the habitat to be navigated. The current algorithm will be troubled if the visual input is identical in two places where headings are in opposite directions. Interestingly, ants learn separate outward and homebound routes and activate one or other route memory dependent on feeding state. Finally, the current algorithm does not tend to draw trajectories back into the route corridor following errors. We are exploring ways that the gradient of familiarity could be used to do this. In addition, we are exploring the use of more sinuous learning paths which capture a wider range of 'correct' headings which might draw the agent back to the route corridor in an analogous way to the way that multiple snapshots can be used to home to a place.

Acknowledgements. Thanks to the anonymous reviewers. AP and BB were funded by the EPSRC. PG was funded by the BBSRC and Leverhulme.

References

1. Schwarz, S., Narendra, A., Zeil, J.: The properties of the visual system in the Australian desert ant Melophorus bagoti. Arthropod Structure & Development 40, 128–134 (2011)
2. Wehner, R., Räber, F.: Visual Spatial memory in desert ants, Cataglyphis bicolor. Experientia. 35, 1569–1571 (1979)
3. Durier, V., Graham, P., Collett, T.S.: Snapshot memories and landmark guidance in wood ants. Current Biology 13, 1614–1618 (2003)
4. Collett, T.S., Dillmann, E., Giger, A., Wehner, R.: Visual Landmarks and Route Following in Desert Ants. Journal of Comparative Physiology a-Sensory Neural and Behavioral Physiology 170, 435–442 (1992)
5. Kohler, M., Wehner, R.: Idiosyncratic route-based memories in desert ants, Melophorus bagoti: How do they interact with path-integration vectors? Neurobiology of Learning and Memory 83, 1–12 (2005)
6. Narendra, A.: Homing strategies of the Australian desert ant Melophorus bagoti - II. Interaction of the path integrator with visual cue information. Journal of Experimental Biology 210, 1804–1812 (2007)
7. Collett, M.: How desert ants use a visual landmark for guidance along a habitual route. Proceedings of the National Academy of Sciences of the United States of America 107, 11638–11643 (2010)
8. Wystrach, A., Beugnon, G., Cheng, K.: Ants might use different view-matching strategies on and off the route. Journal of Experimental Biology 215, 44–55 (2012)
9. Wehner, R.: Desert ant navigation: How miniature brains solve complex tasks. Karl von Frisch lecture. J. Comp. Physiol. A 189, 579–588 (2003)
10. Wehner, R.: The architecture of the desert ant's navigational toolkit (Hymenoptera: Formicidae). Myrmecol. News 12, 85–96 (2009)
11. Shettleworth, S.J.: Cognition, Evolution, and Behavior, 2nd edn. Oxford University Press, New York (2010)
12. Graham, P., Philippides, A.: Insect-Inspired Vision and Visually Guided Behavior. In: Bhushan, B., Winbigler, H.D. (eds.) Encyclopedia of Nanotechnology. Springer (2012)
13. Cartwright, B.A., Collett, T.S.: Landmark Learning in Bees - Experiments and Models. Journal of Comparative Physiology 151, 521–543 (1983)
14. Collett, T.S., Land, M.F.: Visual spatial memory in a hoverfly. J. Comp. Physiol. A 100, 59–84 (1975)
15. Junger, W.: Waterstriders (Gerris-Paludum F) Compensate for Drift with a Discontinuously Working Visual Position Servo. Journal of Comparative Physiology a-Sensory Neural and Behavioral Physiology 169, 633–639 (1991)
16. Collett, T.S., Graham, P., Harris, R.A., Hempel-De-Ibarra, N.: Navigational memories in ants and bees: Memory retrieval when selecting and following routes. Advances in the Study of Behavior 36, 123–172 (2006)

17. Möller, R., Vardy, A.: Local visual homing by matched-filter descent in image distances. Biol. Cybern. 95, 413–430 (2006)
18. Smith, L., Philippides, A., Graham, P., Baddeley, B., Husbands, P.: Linked local navigation for visual route guidance. Adapt. Behav. 15, 257–271 (2007)
19. Smith, L., Philippides, A., Graham, P., Husbands, P.: Linked Local Visual Navigation and Robustness to Motor Noise and Route Displacement. In: Asada, M., Hallam, J.C.T., Meyer, J.-A., Tani, J. (eds.) SAB 2008. LNCS (LNAI), vol. 5040, pp. 179–188. Springer, Heidelberg (2008)
20. Zeil, J., Hofmann, M., Chahl, J.: Catchment areas of panoramic snapshots in outdoor scenes. J. Opt. Soc. Am. A 20, 450–469 (2003)
21. Stürzl, W., Zeil, J.: Depth, contrast and view-based homing in outdoor scenes. Biol. Cybern. 96, 519–531 (2007)
22. Philippides, A., Baddeley, B., Cheng, K., Graham, P.: How might ants use panoramic views for route navigation? J. Exp. Biol. 214, 445–451 (2011)
23. Graham, P., Cheng, K.: Ants use the panoramic skyline as a visual cue during navigation. Curr. Biol. 19, R935–R937 (2009)
24. Baddeley, B., Graham, P., Philippides, A., Husbands, P.: Holistic visual encoding of ant-like routes: Navigation without waypoints. Adaptive Behaviour 19, 3–15 (2011)
25. Baddeley, B., Graham, P., Philippides, A., Husbands, P.: Models of Visually Guided Routes in Ants: Embodiment Simplifies Route Acquisition. In: Jeschke, S., Liu, H., Schilberg, D. (eds.) ICIRA 2011, Part II. LNCS, vol. 7102, pp. 75–84. Springer, Heidelberg (2011)
26. Baddeley, B., Graham, P., Husbands, P., Philippides, A.: A Model of Ant Route Navigation Driven by Scene Familiarity. PLoS Comput. Biol. 8(1), e1002336 (2012)
27. Lulham, A., Bogacz, R., Vogt, S., Brown, M.W.: An infomax algorithm can perform both familiarity discrimination and feature extraction in a single network. Neural Comput. 23, 909–926 (2011)
28. Sommer, S., von Beeren, C., Wehner, R.: Multiroute memories in desert ants. P. Natl. Acad. Sci. USA 105, 317–322 (2008)
29. Müller, M., Wehner, R.: Path integration provides a scaffold for landmark learning in desert ants. Curr. Biol. 20, 1368–1371 (2010)
30. Graham, P., Philippides, A., Baddeley, B.: Animal cognition: Multi-modal interactions in ant learning. Curr. Biol. 20, R639–R640 (2010)
31. Muser, B., Sommer, S., Wolf, H., Wehner, R.: Foraging ecology of the thermophilic Australian desert ant, Melophorus bagoti. Austral. J. Zool. 53, 301–311 (2005)
32. Wehner, R., Meier, C., Zollikofer, C.: The ontogeny of foraging behaviour in desert ants, Cataglyphis bicolor. Ecol. Entomol. 29, 240–250 (2004)

The Dynamical Modeling of Cognitive Robot-Human Centered Interaction

Mikhail I. Rabinovich[1] and Pablo Varona[2]

[1] BioCircuits Institute, University of California San Diego, 9500 Gilman Drive #0328,
La Jolla, CA 92093-0328, USA
mrabinovich@ucsd.edu
[2] Grupo de Neurocomputación Biológica, Dpto. de Ingeniería Informática,
Escuela Politécnica Superior, Universidad Autónoma de Madrid, 28049 Madrid, Spain
pablo.varona@uam.es

Abstract. In this paper we formulate basic principles of cognitive human-robot team dynamics following lessons from experimental neuroscience: 1) the cognitive team dynamics in a changing complex environment is transient and can be considered as a temporal sequence of metastable states; 2) the human mental resources –attention and working memory capacity that are available for the processing of sensory and robot generated information in relation to a specific goal– are finite; 3) the interactive cognitive team activity is robust against noise and at the same time sensitive to information from the environment. We suggest a basic dynamical model that describes the evolution of human cognitive and emotion modes and robot information modes together with the dynamics of mental resources. Using this model we have analyzed the team's dynamical instability, introduced the dynamical description of the information flow capacity, and analyzed the features of the binding dynamics of information flows.

Keywords: sequential transient dynamics, internal representation, human-robot interaction.

1 Introduction

The effectiveness of cognitive activity of human-agent teamwork has been widely investigated during the last decade, both theoretically, experimentally and from the technological point of view (for a review see [1, 2]). Most attention in these works is focused on the coactive design of the participants' interdependence during joint activity [3]. In this article we consider the cognitive robot-human centered interaction from the informational and dynamical points of view. In this perspective, we formulate several fundamental principles and build a basic dynamical model in the form of a set of kinetic differential equations.

Many types of neural activity involved in cognitive processes, i.e., perception, cognition, and emotion, are transient and sequential. When we think about effective interaction between cognitive robots and humans, it makes sense to build models for

T.J. Prescott et al. (Eds.): Living Machines 2012, LNAI 7375, pp. 228–237, 2012.

the internal representation of this interaction based on principles that describe crucial aspects of sequential transient neural dynamics and effective information exchange between members of this team.

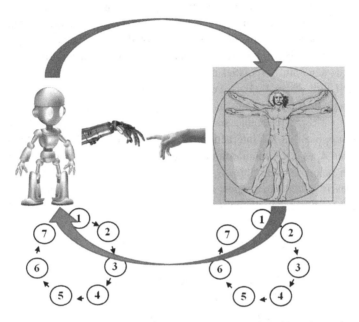

Fig. 1. Schematic illustration of sequential robot-human interaction driven by an internal representation of their joint cognitive dynamics

2   Variables. Common Phase Space. Basic Dynamical Model

We have recently proposed a theoretical formalism to describe brain transient sequential dynamics [4–6]. This formalism is based on two fundamental principles that are also appropriate for the characterization of cognitive activities of human-robot teams. These principles are: 1) finite human mental recourses – attention and working memory, and 2) request for the joint human-robot transient dynamics to be robust against small perturbations but at the same time sensitive to external information signals from the environment and internal signals that inform about the state of the team itself. The corresponding mathematical image representing the self-organized activity of the cognitive team is a stable heteroclinic channel (SHC) [7, 8]. Stable transients are a trajectory that is formed in the vicinity of a sequence of metastable states that are connected by separatrices as we illustrate in Fig. 2. Under proper conditions, all trajectories in the neighborhood of the metastable states that form the chain remain in their vicinity, ensuring robustness and reproducibility over a wide range of control parameters. This vicinity is called Stable Heteroclinic Channel. The robustness, i.e., the structural stability of the transient sequential dynamics implemented by this mathematical object, provides the possibility to characterize phenomena at different description levels, including sensory encoding, working memory and decision making [4]. Following this scheme, our interest in this paper aims to describe the cooperative cognitive robot-human dynamics in terms of a sequence of metastable states in the common phase space. Each metastable state denoted by the index i is represented in physical space by a distributed set of human brain modes and robot artificial brain modes. Let us separate the spatial and temporal variables that characterize the i-th mode as $x_i(t)U_i(k)$, where $U_i(k)$ is the normalized ratio of activity of the k-th member of the i-th mode averaged in time and $x_i(t) \geq 0$ represents the level of the activity of the i-th mode at time t. Metastable states on the x_i-axes, x_i=const$\neq$0, are saddles (see Fig.2). We suppose that the interaction between different modalities – human cognition (variables A_i as x_i), human emotion (variables B_i), robot cognition (variables P_i) and common resources (variables R_i) modulate each other. By keeping just the simplest nonlinearities, we can suggest a phenomenological model of human-robot interaction in the form of kinetic equations (c.f. [5, 9]):

$$\tau_{A_i} \frac{d}{dt} A_i(t) = A_i(t) \cdot F_i(\mathbf{A}, \mathbf{B}, \mathbf{P}, \mathbf{R}, \mathbf{S})$$

$$\tau_{B_i} \frac{d}{dt} B_i(t) = B_i(t) \cdot \Phi_i(\mathbf{A}, \mathbf{B}, \mathbf{P}, \mathbf{R}, \mathbf{S})$$

$$\tau_{P_i} \frac{d}{dt} P_i(t) = P_i(t) \cdot \Psi_i(\mathbf{A}, \mathbf{B}, \mathbf{P}, \mathbf{R}, \mathbf{S}) \tag{1}$$

$$\theta_i \frac{d}{dt} R_i(t) = R_i(t) \cdot Q_i(\mathbf{A}, \mathbf{B}, \mathbf{P}, \mathbf{R}, \mathbf{S})$$

where A_i, B_i, P_i, $R_i \geq 0$ (i = 1,...,N). F_i, Φ_i, Ψ_i and Q_i are functions of A_i, B_i, P_i and R_i, respectively. When initiated properly, this set of equations ensures that all the variables remain non-negative. The vector $\mathbf{S}$ represents the external or/and internal inputs to the system and τ_A, τ_B, τ_P and θ are time constants. In particular cases this model can be written in the form of the generalized Lotka–Volterra model [4]:

$$\tau_i \frac{dx_i^m}{dt} = x_i^m \left[\sigma_i^m - \sum_{j=1}^{N} \rho_{ij}^m x_j^m - \sum_{k=1}^{M} \sum_{j=1}^{N} \xi_{ij}^{mk} x_j^k \right] \tag{2}$$

$$i, j = 1...N \quad , \quad k = 1...M$$

where $x_i^m \geq 0$ represents the instantaneous amplitude of the (i^m)-mode, τ_i is the time constant, $\sigma_i^m \geq 0$ is the growth rate for the mode depending on an external stimulus, $\rho_{ij}^m \geq 0$ and $\xi_{ij}^{mk} \geq 0$ are the interaction strengths between the modes. Here m,k indicate different modalities and i,j indicate different modes within the same modality. The parameters ρ_{ij}^m and ξ_{ij}^{mk} can depend on the stimuli.

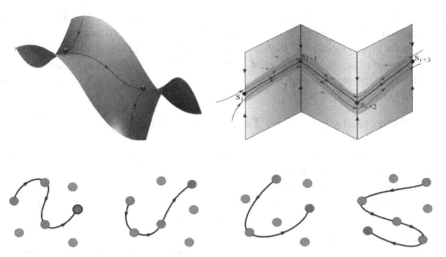

Fig. 2. (Top **panels**) Representation of a simple heteroclinic chain (left) and a robust sequence of metastable states (right). In the phase space of a dynamical model, a temporal winner (metastable state) is represented by a saddle fixed point. Based on this landscape metaphor it is easy to see that two saddles can be connected by an unstable one-dimensional saddle separatrix (see the left panel). This is the simplest heteroclinic sequence. In a many- dimensional phase space (multiple interacting modes) heteroclinic sequences with many connected saddles can exist and form, in a wide area of the control parameter space, a stable heteroclinic channel – a stable heteroclinic flow (see right panel). (**Bottom panels**) Different information inputs (external or internal stimuli) are represented in the phase space by different sequences of global modes activity – different chains of metastable states. The specific topology of the signal-dependent information flow is a key feature that helps to solve the problem of information flow stability against stationary noise.

There is a large area in the control parameter space that corresponds to the existence in the phase space of a heteroclinic channel (see Fig.2). In the general case, the conditions for the stability of the heteroclinic channel – the existence of the SHC can be formulated as:

$$\lambda_1^{(i)} > 0 > \operatorname{Re}\lambda_2^{(i)} \geq \operatorname{Re}\lambda_3^{(i)} \geq \ldots \geq \operatorname{Re}\lambda_d^{(i)} \tag{3}$$

Here the eigenvalues $\lambda_j^{(i)}$ of the saddles are ordered and we introduce the saddle values as: $v_i = -\operatorname{Re}\lambda_2^{(i)}/\lambda_1^{(i)}$.

If

$$\prod_i^g v_i > 1, \tag{4}$$

where g is the number of metastable states along the heteroclinic chain, the compressing is larger than the stretching and the trajectories from the vicinity of a heteroclinic chain cannot escape from the channel, providing its robustness. A rigorous analysis of the structural stability of the heteroclinic channel supports this intuition [7]. The temporal characteristics of transients are related to the exit problem for small random perturbations of dynamical systems with saddle sets. A local stability analysis in the vicinity of a saddle fixed point allows to estimate the characteristic time that the system spends in the vicinity of the saddle as $\tau\,(p) = 1/\lambda_1^{(i)}\ln(1/|\eta|)$, where $\tau\,(p)$ is the mean passage time, $|\eta|$ is the level of noise, and $\lambda_1^{(i)}$ is the maximum eigenvalue corresponding to the unstable separatrices of the saddle.

3   Representation of the Team Information Flow

Sensory information is processed according to certain learning and association rules and then bound into a representation, which is stored, retrieved and matched with new incoming representations. In human-robot interactions that can be considered as active dynamical systems, most of the information perception begins with the emergence of a goal that is implemented by the search for information. The processing channel itself can be a complex, even chaotic, dynamical system [10]. The only input accepted is that which is consistent with the goal and anticipated as a consequence of the searching actions [11, 12]. Active systems demonstrate a lot of interesting information processing phenomena which sometimes appear to be in contradiction with the traditional view or with the intuition.

The classical theory of information transmission, from the perspective of communication between partners of a team, has to be developed in several directions. We remind here just two of them. First, we have to describe the problem when information comes from different sources and then this information is unified [13]. Another general perspective is related to the quantitative description of the information value or its importance: $V = \log(P'/P\,)$, where P is the probability to achieve a goal without the information and P' is the probability to achieve the same goal based on the incoming information [14, 15]. Evidently, V can be negative – when the incoming information is "misinformation" (e.g. working against the team goal). Both directions are very important for the understanding of information processes, especially for the perception of the environment, but these generalizations are still purely algebraic – they do not include the timing or the information dynamics. However, time and dynamics are a key dimension for cognitive team activity. It is important to emphasize that the analysis of the dynamical features of the robot-human team interaction (such as the information flow stability and capacity) can be better done in phase space (along the heteroclinic channel) rather than in physical space (see section 5).

4   Timing of the Cognitive Cycle and Synchronization of Human-Robot Interactions

Be it human, robot, or an autonomous human-robot team, every such agent interacting within a complex dynamical environment must cyclically sample (sense) its environment and act on it, iteratively, in what in [16] is called a cognitive cycle. Cognitive cycles can occur several times a second or be much slower depending on the specific cognitive task. We wish to present here shortly one of the simplest dynamical mechanism of such cycle generation that describes sequential switching in a three cognitive mode competition [9]. In Fig. 3 (upper panel) one can see the architecture of the corresponding mode network and phase portrait of the cyclic dynamics – the heteroclinic cycle. The period of this cycle is controlled by the strength of the competition between modes and the level of noise – exit time (see above).

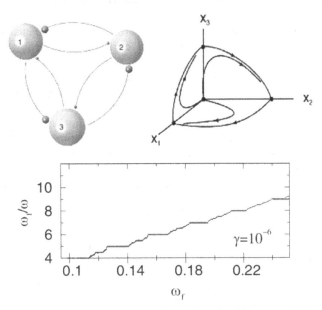

Fig. 3. (Top panels): simplest 3-mode competitive network with sequential dynamics (left) and the mathematical image of the cyclic sequential dynamic – heteroclinic cycle (right). **(Bottom panel):** Heteroclinic synchronization of the three modes rhythmic activity - Devil's staircase in a model (2) with periodic forcing (ω_f is the forcing frequency, ω is the intrinsic frequency and γ is the amplitude of the forcing) – there are zones of synchronization where low frequency harmonics do not depend on the ratio of the intrinsic and external frequency (adapted from [17]).

One of the most important problems for the successive cooperation of a human-robot cognitive team is the temporal coordination or synchronization with the partner's cognitive cycles. To analyze such process one can use model (2) in the case of a three modes interaction with an additional term in the right side of the equation – $\gamma F(\omega_f t)$ – representing a periodic forcing with frequency ω_f and amplitude γ. According to the traditional view of synchronization, a weak periodic input is able to lock a nonlinear oscillator at a frequency close to that of the input (1:1 zone). If the forcing increases, it is possible to achieve synchronization at subharmonic bands also. Using a competitive dynamical system we showed in [17] the inverse phenomenon: with a weak signal the synchronization of ultra-subharmonics is dominant (see Fig.3 bottom panel). In the system's phase space, there exists a heteroclinic contour in the autonomous regime, which is the image of cyclic sequential dynamics. Under the action of a weak periodic forcing, in the vicinity of the contour a stable limit cycle with long period appears. These results in the locking of very low-frequency oscillations with the finite frequency of the forcing. We hypothesize that this phenomenon can be the origin for the synchronization not only of similar time scales but also of slow and fast cognitive cycles of the team partners.

5   Binding and Communication Inside the Team

Suppose we communicate with our artificial partner through tree channels - somatosensory (like in Fig.1), speech stream and eye-tracking (gaze), for example. Such interaction can be described by the generalized Lotka-Volterra model (2). The main results of our previous theoretical and computational analyses [18] can be summarized as: (i) for a wide range of control parameters in the phase space of model (2) there exists an object that we name a multimodality heteroclinic channel –binded heteroclinic channels (see Fig. 4), and the trajectories in the vicinity/inside of this dynamical object represent an integrated (binded) information flow of different modalities; (ii) the time series and spectrum of these multimodality-trajectories demonstrate new features – mutual modulation and regularization of the different modalities and, correspondingly, the appearance of new components in the power spectrum. The properties displayed by the model can be key features for the next step of multimodality information processing of the team. The proposed dynamical mechanism for binding can account for different levels of temporal hierarchy, from milliseconds to minutes.

To quantitatively characterize the effectiveness of the heteroclinic binding, let us introduce a new function called Information Flow Capacity (C_{IF}) as

$$C_{IF}(L) = \sum_{l}^{L} (\Delta C_{IF}(l)) \qquad (5)$$

where

$$\Delta C_{IF}(l) = J_l + \sum_{j=1}^{J_l} \frac{\operatorname{Re} \lambda_j^l}{\left| \lambda_{J_l+1}^l \right|} \qquad (6)$$

Here l is the index of the metastable state (saddle) along a channel, L is the number of saddles that the system passes until time t_L, and the integer J is determined by the following conditions:

$$\sum_{j=1}^{J} \operatorname{Re} \lambda_j > 0, \quad \sum_{j=1}^{J+1} \operatorname{Re} \lambda_j < 0 \qquad (7)$$

In contrast to (3), here we took into account that the dimension of the stable manifold can be larger than 1. For example in Fig. 4, each saddle along the binded heteroclinic channels has two unstable separatrices. That means all J_l=2. Thus, the estimation of the C_{IF} described above tells us that the flow capacity for a binding channel is at least two times larger than the C_{IF} of three independent channels. We can interpret this result in the following way: the information flow capacity characterizes the complexity level of the trajectories within the binded heteroclinic channels. We can hypothesize that such complexity supports for fast and rich information transduction between members of the team.

It would be very interesting to connect the function C_{IF} with Shannon information and the capacity dimension of chaotic sets (see [19]). However there are two main steps that have to be done first in order to build a bridge between the description of transient trajectories in heteroclinic networks and asymptotic dynamics on chaotic attractors: (i) it is necessary to introduce a specific non-invariant measure, and (ii) to consider not continuous flows but maps. We are sure that this is doable in the near future.

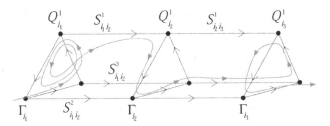

Fig. 4. Illustration of a multimodality heteroclinic sequence and a trajectory corresponding to the binding activity (adapted from [18])

Fortunately, there is no information conservation law in a cognitive system. Because the human-robot team is an active – non-equilibrium system, it can create or generate information as a result of sequential instabilities. The quantity of such information can be characterized by the Kolmogorov–Sinai entropy and the information quality depends on how adequate is for the cognitive goal, i.e., what is the distance between the generated information and the optimal one in some sense. This is a very challenging and provocative problem. In fact, information in cognitive science is determined differently depending on the problem. For example, the quantity of consciousness corresponds to the amount of integrated information generated by the team partners (c.f. [20]). The same approach can be used to understand creativity and imagination, which are the ability to combine together in many ways local instabilities in larger coherent patterns. The results have to be filtered, granting access to the working memory of only the most stable products of imagination and intuition (see also [21]). One of the most interesting examples of relationship between local space–time instability, uncertainty and creativity is our everyday language. Suppose we are going to describe some subject – idea, situation or person. This can be done in many different ways and by using different words and linguistic structures. The chosen way depends on the personality, emotion, memory, etc. (see [21, 22]). These ideas can be compactly integrated in the models discussed above which can in turn be used as internal representations of the sequential cognitive human-robot interaction to drive the robot interactive behavior.

6   Discussion

Bio-inspired sequence generation in robotics if often limited to specific aspects of mechanics, locomotion and associated control paradigms [23–25]. In this paper, we have argued that models proposed to describe neural sequential transient activity can also be used to build internal representations to drive *robot-human cognitive interactions*.

In particular, we have shown that rate equations implementing stable heteroclinic channels can describe the team's information flow, the synchronization and the binding of the activity. As we have discussed before, these models could be used to describe cognitive tasks such as working memory [5] and decision making [7]. The combination of robustness and sensibility to inputs inherent in this type of models makes them good candidates to drive cycles of perception-action. The parameters of the models could be refined during the interaction once the goal and performance measurements are specified for the team activity. It is important to note that the proposed paradigms can be used to describe interactions that range from sequential movements [25] and perceptual tasks involving navigation and obstacle avoidance [26] to decision making and speech interactions.

In this scheme, it is important to remind that the cognitive activity of the intelligent robot – human centered team may be able to create new information that also has to be represented in the model. The environment can initiate and modulate team cognitive activity. After the learning, the human and the robot brains are organized as an interactive functional network reflected in the internal representation. The sequential spatiotemporal activity of the tasks involved for the interaction goal can be described by the corresponding dynamical model of human and robot mode interaction. As we have discussed, the informational flow, in fact, can be described and analyzed as a heteroclinic flow in a *common robot-human cognitive phase space*.

We realize that the problem of integrating an internal representation built with a dynamical model based on the principles that we have discussed to participate in the interaction with a living brain is very challenging. Nevertheless the above discussed models provide essential mechanisms that can be used as control systems to drive the sequential robot-human closed-loop in changing environments. In particular, the proposed models can represent and predict events in complex tasks such as consensus decision making through information flow transient coherence and binding.

Acknowledgements. Mikhail I. Rabinovich acknowledges support from ONR grant N00014-07-1-074. Pablo Varona was supported by MICINN BFU2009-08473 and IPT-2011-0727-020000.

References

1. Goodrich, M.A., Schultz, A.C.: Human-robot interaction: a survey. Found. Trends Hum.-Comput. Interact. 1, 203–275 (2007)
2. Goodrich, M.A., Pendleton, B., Sujit, P.B., Pinto, J.: Toward human interaction with bio-inspired robot teams. In: 2011 IEEE International Conference on Systems, Man, and Cybernetics (SMC), pp. 2859–2864 (2011)
3. Johnson, M., Bradshaw, J.M., Feltovich, P.J., Jonker, C.M., van Riemsdijk, B., Sierhuis, M.: The Fundamental Principle of Coactive Design: Interdependence Must Shape Autonomy. In: De Vos, M., Fornara, N., Pitt, J.V., Vouros, G. (eds.) COIN 2010. LNCS, vol. 6541, pp. 172–191. Springer, Heidelberg (2011)
4. Rabinovich, M.I., Varona, P.: Robust transient dynamics and brain functions. Front Comput. Neurosci. 5, 24 (2011)

5. Rabinovich, M.I., Afraimovich, V.S., Bick, C., Varona, P.: Information flow dynamics in the brain. Physics of Life Reviews 9, 51–73 (2012)

6. Rabinovich, M.I., Afraimovich, V.S., Bick, C., Varona, P.: Instability, semantic dynamics and modeling brain data. Physics of Life Reviews 9, 80–83 (2012)

7. Rabinovich, M.I., Huerta, R., Varona, P., Afraimovich, V.S.: Transient cognitive dynamics, metastability, and decision making. PLoS Comput. Biol. 4, e1000072 (2008)

8. Rabinovich, M., Huerta, R., Laurent, G.: Neuroscience. Transient dynamics for neural processing. Science 321, 48–50 (2008)

9. Rabinovich, M.I., Friston, K., Varona, P. (eds.): Principles of brain dynamics: global state interactions. MIT Press, Cambridge (2012)

10. Pecora, L.M., Carroll, T.L., Johnson, G.A., Mar, D.J., Heagy, J.F.: Fundamentals of synchronization in chaotic systems, concepts, and applications. Chaos 7, 520–543 (1997)

11. Freeman, W.J.: Comparison of Brain Models for Active vs. Passive Perception. Information Sciences 116, 97–107 (1999)

12. Jirsa, V.K., Kelso, J.A.S. (eds.): Coordination Dynamics: Issues and Trends. Springer (2004)

13. Massaro, D.W., Friedman, D.: Models of integration given multiple sources of information. Psychol. Rev. 97, 225–252 (1990)

14. Bongard, M.M.: Pattern Recognition (Original publication: Problema uznavania, Nauka Press, Moscow, 1967), Rochelle Park, N.J (1970)

15. Howard, R.A.: Information Value Theory. IEEE Transactions on Systems, Science and Cybernetics 2, 22–26 (1966)

16. Madl, T., Baars, B.J., Franklin, S.: The timing of the cognitive cycle. PLoS One 6, e14803 (2011)

17. Rabinovich, M.I., Huerta, R., Varona, P.: Heteroclinic synchronization: ultrasubharmonic locking. Phys. Rev. Lett. 96, 141001 (2006)

18. Rabinovich, M.I., Afraimovich, V.S., Varona, P.: Heteroclinic Binding. Dynamical Systems: An International Journal 25, 433–442 (2010)

19. Baker, G.L., Gollub, J.B.: Chaotic Dynamics: An Introduction. Cambridge University Press (1996)

20. Tononi, G.: Consciousness as integrated information: a provisional manifesto. Biol. Bull. 215, 216–242 (2008)

21. Perlovsky, I.L.: Toward physics of the mind: Concepts, emotions, consciousness, and symbols. Physics of Life Reviews 3, 23–55 (2006)

22. Maybin, J., Swann, J.: Everyday Creativity in Language: Textuality, Contextuality, and Critique. Applied Linguistics 28, 497–517 (2007)

23. Ijspeert, A.J.: Central pattern generators for locomotion control in animals and robots: a review. Neural Netw. 21, 642–653 (2008)

24. Liu, C., Chen, Q., Wang, D.: CPG-inspired workspace trajectory generation and adaptive locomotion control for quadruped robots. IEEE Trans. Syst. Man Cybern. B Cybern. 41, 867–880 (2011)

25. Herrero-Carrón, F., Rodríguez, F.B., Varona, P.: Bio-inspired design strategies for central pattern generator control in modular robotics. Bioinspir. Biomim. 6, 16006 (2011)

26. Arena, P., Fortuna, L., Lombardo, D., Patanè, L., Velarde, M.G.: The winnerless competition paradigm in cellular nonlinear networks: Models and applications. Int. J. Circuit Theory Appl. 37, 505–528 (2009)

Internal Drive Regulation of Sensorimotor Reflexes in the Control of a Catering Assistant Autonomous Robot

César Rennó-Costa[1], André Luvizotto[1], Alberto Betella[1],
Martí Sánchez-Fibla[1], and Paul F.M.J. Verschure[1,2]

[1] SPECS, Technology Department, Universitat Pompeu Fabra, Barcelona, Spain
[2] ICREA, Barcelona, Spain
paul.verschure@upf.edu

Abstract. We present an autonomous waiter robot control system based on the reactive layer of the Distributed Adaptive Control (DAC) architecture. The waiterbot has to explore the space where catering is set and invite the guests to serve themselves with chocolate or candies. The control model is taking advantage of DAC's allostatic control system that allows the selection of actions through the modulation of drive states. In the robot´s control system two independent behavioral loops are implemented serving specific goals: a navigation system to explore the space and a gazing behavior that invites human users to serve themselves. By approaching and gazing at a potential consumer the robot performs its serving behavior. The system was tested in a simulated environment and during a public event where it successfully delivered its wares. From the observed interactions the effect of drive based self-regulated action in living machines is discussed.

Keywords: biomimetic robot control, autonomous robot, human-robot interaction, autonomous control.

1   Introduction

Many day-by-day human tasks require the ability to interact with others. That is the case of serving chocolate and candies in a social event. The waiter has to navigate through the hall, approach the guests and invite them to try what is on his plate. An optimal waiter would be able to map the space and program the visit to every guest remembering the time constraints involved. Those are not simple tasks for an autonomous robot. In this paper, we propose a purely reactive controller aimed to allow robots to assist catering services. Although it does not envisage a performance comparable to a human waiter, we suggest that the complexity added to the agent's behavior by the internal regulation of the sensorimotor reflexes can establish a non-verbal communication channel and create a rich and effective serving experience to the user.

The approach taken is to communicate with the guest through gaze (Knapp and Hall 2009) and smooth reduction of interpersonal distance (Inderbitzin et al. 2009; Lawson 2001). This is accomplished by the control of a camera mounted on a pan-tilt unit and the navigation of a mobile robot. The theory we pursue is that complex

T.J. Prescott et al. (Eds.): Living Machines 2012, LNAI 7375, pp. 238–249, 2012.

behavior, such as effective serving of food might emerge from simple and limited constructions without relying on complex processes such as representations, memory or inference skills (Braitenberg 1984). The apparent complexity of behavior over time is a reflection of the complexity of the environment in which the agent finds itself (Simon 1969). By allowing basic behaviors of gaze and motion to be coordinated through the interaction with the environment we aim to achieve emergent behavioral regularities: designing for emergence. Our approach is essentially different from the traditional top-down robot design methodology in which the environment and the possible interactions are parameterized and behavior follows specific and declarative predefined schemes (Pfeifer and Verschure 1994; Pfeifer and Bongard 2006). Designing for emergence might appear more of an art than a science but we base our methods on the fact that the synergy between perception and behavior is mediated by the environment in controllable ways (Verschure et al. 2003). This supports a bottom-up robot design methodology grounded on a more generic and also simple control system. The complexity of the interaction is expected to emerge naturally from the immersion in the environment and from the contact with other agents such as animals (Lund et al. 1997), humans (Eng et al. 2003) or other robots (Asama et al. 1994).

Robots that can establish a closed loop in the sensorimotor interaction with the environment and generate emergent complex behavior have been called *living machines* (Hasslacher and Tilden 1995), which established a significant field inside the autonomous robotics community (Bekey 2005). The objective of the proposed control system is to implement a living machine with behavioral complexity that matches the environmental requirements and human expectations of the catering task, without the need for explicitly and centrally declaring the task in the robot control system.

The main source of behavioral complexity in the control system we propose is the dynamic regulation of the sensorimotor loops through internal drives. Drives are defined as internal states that are defined by the fundamental homeostatic needs of the agent and that adjust the link between perception and action. Drives are set dynamically by the activity of the system and may follow different time scales than the time course of a motor action or the sensorimotor cycle. These two properties together allow a repertoire of multiple responses to the same stimuli with a varying time frame that might not be perceived as a change in perceptual reaction (Simons and Levin 1998).

The implementation of the control system is based on the Distributed Adaptive Control (DAC) architecture (Duff and Verschure 2010; Verschure and Althaus 2003; Verschure et al. 1992; Verschure et al. 2003). DAC is a sensory-action closed loop embodied control system based on the theories of conditioning and grounded on its neurophysiological constraints. Its vertical multi-layered scheme allows the acquisition of internal representations of conditional stimuli with crescent complexity supporting adaptive environmental-mediated behavior (Verschure et al. 2003). The implementation of internal drive in DAC's reactive layer allows the low-level regulation of vital needs in an allostatic control framework (Sanchez-Fibla et al. 2010) and supports goal-oriented behavior by acquiring higher cognitive representations such as rules and plans (Duff et al. 2011).

Although DAC provides the tools for learning and adaptation, we limit the robot´s controller to an implementation of the reactive layer. The reactive layer offers a basic repertoire of actions, sensations, reflexes and drives. Through this basic control system only non-adaptive behavior is generated since no persistent memory of any kind is implemented. The emerging sensorimotor dynamics might ultimately form the basis for the acquisition of internal representations and adaptive behavior (Duff and Verschure 2010), so this structure might work as a base for adaptive behaving system working in the same environmental context.

Another important aspect is that DAC has already been used in the control of living machines that interact with humans. It is the case of the Ada interactive space (Eng et al. 2003). Ada is a room that interacts with the user using light and dynamically composed music (Manzolli and Verschure 2005). A key element for evoking the sense of interaction in humans is the ability to generate complex and unpredictable behavior, which does not look (or sound) like a pure reactive response (Michaud et al. 2005). In this respect, as a benchmark the control system, it was used in the control of an unmanned mobile vehicle in a real catering situation during the Future and Emergent Technologies (FET'11) meeting in Warsaw.

In the following sections we describe the control architecture, the results of computer simulations and provide a brief report on the public demonstration of the system. We further discuss the effect of drive regulation and the limitations imposed by the use of purely reactive systems.

2   The Control Architecture

2.1   The Hardware

The robotic platform used to implement the living machine is a 50 kg unmanned mobile vehicle (124x80x67 cm) with 6 wheels and an articulated body with 3 subsections (Fig. 1). Its sensory system is composed by an array of 16 ultrasound and 16 infrared proximity sensors and a Firewire color camera mounted over a pan-tilt unit in the front body subsection. A notebook (MacBook Pro running custom application under Linux Ubuntu) that is docked in top of the middle body subsection performs computer processing. The load to be distributed is placed in the back body subsection. Robosoft (Bidart, France) developed the platform in cooperation with Universitat Pompeu Fabra (Barcelona, Spain). As an external interface, an iPad with a custom application allowed the visualization of all variables in real time, the override of the navigation loop and emergency locking. The simulation environment used to optimize the parameters of the reactive controller (Sanchez-Fibla et al. 2010) allows a control that is similar to that of the real robot. One of the main features of the simulator is the possibility to customize different virtual environments and realize experiments in computational time rather than real time.

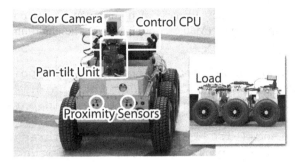

Fig. 1. Robot overview. The color camera is mounted over the pan-tilt unit in the front segment of the robot. The control CPU is placed over the middle segment. The load is placed over the rear segment. The proximity sensors are placed all around the robot, but only the ones in the front and in the back are used for the experiment. Robot produced by Robosoft.

2.2   The Autonomous Control System

The autonomous control system is divided in two main sensorimotor loops: the vision loop and the navigation loop. This is an approximation of the subcortical loops formed by the basal ganglia and brainstem sensorimotor structures comprise sensing, internal motivational states and action (McHaffie, Stanford et al. 2005). While the navigation loop allows the robot to explore the space searching for unattended guests, the vision loop will permit "eye contact" through the gaze system, showing to the user the "intention" of the robot to serve that specific person. Both sensorimotor loops conform to structures with the same components: perception, action, drive and reflexes. Perception and action are the sensor/actuator interfaces to the environment. Drive is the internal state of the agent. Reflexes are the hardwired connections among these components. Each control loop is independent and no information or signals are interchanged. Moreover, they differ with respect to the types of perception, action, drive and the specific organization of the reflexes.

The vision loop receives sensory input from a color camera and acts through the motorized pan-tilt unit. The expected behavior is that the robot tracks faces using the gaze system. Faces are detected by a cascade of boosted classifiers working with Haar-like features and trained with face samples (Lienhart et al. 2003). The output of the visual processing is a salience map that is aligned to the Cartesian representation of the retinotopic input (Fig. 2ab) and can be seen as an approximation of the sensorimotor mappings found in the superior colliculus (Song, Rafal et al. 2010; Gandhi and Katnani 2011). We interpret the active control of the gaze unit in terms of a cerebellum-like saccadic control system acquired through conditioning-like learning mechanisms (Blazquez et al. 2003; Schweighofer et al. 1996; Hofstötter, Mintz et al. 2002). From this salience map it is possible to determine whether there are relevant eye movement targets, i.e. faces, in specific areas of the visual field. An attention mechanism based on competition and predictive anticipation from the recent response history is used to select one single salient face in the visual field. The algorithm consists of searching for a peak in the neighboring area of the last salient point. If no peak is found (highest value in the neighborhood is below the salience threshold, Ω), it searches for the highest peak in the whole visual field. The neighborhood is defined as the predicted area of the highest likeness of finding new salient point given recent history, or an anticipatory gate (Mathews et al. 2009). The algorithm that define the salient points follows:

```
neighbor = compute_neighbor(current_salient_point)
if(max_salience(neighbor)>salience_threshold)
  current_salient_point = max_salience(neighbor)
else
  if(max_salience(visual_field)>salience_threshold)
    current_salient_point = max_salience(visual_field)
  else
    current_salient_point = center(visual_field)
  endif
endif
```

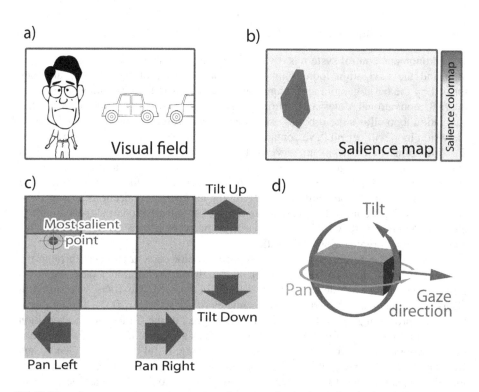

Fig. 2. From face perception to gaze action: the perception/action reflex in the visuomotor loop (a) Illustration of the visual field of the robot and (b) the associated salience map emitted by the face detection and salience map system. (c) Through a competitive process the most salient point in the visual field is selected. If the most salient point is active in a zone associated with a gaze action, a saccade is activated. In this specific case, the action "Pan Left" is triggered, moving the detected face to the center of the visual field. (d) Illustration of the two possible action types ("tilt" and "pan") in relation to the gaze direction.

Reflexes were set connecting the face detection output to the movements of the pan-tilt unit (Fig. 2c and d). The visual field was divided in five horizontal and five vertical zones. To each zone two reflexes are assigned, one horizontal and another vertical. The movement triggered by a certain reflex is capable of centering the camera in the active zones.

In the vision sensorimotor loop, the drive variable is analog to the concept of "curiosity" or "novelty" (Fig 3). Higher levels of curiosity make the gaze system search for new faces, whilst lower levels of curiosity make the gaze system stick to a certain face or remain still. The level of motor action regulates the curiosity level. This is an indirect measure of the variability of the visual input, which allows an environmental mediation of the drive regulation. This is an visual analog of exploration behavior (Sanchez-Fibla, Bernardet et al. 2010).

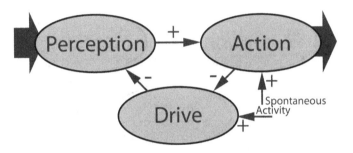

Fig. 3. Visual loop organization. Perception excites action. Action inhibits the drive. The drive inhibits perception. Action and drive have spontaneous activity. Drives regulates the system by allowing action spontaneous activity to take place when few actions are triggered.

The perceptual activity is modulated by the curiosity drive through an inhibitory reflex (Equation 1). The curiosity drive is set accordingly to the number of actions triggered in a specific time window (δ) in relation to an activation threshold (T_{Action}) implemented as a spontaneous activity (Equation 2). Positive curiosity levels, caused by low action activity, lead to the inhibition of perception by a combination of spontaneous activity and an inhibitory reflex from the action set. The action network also has spontaneous activity, so that when all input is extinguished the camera will make some random movement. When perception is active it overrides the spontaneous activity (Equation 3). This intrinsic action by itself constitutes a kind of exploratory searching behavior set when no perception is available to lead the action. Random search is also regulated by the curiosity drive since the intrinsic actions will lower the curiosity level, which will allow perception to rule again. This system allows the gaze control to continuously track for faces and avoids getting stuck in misclassified points. In pseudo code the visuamotor gaze system could be described as:

$$Perception = \begin{cases} Perception & Drive \leq 0 \\ 0 & Drive > 0 \end{cases} \tag{1}$$

$$Drive = T_{Action} - \sum_{t=now}^{now-\delta} Action \tag{2}$$

$$Action = \begin{cases} Perception\ control & Perception > 0 \\ Random\ control & Perception = 0 \end{cases} \tag{3}$$

The navigation loop follows a different architecture when compared to the visual loop (Fig. 4). It receives sensory input from an array of proximity sensors positioned around the robot body and acts through motor commands that spin the wheels. The expected behavior is that the robot runs around the location, avoiding collisions of any kind. Differently to the vision loop, in the navigation loop the reflex from the perception to the action in inhibitory. Whenever an obstacle is perceived the motor control is inhibited. Motor action has spontaneous activity, which when is not inhibited causes the robot to follow a random direction (Equation 4). Direction is changed randomly and periodically with a time constant (σ) defined as 10 seconds.

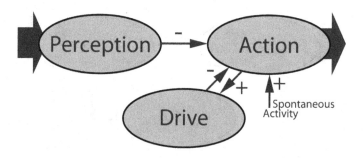

Fig. 4. Navigation loop organization. Action has spontaneous activity. Action can be inhibited by the perception and by the drive. The drive is activated by the action.

The drive in the navigation task works as a negative feedback. It is set in a way that the robot periodically stops at a certain position so that the guests have time to collect the load. The combination between the time constant (δ) and the drive activation threshold (T_{Action}) sets the time the robot takes in the stop phase and the time it stays in the moving phase (Equation 5).

$$Action = \begin{cases} 0 & Perception\ or\ Drive > 0 \\ Random\ control & otherwise \end{cases} \tag{4}$$

$$Drive = \sum_{t=now}^{now-\delta} Action - T_{Action} \tag{5}$$

3   Results

A full demonstration of the living machine was performed during the Future and Emergent Technologies (FET'11) at the Polytechnic University of Warsaw in November of 2011 (Fig. 5). The robot was used to distribute chocolates to the attendees of the event during the coffee breaks. The venue consisted of a large hall of approximately 4000 square meter filled with tables and other sorts of obstacles.

Fig. 5. Demonstration venue (right) and the robot (top-left), chocolates and candies are placed in a plate located in the back part of the robot body

Before the demonstration, in order to profile the behavioral capability of the navigation loop and validate the parameters we used the control mechanism to steer a simulated robot in a virtual environment. For this experiment we used a drive time constant of 20 seconds and a T_{Action} of on average 2.5 kilometers/hour or 5 degrees per second. The aim of the simulations was to verify the ability of the robot to cover the whole catering space and quantify the quality of its delivery service. We run five trials of three hours with different areas from 1600 m^2 to 7000 m^2 (Fig. 6). A specific area was considered "visited" when it was closer than 3.75 meters from a position where the robot was stationary (drive above zero). The control paradigm was able to deliver to more than 90% of the locations in all the arena sizes. Moreover, after 1 hour it was able to visit more than 60% of the arena with an area similar to the experimental site.

The gaze system was also verified in lab conditions. T_{Action} was set to .1 movements/second in average and the time constant to 10 seconds. This would imply that if no movement was done in 10 seconds, the drive would be active and gaze would follow a random direction. With controlled conditions - single face with white background – the vision loop was able to fix the gaze correctly in all the trials (n=10) and with the subject stationary it took an average time of 10.5 seconds to gaze in another direction. With the subject moving slowly (< 1 steps per second or 0.76 m/sec), the gaze system never lost track of the subject (n=10). In the case in which the subject was moving fast (> 1 steps per second), the gaze system always lost track of the user when the movement followed continuously in the same direction for more than 2 seconds (n=10). Movements confined to the view range were successfully tracked by the vision loop. Thus, the system is highly reliable in localizing stationary or slowly moving human faces with a speed of up to 3 km/h.

In the real demonstration, the robot was successful in the task of distributing chocolates. In total, 213 pieces were distributed in a time course of 10 hours or on the average about 1 piece every 3 minutes. The gaze system worked for the whole time course of the demonstration and was effective in calling attention and in creating a connection with the guests (Fig. 7).

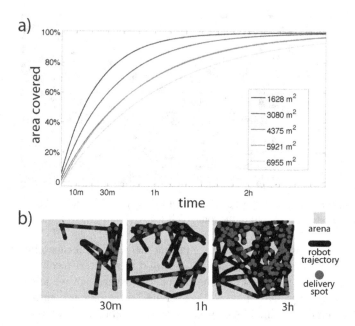

Fig. 6. Simulated navigation data. (a) Time evolution (exponential fit) for the percentage of the area covered by delivery stops for five different arena sizes. (b) Sample of robot trajectories and delivery spot spatial distributions at different time windows for the 5921 m^2 square arena.

The robot was able to cover autonomously a large part of the hall´s surface, confirming the simulation results. The variability in the movements caused by the drive regulation showed to be effective since most of the guest would only approach the robot when it stopped or was moving slowly.

We could observe situations in which it got stuck (N=X in 10 hours). This would happen mostly in corners and in places with high desk and tables density. In these cases the navigation system was overridden by a remote control. Moreover, due to the different kind of materials used in the room, the proximity sensor could not detect a few obstacles causing the robot to crash. In moments with high guest density the navigation system was overridden for security reasons.

Although both control loops were set independently and could not exchange any kind of messages through the computer, the influence of navigation loop over the gaze loop is striking. When the robot was in movement, gaze would keep directed to the same face for a longer time if compared to when the robot is still. The reason is that the movement of the robot forced the gaze system to activate the saccade movements more frequently, extending the time of drive integration. With the parameters used in the demonstration, the robot would keep the camera focused for about 10 seconds if the target face remained still. With the robot in movement (> 0.5 meters per second) the robot would only change the target face in case it lost the target face for a period greater than 1 second (caused mainly by visual obstruction).

Fig. 7. Demonstration of the gazing behavior by the visuomotor loop in a sequence with a moving subject

4   Discussion

We addressed the question whether designing a controller that achieves its coherence through dynamic interaction with the environment can render robust and effective behavior. In this example of designing for emergence we targeted a delivery service task where a mobile robot had to deliver chocolate to a naïve audience in a public event. The public demonstration of the presented control system is an example of how enhanced drive based reactive control can lead to emergent behavioral skills sufficient for permitting effective human-robots interaction. More specifically to the waiter task, it generates gaze and interpersonal distance regulation behaviors. The approach used is grounded on the regulation of reactive sensorimotor loops through internal drives that adds complexity to the performed activity and modulates the reactive response on an uncorrelated time-scale. Most importantly, the robot accomplishes its mission without the need of any declarative representation of the task or the other agents involved.

Nevertheless, the demonstration showed that the system is sufficient but not optimal to reproduce a waiter performance. This is somehow expected since the system can be enhanced in many ways. But it is important to highlight the sufficiency of the reactive control since it gives a safe behaving procedure for any system eventually built on top of it. This form of environmentally mediated allostatic control can be seen as an hypothesis on how biological systems ultimately support survival in potentially harmful situations or to satisfy a range of needs (Sanchez-Fibla et al. 2010). Moreover, it would be interesting to observe how multiple robots would interact in the performance of the task. Since the covering of space tended to be confined to a certain region for a limited time, a simple communication channel in which robots avoid other machines might help to establish zones of action, reducing the active area of each robot and diminishing the time needed to cover the space through emergent collaboration.

Regarding the extension of the control system, it is possible to keep the reactive idea by establishing direct communication between the two control loops. One example would be to use the face detection component to modulate the drive of the navigation, enhancing the sensation of approaching a guest. However, the clearest possibility is the addition of more cognitive skills such as memory and planning. The reactive system can support the construction of highly structured representations when considering the other two layers of the DAC architecture (Duff et al. 2010; Duff and Verschure 2010). The same platform has been controlled by a full DAC architecture in non-interactive tasks in which we included features such as mapping, sequence learning, object recognition and spatial memory (Rennó-Costa et al. 2011).

More specifically to the control loops, the visual-motor loop can be enhanced by face recognition itself (Luvizotto et al. 2010). This would allow the robot to focus on guests who have not yet been served. Regarding the navigation loop, the use of a spatial memory system would allow a homogeneous covering of the space since it would be possible to remember where it was before and avoid recently visited locations. Another possibility is to integrate both loops in the mapping of the space (Verschure et al. 2006).

Acknowledgements. This work was supported by the projects: GOAL-LEADERS (FP7-ICT-97732), EFAA (FP7-ICT-270490), eSMCs (FP7-ICT-270212) and CEEDS (FP7-ICT-258749).

References

1. Asama, H., Fukuda, T., Arai, T.: Distributed Autonomous Robotic Systems. Springer (1994)
2. Bekey, G.A.: Autonomous Robots: From Biological Inspiration to Implementation and Control. MIT Press (2005)
3. Blazquez, P.M., Hirata, Y., Heiney, S.A., et al.: Cerebellar Signatures of Vestibulo-Ocular Reflex Motor Learning. J. Neurosci. 23, 9742–9751 (2003)
4. Braitenberg, V.: Vehicles: Experiments in synthetic psychology. MIT Press (1984)
5. Duff, A., Rennó-Costa, C., Marcos, E., et al.: From Motor Learning to Interaction Learning in Robots. From Motor Learning to Interaction Learning in Robots 264, 15–41 (2010)
6. Duff, A., Sanchez Fibla, M., Verschure, P.F.M.J.: A biologically based model for the integration of sensory-motor contingencies in rules and plans: A prefrontal cortex based extension of the DAC architecture. Brain Research Bulletin 85, 289–304 (2011)
7. Duff, A., Verschure, P.F.M.J.: Unifying perceptual and behavioral learning with a correlative subspace learning rule. Neurocomputing 73, 1818–1830 (2010)
8. Eng, K., Babler, A., et al.: Ada - intelligent space: an artificial creature for the SwissExpo.02. In: IEEE International Conference on Robotics and Automation ICRA, pp. 4154–4159 (2003)
9. Hasslacher, B., Tilden, M.W.: Living machines. Robotics and Autonomous Systems 15, 143–169 (1995)
10. Inderbitzin, M., Wierenga, S., et al.: Social cooperation and competition in the mixed reality space eXperience Induction Machine XIM. Virtual Reality 13, 153–158 (2009)
11. Knapp, M.L., Hall, J.A.: Nonverbal Communication in Human Interaction. Cengage Learning (2009)
12. Lawson, B.: The Language of Space. Routledge Chapman & Hall (2001)
13. Lienhart, R., Kuranov, E., Pisarevsky, V.: Empirical Analysis of Detection Cascades of Boosted Classifiers for Rapid Object Detection. In: DAGM 25th Pattern Recognition Symposium, pp. 297–304 (2003)

14. Lund, H., Gerstner, W., Germond, A., et al.: Something. In: Gerstner, W., Hasler, M., Germond, A., Nicoud, J.-D. (eds.) ICANN 1997. LNCS, vol. 1327, pp. 745–750. Springer, Heidelberg (1997)
15. Luvizotto, A., Rennó-Costa, C., Pattacini, U., Verschure, P.F.M.J.: The Encoding of Complex Visual Stimuli by a Canonical Model of the Primary Visual Cortex: Temporal Population Code for Face Recognition on the iCub Robot. In: Proceedings of the 2011 IEEE International Conference on Robotics and Biomimetics, pp. 313–318 (2010)
16. Manzolli, J., Verschure, P.F.M.J.: Roboser: A Real-World Composition System. Computer Music Journal 29, 55–74 (2005)
17. Mathews, Z., Lechon, M., Calvo, J.M.B., et al.: Insect-Like mapless navigation based on head direction cells and contextual learning using chemo-visual sensors. In: IEEE/RSJ International Conference on Intelligent Robots and Systems, pp. 2243–2250. IEEE (2009)
18. Michaud, F., Brosseau, Y., Cote, C., et al.: Modularity and integration in the design of a socially interactive robot. In: IEEE International Workshop on Robot and Human Interactive Communication, pp. 172–177. IEEE (2005)
19. Rennó-Costa, C., Luvizotto, A.L., Marcos, E., et al.: Integrating Neuroscience-based Models Towards an Autonomous Biomimetic Synthetic Forager. In: Proceedings of the 2011 IEEE International Conference on Robotics and Biomimetics, pp. 210–215 (2011)
20. Sanchez-Fibla, M., Bernardet, U., et al.: Allostatic Control for Robot Behavior Regulation: a Comparative Rodent-Robot Study. Advances in Complex Systems 13, 377 (2010)
21. Schweighofer, N., Arbib, M.A., Dominey, P.F.: A model of the cerebellum in adaptive control of saccadic gain. Biological Cybernetics 75, 29–36 (1996)
22. Simon, H.A.: The architecture of complexity. The Sciences of the Artificial, pp. 192–229. MIT Press, Cambridge (1969)
23. Simons, D.J., Levin, D.T.: Failure to detect changes to people during a real-world interaction. Psychonomic Bulletin & Review 5, 644–649 (1998)
24. Verschure, P.F.M.J., Althaus, P.: A real-world rational agent: unifying old and new AI. Cognitive Science 27, 30 (2003)
25. Verschure, P.F.M.J., Kröse, B., Pfeifer, R.: Distributed Adaptive Control: The self-organization of structured behavior. Robotics and Autonomous Systems 9, 181–196 (1992)
26. Verschure, P.F.M.J., Voegtlin, T., Douglas, R.J.: Environmentally mediated synergy between perception and behaviour in mobile robots. Nature 425, 620–624 (2003)
27. Verschure, P.F.M.J., Wyss, R., König, P.: A Model of the Ventral Visual System Based on Temporal Stability and Local Memory. PLOS Biology 4, 1–8 (2006)
28. Gandhi, N.J., Katnani, H.A.: Motor functions of the superior colliculus. Annual Review of Neuroscience 34, 205–231 (2011)
29. Hofstötter, C., Mintz, M., et al.: The cerebellum in action: a simulation and robotics study. European Journal of Neuroscience 16, 1361–1376 (2002)
30. McHaffie, J.G., Stanford, T.R., et al.: Subcortical loops through the basal ganglia. Trends in Neurosciences 28(8), 401–407 (2005)
31. Pfeifer, R., Bongard, J.C.: How the Body Shapes the Way We Think: A New View of Intelligence. MIT Press, Cambridge (2006)
32. Pfeifer, R., Verschure, P.F.M.J.: The challenge of autonomous agents: Pitfalls and how to avoid them. In: Steels, L., Brooks, R. (eds.) Autonomous Agents, pp. 237–263. Springer (1994)
33. Sanchez-Fibla, M., Bernardet, U., et al.: Allostatic control for robot behavior regulation: a comparative rodent-robot study. In: Advances In Complex Systems (2010)
34. Song, J.H., Rafal, R., et al.: Neural substrates of target selection for reaching movements in superior colliculus. Journal of Vision 10(7), 1082–1082 (2010)

Incremental Learning in a 14 DOF Simulated iCub Robot: Modeling Infant Reach/Grasp Development

Piero Savastano and Stefano Nolfi

ISTC-CNR
Via San Martino della Battaglia 44, 00185 Rome, Italy
{piero.savastano,stefano.nolfi}@istc.cnr.it

Abstract. We present a neurorobotic model that develops reaching and grasping skills analogous to those displayed by infants during their early developmental stages. The learning process is realized in an incremental manner, taking into account the reflex behaviors initially possessed by infants and the neurophysiological and cognitive maturations occurring during the relevant developmental period. The behavioral skills acquired by the robots closely match those displayed by children. Moreover, the comparison of the results obtained in a control non-incremental experiment demonstrates how the limitations characterizing the initial developmental phase channel the learning process toward better solutions.

1 Introduction

The control of arm and hand movements in humans represents a fascinating research area for researchers in psychology, neurosciences and robotics. The development of reaching and grasping behaviors in humans, in particular, constitutes one of the most deeply studied area of motor control. Despite of that, the attempt to replicate the development of reaching and grasping skills comparable to those acquired by humans still represents a challenging objective and an open issue [24]. We present a neurorobotic model that develops reaching and grasping skills analogous to those displayed by infants during their early developmental stages. The model is designed by using a holistic approach that aims to identify and model all the key characteristics of the natural phenomena, while abstracting and simplifying all the aspects that do not play a key causal role. In the context of infant reaching and grasping development, we hypothesize that the following aspects and modeling choices constitute essential prerequisites:

Embodiment. With the term embodiment we refer to the fact that the morphological and sensory-motor characteristics of the agent play an essential role in adaptive behavior [21]. For this reason we carry out our experiments by using a humanoid robot (the iCub) that matches to a good extent the characteristics of human infants in term of morphology, kinematic structure, and DoFs. Moreover, we design the sensory-motor system of the robot by taking into account the empirical evidences about infants' development.

T.J. Prescott et al. (Eds.): Living Machines 2012, LNAI 7375, pp. 250–261, 2012.

Situatedness. Behavior is not only the result of the agent's characteristics but also of the agent/ environmental interactions. This aspect is accounted for in our experiments by simulating the characteristics of the physical environment and of the robot/ environmental interaction in detail, and by using a learning process and a control architecture that allow the robot to exploit sensory-motor coordination and more generally properties emerging from the agent/ environmental interaction. Moreover, we replicate as much as possible the characteristics of the experimental settings in which the behavior of infants was studied [27,33], see Fig. 1. This allows us to generate data more easily comparable with experimental data, and to produce testable predications for infant motor learning.

Fig. 1. The simulated setting (Left) is derived from experiments on real children (Center and Right, adapted respectively from references [33] and [27])

Nervous System and Learning Process. Here we refer to the formalism used to specify the agent's nervous system (or robot's controller) and its plasticity. In the context of infant reach/grasp development modeling addressed in this paper, we implement the robot's controller with an artificial neural network and the learning process through a simple trial and error learning algorithm that is driven by the observed consequences of the robot's action (visual and tactile feedback). The neuromimetic controller is not intended to reproduce the detailed characteristics of the infants nervous system (at the level of the single neurons or at the level of the nervous system architecture), but to capture its essential features. The neural network formalism encodes and processes quantitative information, operates over time, displays generalization properties, and is a suitable and biologically plausible media for the learning process. A learning algorithm operating on the basis of distal somatosensory feedback complies with empirical evidences suggesting that young infants overcome problems associated with reaching and grasping by a self-learning trial and error process [28]. This form of learning allows the exploitation of sensory-motor coordination and overcomes the initial limited visual capabilities of young infants. Those perceptual limitations may prevent alternative form of learning, like imitation learning [17].

Incrementality. The fourth and last key aspect is constituted by the incremental nature of the developmental process. Action development in newborn infants does not start from scratch, as it is strongly influenced by pre-existing behavioral skills and by concurrent maturational and developmental processes. To take this aspects into account we provide the robot, before learning takes place, with few simple reflexes homologous to some of the reflexes initially possessed by infants. Moreover, we model the developmental process in a series of cumulative phases subjected to physiological modifications originating from tissues maturation [31] and cognitive modifications (e.g. increased ability to process visual information [2]).

In the next Section we describe the robotic model and the experimental scenario in detail. In Section 3 we present the results. Finally, in Section 4 we discuss the implication of the results and plans for the future.

2   Robotic Model and Experimental Scenario

A simulated iCub robot [23] is trained for the ability to reach and grasp a colored ball located in its peripersonal space. The experimental scenario in which we train the robot is derived from the experiments carried on with children of about 4 months of age by Spencer and Thelen [27] and von Hofsten [33] (see Fig. 1). The robot is suspended vertically over a stick attached to the pelvis. In each trial the ball is placed in a randomly selected point located within one of the 9 sectors of the spherical surface centered on the iCub neck (Fig. 2). The ball is attached to a pendulum. The robot is provided with a neural controller that is trained through a simple incremental trial and error process (Par. 2.4).

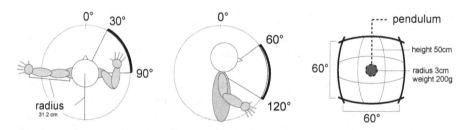

Fig. 2. The thick line in the three pictures shows the portion of the spherical surface in which the ball can be placed. To ensure a good distribution over space of the target objects, the surface is virtually divided into 9 sectors.

2.1   The Robot

The iCub is a humanoid robot developed at IIT as part of the EU project RobotCub [23,30]. It has 53 motors that move the head, arms and hands, waist, and legs. From the sensory point of view, the iCub is equipped with digital cameras, gyroscopes and accelerometers, microphones, force/torque sensors, tactile sensors. In the experiment reported in this paper, the sensors and actuators located on the left arm and on the legs have not been used. The experiments have been carried out by using the simulator developed at our lab by Gianluca Massera and collaborators (freely available from *http://laral.istc.cnr.it/laral++/farsa*).

The simulator reproduces as accurately as possible the physics and the dynamics of the robot and robot/environment interaction, and is based on the Newton Game Dynamics open-source physics engine (*http://newtondynamics.com*).

2.2   The Robot's Neural Controller and Sensory-Motor System

The robot's neural controller is constituted by a recurrent neural network that receives proprioceptive input from the right arm, torso, and head, exteroceptive input from the camera and the tactile sensors located on the right hand, and controls the motors of the torso, head, and of the right arm/hand (Figure 3). As can be seen from the figure, the sensory layer is connected to the motor layer either directly, to take into account the fact that the initial pre-reaching behavior observed in children is highly reflexive and oriented to sensory-motor exploration [1,20], or through 8 internal neurons to allow the robot to develop more elaborated and effective motor strategies.

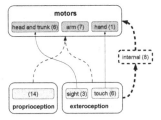

Fig. 3. The architecture of the robot's neural controller. Numbers between parenthesis represent the number of neurons, arrows indicate connections. Full arrows indicate hand-designed connection weights used to implement motor reflexes. Dashed thin and thick arrows indicate connection subjected to plasticity during the first and the second training phases, respectively. Internal neurons are added in the second phase. Notice that dashed arrows pointing to the motor layer indicate connections toward all motor neurons.

Internal and motor neurons consist of integrator units (i.e. neurons whose current state also depends on their previous state) that are updated as follows:

$$x_i^{(t)} = \tau_i x_i^{(t-1)} + (1 - \tau_i)s_i^{(t)}$$

Where $x_i^{(t)}$ is the state of the i-th neuron at timestep t and $0 \leq \tau_i \leq 1$ is a time constant associated to each neuron [16,18]. $s_i^{(t)}$ is computed as:

$$s_i^{(t)} = \sigma \left(\sum_{j}^{n} (w_{ij}x_j^{(t)}) - \theta_i \right)$$

Where w_{ij} is the connection weight between the j-th and the i-th neuron, θ_i is the neuron threshold and $\sigma(z)$ is the sigmoidal function $= 1/(1 + e^{-z})$.

The state of the sensors, the network, and the motors is updated every timestep (0.1 seconds). The motor neurons set the desired angular position (scaled within the robot's joint limits) of 14 actuators controlling the following DoFs: head (3), torso (3), right arm (7), right hand (1). Each motor neuron controls a DoF of the robot with the exception of the hand, in which a single motor neuron controls the extension/flexion of all the fingers. The proprioceptors encode the current angular position of the corresponding joints (the average extension/flexion of the fingers' joints, in the case of the hand), scaled from -1 to 1.

A set of 6 tactile neurons binarily encode (-1 or 1) whether the corresponding touch sensor located in the right hand palm and fingertips (Fig. 4, Left) detects an obstacle or not. The 3 sight sensors encode pre-elaborated information extracted from the cameras through a simple color blob identification software routine. These sensors thus provide only a limited visual analysis of the object, its position (or approximate position, see below) and a crude assessment of grasp affordance. More precisely the first two encode the relative position of the color blob corresponding to the ball in the robot's visual field (Eq. 1 and 2) and the third encodes the estimated ball distance up to 50cm (Eq. 3).

$$x_{\text{sight1}} = \text{sgn}(c_x) \cdot |c_x|^a \tag{1}$$

$$x_{\text{sight2}} = \text{sgn}(c_y) \cdot |c_y|^a \tag{2}$$

$$x_{\text{sight3}} = \begin{cases} 1 - 2l, & \text{if } l < 0.5 \\ 0, & \text{otherwise.} \end{cases} \tag{3}$$

Here c_x and c_y represent the coordinates of the detected color blob in the camera image and $\text{sgn}(x)$ is the sign of x. In accordance with experimental findings on sight development [7,8], we vary the visual acuity/peripherality of the robot during the first and second training phases by setting the value of a to 3 and to 1, respectively (see Fig. 4, Right). The third sensor encodes l, the eye-object distance.

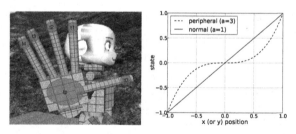

Fig. 4. Left: Location of the touch sensors in the robot's hand. Right: Dashed and filled lines indicate the state assumed by sight sensors for different positions of the colored blob in the camera image, for low and high acuity vision respectively.

2.3 Learning Process

In accordance with empirical evidences indicating that early reaching and grasping skills in infants are acquired through self-learning mechanisms rather than by imitation [17], the robot's training is realized through a form of trial and error learning during which the robot is rewarded for sensorial exploration and multimodal perception (seeing and touching [22]). More specifically, we evaluate the performance level of the robot at each time step by taking the smaller score between the perceptual modalities:

$$p_{\text{multimodal}} = \min\left(p_{\text{sight}}, p_{\text{touch}}\right)$$

The value is averaged over 18 trials each lasting 20 seconds. p_{sight} measures the distance between the barycenter of the object and the center of the robot's visual field, p_{touch} measures the number of inner hand/fingers segments in contact with the object. Both factors are scaled between 0 and 1.

The agents are trained through a trial and error process in which the free parameters are varied randomly and variations are retained or discarded depending on whether they lead to maximization of $p_{\text{multimodal}}$ at the end of the 18 trials. This is realized by using an evolutionary method [15]. The initial population consists of 20 randomly generated genotypes encoding the connection weights, the biases, and the time constants of 20 corresponding neural controllers (each parameter is encoded by eight bits and mutated with probability 0.02). The training process intends to represent ontogenetic learning. The reason behind the choice of this algorithm is that is one of the simpler yet effective ways to train an embodied neural network through a trial and error process based on a distal reward [25].

2.4 Incremental Training

The robot is subjected to an incremental training process organized into the following three phases, inpired to those used to describe the development of reaching/grasping in infants [10]:

1. The pre-reaching phase, that in infants extends from birth to approximately 4 months of age, is characterized by the presence of simple head orientation [26] and grasping reflex behaviors [9], by a low involvement of cortical areas [13], and by a low visual acuity [7,8]. From the behavioral point of view this phase is characterized by a primitive orientation behavior of the arm [32], by the freezing of certain DoFs (i.e. by the reduced use of the distal DoFs [3]), and by the emergence of a form of motor babbling (i.e. a quasi-periodic behavior of the arm/hand leading to a form of exploration of the area in which the object is located) [28,34]. To subject the robot to a similar process we initially provide it with two simple motor reflexes: an orienting response that makes the robot turn its head toward the colored object [26] and a grasp reflex that makes the robot close its fingers when its right palm touch sensor becomes activated [9]. These reflexes are realized by manually setting the

connection weights indicated with full lines in Fig. 3. The immature visual system is simulated by degrading visual acuity (see Section 2.2). Finally, the limited role of cortical areas during this phase is realized by freezing the connection weights to and from internal neurons to a null value (i.e. by subjecting to plasticity only direct sensory-motor areas).

2. A gross-reaching phase, that extends approximately from month 4 to the first year of age, is characterized by an improved visual acuity [7,8] and by an higher involvement of cortical areas [13]. This phase, that leads to an improved reaching and grasping ability, is characterized by a initial motor suppression [33], by a reduced use of motor babbling [34] and by de-freezing of the distal DoFs [4,9]. The variations occurring during this phase have been modeled in the robotic experiment by increasing the visual acuity (see Section 2.2) and by subjecting to plasticity also the internal neurons' incoming and outgoing connection weights. This loosely simulates the intervention of cortical centers to mediate the sensori-motor reflexive behavior [13].

3. A fine-reaching phase not yet modeled in the experiment reported in this paper, that follows the first year of life. From the behavioral point of view this phase is characterized by a more reliable, faster and smoother reaching and grasping behavior [4,12,14,29]. In future experiments the variations occurring during this phase will be modeled providing the robot's neural controller with additional sensory neurons encoding the current hand/object spatial relation [6]. For more details and for videos of the trained robots see *http://laral.istc.cnr.it/esm/reach/*.

3   Results and Discussion

The first objective of the work is to verify whether the robotic model proposed could effectively lead to the acquisition of reaching and grasping skills analogous to those developed by infants. The analysis of the performance (Fig. 5) shows indeed that at the end of the pre-reaching phase robots manage to reach (i.e. touch the object) through the exhibition of an exploratory behaviour (more details below) in about half of the trials.

During the gross-reaching phase the robots develop an ability to orient their arm toward the area in which the object is located and improve their ability to grasp the object (i.e. touch the object with the palm and at least one of the fingers). The comparison of the robots' performance at the end of the pre-reaching and gross-reaching phase reveals a significant improvement in the frequency of successful grasps and reach straightness (both with $p < 0.001$, two-tailed Mann-Whitney U test). Surprisingly, the frequency of successful reaches is lower in the gross-reaching phase ($p < 0.001$). This indicates that the robots specialize on certain regions of their peripersonal space in which they can robustly grasp an object, instead of trying to reach it in each position. For videos of robots behavior see *http://laral.istc.cnr.it/esm/reach/*.

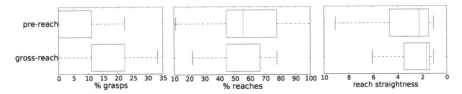

Fig. 5. Boxplot of the performances observed at the end of the pre-reaching and gross-reaching phases. Data computed by post-evaluating the best 5 robots of each replication of the experiment for 45 trials. The straightness indicates the ratio between the length of the trajectory of the hand during successfully reaching actions and the hand/object initial distance (lower values correspond to more efficient reaches).

The robots' behavior at the end of the pre-reaching phase (Fig. 6, Top and Bottom respectively) is characterized by an exploratory motor babbling behavior that is realized by extending the arm and by producing circular movements [19,20,28,34] around the area in which the object can be located (Fig. 6, Top). We compared this behavior (Fig. 6, Top Left) with a control condition in which tactile stimulation is impaired (Fig. 6, Top Right). The confrontation indicates that tactile stimulation plays almost no role in this phase. In the gross-reaching phase instead, the robots keep producing an exploratory motor babbling behavior which is now restricted in the area in which the object is located and regulate their movement on the basis of tactile information so to keep touching and to grasp the object (Fig. 6, Bottom Left and Right pictures).

Moreover, the behavior displayed by the robots at the end of the pre-reaching phase is characterized by a large use of the DoFs of the trunk and of the shoulder and by a reduced use (locking) of the elbow DoF. This is demonstrated by the fact that, as in the case of real infants [3,27], the distance between the shoulder and the hand remains almost constant during reaching attempts (Fig. 7). Note that after touching the ball (distance< 0.1 meters in Fig. 7) the robot is not yet able to remain near the ball at the end of the pre-reaching phase. This skill starts to be developed during the gross-reaching phase.

The second objective of the experiments is to verify whether the realization of an incremental process analogous to those occurring in humans facilitates the development of the required skills and/or channels the developmental process toward specific solutions (Fig. 8). To study this aspect we ran a non-incremental control experiment involving a single developmental phase (lasting the sum of the pre-reaching and gross-reaching phases) in which the robots are not provided with reflexes and are have from the beginning both high visual acuity and internal processing resources (i.e. plastic internal neurons). The performance in this control condition is significantly lower than the performance of the gross-reaching condition ($p < 0.001$, two-tailed Mann-Whitney U test), indicating that indeed the incremental process enables the development of more effective solutions. The fact that robots trained in the non-incremental condition do not display motor babbling suggests that the development of motor babbling plays an important

Touch   No touch

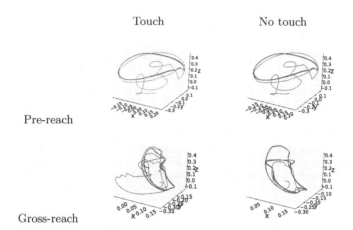

Pre-reach

Gross-reach

Fig. 6. Typical trajectories of the hand in 3D during a trial produced by a robot at the end of the pre-reaching (Top) and gross-reaching (Bottom) phase. Tests performed by placing the target object in the central position. Results produced in a normal (Left) and control (Right) condition in which robots are deprived from tactile stimulation.

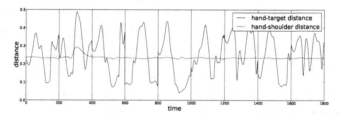

Fig. 7. Hand-shoulder and hand-target distance (in meters) during 9 trials. At the beginning of each trial the posture of the robot is reset and the target object is placed in a different position.

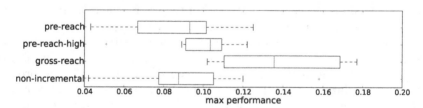

Fig. 8. Performance of the different developmental phases and experimental conditions. Each bar represents the best performances obtained in each condition over ten different replications. Pre-reaching-high inidicates the pre-reaching control condition with high visual acuity.

functional role. An additional control experiment performed in a pre-reaching condition with high visual acuity did not lead to significantly different performance respect the low-acuity pre-reaching condition ($p = 0.14$), indicating that the main factor channeling the developmental process toward effective human-like behavior is constituted by the addition of the internal neurons.

4 Conclusions

We illustrated how the design of humanoid robots able to develop relatively complex action capabilities can be successfully approached by using robotic models that incorporate the following fundamental aspects: embodiment, situatedness, brain-like control/learning, incremental development. The use of an experimental setting similar to those employed by experimental psychologists to study children's behavior allowed us to closely compare human and robot data.

The analysis of the behavioral skills acquired by the robots indicate they closely match those displayed by infants. More specifically, during the pre-reaching phase robots develop a motor babbling strategy similar to that observed in infants [20,28,34]. The development of such strategy is channelled primarily by the availability of limited internal processing capabilities. This maturational constraint [31], possibly coupled with the presence of reflexes, constitutes a prerequisite for the development of better capability later on since allows robots initially provided with limited internal resources to later outperform (when internal resources become available) robots not affected by this initial limitation. As in the case of infants, the developmental process in robots leads to a form of proximo-distal maturation of the DoFs [3]. Indeed, during the first developmental phase, the robots immediately extend the arm and try to reach and grasp the object by exploiting the DoFs of the trunk and of the shoulder while locking the DoF of the elbow. The spontaneous freezing of the elbow joint in our experiments suggests that proximo-distal maturation might result as a side effect of the need to start from simple control policies rather than a maturational constraint [5]. Finally, the results indicating that the incremental version of the model leads to better performing robots than the non-incremental control experiment demonstrates that the incremental process might represent a key factor not only from a modeling but also from an engineering point of view.

In future works we will extend the simulations to the fine-reaching phase, by providing the robot with sensors encoding information about the hand/object offset. Finally we will test the trained neural controller in hardware. The long term challenge we propose is to identify the scalability of this approach and its applications to different domains in robotics and control.

Acknowledgement. This research work was supported by the ITALK project (EU, ICT, Cognitive Systems and Robotics Integrated Project, grant n. 214668).

References

1. Angulo-Kinzler, R.: Exploration and selection of intralimb coordination patterns in 3-month-old infants. Journal of Motor Behavior 33, 363–376 (2001)
2. Atkinson, J.: Human visual development over the first 6 months of life. A review and a hypothesis. Human Neurobiology 3(2), 61–74 (1984)

3. Berthier, N.E., Clifton, R.K., McCall, D.D., Robin, D.J.: Proximodistal structure of early reaching in human infants. Experimental Brain Research 127(3), 259–269 (1999)
4. Berthier, N.E., Keen, R.: Development of reaching in infancy. Experimental Brain Research (2005)
5. Berthouze, L., Lungarella, M.: Motor skill acquisition under environmental perturbations: On the necessity of alternate freezing and freeing of degrees of freedom. Adaptive Behavior 12(1), 47–63 (2004)
6. Burnod, Y., Baraduc, P., Battaglia-Mayer, A., Guigon, E., Koechlin, E., Ferraina, S., Lacquaniti, F., Caminiti, R.: Parieto-frontal coding of reaching: an integrated framework. Experimental Brain Research 129(3), 325–346 (1999)
7. Courage, M.L., Adams, R.J.: Infant peripheral vision: the development of monocular visual acuity in the first 3 months of postnatal life. Vision Research 36(8), 1207–1215 (1996)
8. Hendrickson, A., Druker, D.: The development of parafoveal and mid-peripheral human retina. Behavioural Brain Research 49(1), 21–31 (1992)
9. Lantz, C., Meln, K., Forssberg, H.: Early infant grasping involves radial fingers. Developmental Medicine and Child Neurology 38(8), 668–674 (1996)
10. Konczak, J., Borutta, M., Topka, H., Dichgans, J.: The development of goal-directed reaching in infants: hand trajectory formation and joint torque control. Experimental Brain Research 106(1), 156–168 (1995)
11. Konczak, J., Borutta, M., Dichgans, J.: The development of goal-directed reaching in infants. II. Learning to produce task-adequate patterns of joint torque. Experimental Brain Research 113(3), 465–474 (1997)
12. Konczak, J., Dichgans, J.: The development toward stereotypic arm kinematics during reaching in the first 3 years of life. Experimental Brain Research 117(2), 346–354 (1997)
13. Martin, J.H.: The corticospinal system: from development to motor control. The Neuroscientist 11(2), 161–173 (2005)
14. McCarty, M.K., Clifton, R.K., Ashmead, D.H., Lee, P., Goulet, N.: How infants use vision for grasping objects. Child Development 72, 973–987 (2001)
15. Nolfi, S., Floreano, D.: Evolutionary Robotics: The Biology, Intelligence, and Technology of Self-Organizing Machines. MIT Press, Cambridge (2000)
16. Nolfi, S., Marocco, D.: Evolving robots able to integrate sensory-motor information over time. Theory in Biosciences 120(3-4), 287–310 (2001)
17. Oztop, E., Bradley, N.S., Arbib, M.A.: Infant grasp learning: a computational model. Experimental Brain Research 158(4), 480–503 (2004)
18. Paine, R.W., Tani, J.: Motor primitive and sequence self-organization in a hierarchical recurrent neural network. Neural Networks 17(8-9) (2004)
19. Piaget, J.: The origin of intelligence in the child. Routledge and Kegan Paul, London (1953)
20. Piek, J.: The role of variability in early motor development. Infant Behavior and Development 25(4), 452–465 (2002)
21. Pfeifer, R., Iida, F., Bongard, J.: New robotics: design principles for intelligent systems. Artificial Life 11(1-2), 99–120 (2005)
22. Rochat, P.: Object manipulation and exploration in 2- to 5-month-old infants. Developmental Psychology 25(6), 871–884 (1989)
23. Sandini, G., Metta, G., Vernon, D.: Robotcub: An open framework for research in embodied cognition. International Journal of Humanoid Robotics 8(2), 18–31 (2004)

24. Schaal, S.: Arm and hand movement control. In: Arbib, M.A. (ed.) Handbook of Brain Theory and Neural Networks, 2nd edn., pp. 110–113. MIT Press, Cambridge (2002)

25. Schlesinger, M.: Evolving agents as a metaphor for the developing child. Developmental Science 7, 154–168 (2004)

26. Sokolov, Y.N.: Perception and the conditional reflex. Pergamon Press, London (1963)

27. Spencer, J.P., Thelen, E.: Spatially Specific Changes in Infants' Muscle Coactivity as They Learn to Reach. Infancy 1(3), 275–302 (2000)

28. Thelen, E., Corbetta, D., Kamm, K., Spencer, J.P., Schneider, K., Zernicke, R.F.: The Transition to Reaching: Mapping Intention and Intrinsic Dynamics The Transition to Reaching: Mapping Intention and Intrinsic Dynamics. Child Development 64(4), 1058–1098 (1993)

29. Thelen, E., Corbetta, D., Spencer, J.P.: Development of reaching during the first year: role of movement speed. Journal of experimental psychology. Human Perception and Performance 22(5), 1059–1076 (1996)

30. Tsagarakis, N.G., Metta, G., Sandini, G., Vernon, D., Beira, R., Santos-Victor, J., Carrazzo, M.C., Becchi, F., Caldwell, D.G.: iCub The Design and Realization of an Open Humanoid Platform for Cognitive and Neuroscience Research. International Journal of Advanced Robotics 21(10), 1151–1175 (2007)

31. Turkewitz, G., Kenny, P.: Limitations on input as a basis for neural organization and perceptual development: A preliminary theoretical statement. Developmental Psychobiology 1, 357–368 (1982)

32. von Hofsten, C.: Eye-Hand Coordination in the Newborn. Developmental Psychology 18(3), 450–461 (1982)

33. von Hofsten, C.: Developmental changes in the organization of prereaching movements. Developmental Psychology 20(3), 378–388 (1984)

34. von Hofsten, C., Rönnqvist, L.: The structuring of neonatal arm movements. Child Development 64(4), 1046–1057 (1993)

A True-Slime-Mold-Inspired Fluid-Filled Robot Exhibiting Versatile Behavior

Takuya Umedachi[1,3,4], Ryo Idei[2], and Akio Ishiguro[2,4]

[1] Graduate School of Science, Hiroshima University,
1-3-1 Kagamiyama, Higashi Hiroshima 739-8526, Japan
umedachi@hiroshima-u.ac.jp
[2] Research Institute of Electrical Communication, Tohoku University,
2-1-1 Katahira, Aoba-ku, Sendai 980-8577, Japan
{idei,ishiguro}@riec.tohoku.ac.jp
[3] Research Fellow of the Japan Society for the Promotion (PD),
Sumitomo-Ichibancho FS Bldg., 8 Ichibancho, Chiyoda-ku, Tokyo 102-8472, Japan
[4] Japan Science and Technology Agency, CREST,
Sanban-cho, Chiyoda-ku, Tokyo, 102-0075, Japan

Abstract. Behavioral diversity is one essential feature of living systems in order to exhibit adaptive behavior in hostile and dynamically changing environments. However, classical engineering approaches strive to avoid the behavioral diversity of artificial systems to achieve high performance in specific environments for given tasks. The goals of this research include understanding how living systems exhibit behavioral diversity and use these findings to build robots that exhibit truly adaptive behaviors. To this end, we have focused on an amoeba-like unicellular organism, i.e., the plasmodium of true slime mold. Despite the absence of a central nervous system, the plasmodium exhibits versatile spatiotemporal oscillatory patterns and switches spontaneously between the patterns. Inspired by this, we build a real physical robot that exhibits versatile oscillatory patterns and spontaneous transition between the patterns. The results are expected to shed new light on the design scheme for life-like robots that exhibit amazingly versatile and adaptive behavior.

Keywords: Behavioral Diversity, Decentralized Control, Biologically-Inspired Robot, Soft Actuator.

1 Introduction

Classical engineering approaches strive to avoid, or suppress, the behavioral diversity of artificial systems in order to achieve high performance in specific environments for given tasks. Living systems, in contrast, exhibit qualitatively versatile behaviors and spontaneously switch among these behaviors in response to the situation encountered. Owing to this behavioral diversity, living systems are able to exhibit adaptive behavior in hostile and dynamically changing environments. The goals of the present research include understanding how living

T.J. Prescott et al. (Eds.): Living Machines 2012, LNAI 7375, pp. 262–273, 2012.

systems generate this behavioral diversity and use these findings to build life-like robots that exhibit truly adaptive behavior.

To this end, we have employed the so-called back-to-basics approach. More specifically, we have focused on one of the most primitive living organisms: plasmodium of true slime mold (see Fig. 1). The plasmodium is a large amoeba-like multinucleated unicellular organism, whose motion is driven by spatially distributed biochemical oscillators in the body [1]. The oscillators induce rhythmic mechanical contractions, leading to a pressure increase in the protoplasm, which in turn generates protoplasmic streaming according to the pressure gradient [2]. Hence, the interaction between the homogeneous elements (i.e., the biochemical oscillators) induces global behavior in the plasmodium in the absence of a central nervous system or specialized organs. Yet, despite such a decentralized system, the plasmodium exhibits versatile spatiotemporal oscillatory patterns [3] and, more interestingly, switches between the versatile oscillatory patterns spontaneously [4]. By exploiting this behavioral diversity, the plasmodium is thought to be capable of exhibiting adaptive behavior according to the situation encountered. Thus, the plasmodium is one of the most simple model organisms for investigating the key mechanism that induces the underlying versatile behavior in a fully decentralized manner.

Fig. 1. Plasmodium of true slime mold (*Physarum polycephalum*). The plasmodium of true slime mold exhibits amoeboid locomotion, which is controlled in a fully decentralized manner.

There are two important factors that help the plasmodium exhibit such oscillatory patterns: phase modification of the *mechanosensory* oscillators and physical communication (i.e., morphological communication) stemming from protoplasmic streaming. The oscillators of the plasmodium, which are similar to a central pattern generator (CPG) [3], are physically coupled with tubes filled with protoplasm. By generating protoplasmic streaming through the tubes, long-distance physical interaction is induced between the oscillators. This physical interaction leads to phase modification on the basis of the pressure (i.e., mechanosensory information) applied by the protoplasm [5]. In light of biological knowledge, the versatile oscillatory pattern is attributed mainly to the synergistic effect of phase modification and morphological communication stemming from the protoplasmic streaming. Thus far, we have developed a mathematical model of the oscillator system that reproduces the versatile oscillatory patterns and have demonstrated the validity of the model through numerical experiments [6].

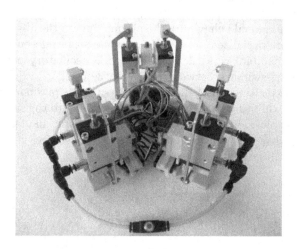

Fig. 2. Image of entire robot composed of three modules

The purpose of this study was to build a real physical robot that exhibits various oscillatory patterns and to understand the design principle behind such intelligence. To this end, we developed a robot inspired by the plasmodium of true slime mold (Fig. 2) that consists of several homogeneous modules filled with air (protoplasm). The significant features of this robot are threefold: (i) the modules are physically connected by tubes so as to induce long-distance physical interaction among the modules, (ii) soft actuators are embedded in the modules in order to aid the flow of protoplasm between the modules, and (iii) phase modification exploits the mechanosensory information of the soft actuator. Experimental results show that the robot is capable of exhibiting surprisingly versatile oscillatory patterns and transitions among them in a fully decentralized manner. The results obtained are expected to shed new light on the design scheme of a new type of life-like robot with the capability to switch behavior (e.g., taxiing, exploratory, and escape behavior) spontaneously according to the situation encountered without modifying any parameters of the robot, as shown in Fig. 3.

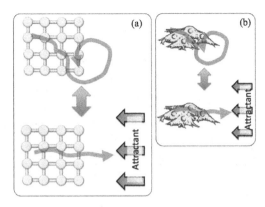

Fig. 3. Conceptual diagram of a modular robot that can switch behaviors (such as taxiing behavior, exploratory, and escape behavior) spontaneously without changing robot parameters according to situation encountered (a), similar to true slime mold (b). Each circle in (a) indicates one module of the robot connected with fluid-filled tubes.

2   Robot Design

2.1   Mechanical System

The robot consists of several homogeneous modules, each of which consists of two cylinders, soft actuators, and control circuit boards (see Fig. 4 (a)). The cylinders are physically connected with a tube, and air is sealed inside the cylinders and tube as protoplasm. One phase oscillator is implemented in the microcomputer in each control circuit board in order to control each actuator according to the phase $\phi_{i,n}(n = r, l)$. As shown in Fig. 4 (b), the motion of the soft actuator is converted into an up-down movement, which aids the flow of protoplasm inside the cylinder.

Soft Actuator. In order to obtain mechanosensory information, we designed a soft actuator composed of a servomotor, an elastic element, and a potentiometer (see Fig. 4 (c)). An important feature of this actuator is that the servomotor and cylinder rod are connected via the elastic element. Owing to this mechanical design, any gdiscrepancyh between the controlled value (the rotation angle of the servomotor) and the actual value (the actual angle sensed by the potentiometer) can be measured as mechanosensory information (as explained in section 2.2).

The rotation angle of the servomotor, $\bar{\theta}_{i,n}$, is controlled according to $\phi_{i,n}$ and is given by

$$\bar{\theta}_{i,n}(\phi_{i,n}) = a \sin \phi_{i,n} \tag{1}$$

where a is a constant in space and time that defines the amplitude of the rotation angle.

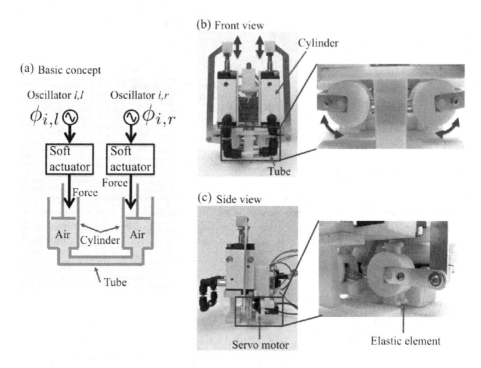

Fig. 4. (a) Schematic of the module proposed. Photographs of (b) front view and (c) side view of constructed robot.

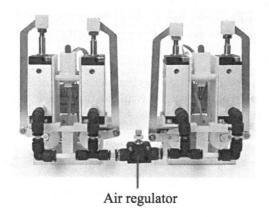

Fig. 5. Photograph of robot composed of two modules

Protoplasmic Streaming. As shown in Figs. 5 and 2, the modules are physically connected via the fluid-filled tube with an air regulator. Through the tube, protoplasmic streaming is generated between the modules according to the pressure gradient. It should be noted that the total volume of the protoplasm inside all the cylinders and tubes is conserved. Because of this, long-distance physical interaction is induced among the modules. In addition, fluid conductance among the modules can be manually adjusted using the air regulator.

2.2 Control System

Here, we introduce the dynamics of the oscillator model to be implemented in each microcomputer. The equation for the oscillator is given as [7,8,9]

$$\dot{\phi}_{i,n} = \omega_n - \frac{\partial I_{i,n}}{\partial \phi_{i,n}}, \tag{2}$$

where ω_n is the intrinsic frequency of the phase oscillator, and the second term is the local sensory feedback that can be calculated from the discrepancy function $I_{i,n}$. This function is based on the mechanosensory information of the soft actuator, as mentioned in 2.1. Note that the phase oscillators only interact with each other through the mechanical system (i.e., the protoplasm).

To define the discrepancy function, let us discuss the possible cytological processes that play a role in the discrepancy function. It is well known that the so-called stretch-activated Ca^{2+} channel is regulated by the mechanical deformation of the cell shape [10]. This means that the channel can detect the local cell curvature and open or close in response. This type of channel can relax rhythmic contractions because a high Ca^{2+} concentration leads to the relaxation of the actomyosin contraction in the plasmodium of true slime mold [5]. When a local oscillator is pushed strongly by the protoplasm under high pressure, it stretches even if it is trying to contract. The amount of stretching is related to the increase in the Ca^{2+} concentration, which leads to the relaxation of the oscillator. This is one possible cytological scenario for the role of the discrepancy function.

On the basis of this biological finding, we define the discrepancy function for this robot as:

$$I_{i,n}(\tau_{i,n}) = \frac{\sigma}{2}\tau_{i,n}^2, \tag{3}$$

where σ is a coefficient that defines the strength of the feedback, and $\tau_{i,n}$ is the tension in the soft actuator. The function is designed to increase in value when the absolute value of $\tau_{i,n}$ increases, and $\tau_{i,n}$ can be described by the following equation:

$$\tau_{i,n} = k(\theta_{i,n} - \bar{\theta}_{i,n}(\phi_{i,n})), \tag{4}$$

where k is the elastic coefficient of the elastic element. It should be noted that this mechanosensory information can only be produced from the mechanical softness of the actuator.

Based on the discrepancy function, (2) can be rewritten as:

$$\dot{\phi}_{i,n} = \omega_n + \sigma k^2 a(\theta_{i,n} - a\sin\phi_{i,n})\cos\phi_{i,n}. \tag{5}$$

The second term on the right-hand side of (5) is the local sensory feedback that reduces the discrepancy function $I_{i,n}$, and it can be calculated using only locally available variables, which include the discrepancy between the controlled value and its actual value.

3   Experimental Results

3.1   Experiment with One Module

To confirm the validity of the local sensory feedback based on the discrepancy function, an experiment with one module was conducted, and the results are presented in Fig. 6. The experiment is started from the nearly in-phase condition. As can be seen in the figure, over time, transitions from the in-phase condition to antiphase coordination inside the module occur with a decreasing total value of the discrepancy function.

The parameters of the robot are as follows: $\omega_n = 0.3$ rad/sec $(n = r, l)$; $\sigma = 0.005$; $\phi_{0,r}(t = 0) = \pi/6$; $\phi_{0,l}(t = 0) = 0.0$.

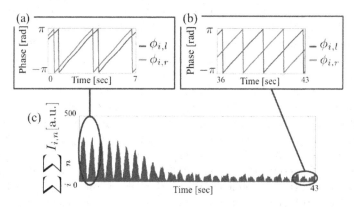

Fig. 6. Representation data of oscillatory pattern transition in one module. (a) and (b) show the time evolution of the phases of the oscillators from 0 to 7 sec and from 36 to 43 sec, respectively. (c) shows the time evolution of $\sum_i \sum_n I_{i,n}$.

3.2   Experiment with Two Modules

Now, let us describe what happens when two modules are connected, as shown in Fig. 5. In this experiment, we increased the fluid conductance during the experiment (at 28 sec); the results are presented in Figs. 7 and 8.

From 0 to 28 sec, the fluid conductance is small. In this condition, the volumes of the two modules are oscillating in antiphase, which means that the two modules are exchanging the protoplasm, as can be seen in Fig. 7. After 28 sec, the fluid conductance increases, and it becomes difficult for the two modules to exchange the protoplasm. In this condition, the modules decrease the total value of the discrepancy function "inside" each module, which results in antiphase oscillation between two phase oscillators in each module, as shown in Fig. 8. Note that this behavioral change is produced as a result of decreasing the total value of the discrepancy function in a fully decentralized manner.

The parameters of the robot are as follows: $\omega_r = 0.305$ rad/sec; $\omega_l = 0.3$ rad/sec; $\sigma = 0.005$; $\phi_{0,n}(t = 0) = 0.0$ $(n = r, l)$; $\phi_{1,n}(t = 0) = \pi$ $(n = r, l)$.

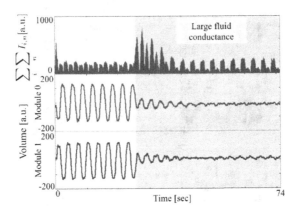

Fig. 7. Representation data of oscillatory pattern transition in two modules. These show the time evolution of $\sum_i \sum_n I_{i,n}$ (top) and the volumes of the modules (middle and bottom) from 0 to 74 sec, respectively.

3.3   Experiment with Three Modules

We now describe what happens when three modules are connected, as shown in Fig. 2; the experimental results are shown in Figs. 9 and 10. As can be seen in Fig. 9, we confirmed the existence of four oscillatory patterns: (a) a rotation mode, (b) a partial in-phase mode, (c) a partial antiphase mode, (d) an intra-oscillation mode[1]. Furthermore, the robot spontaneously switched between these four oscillation modes during one observation, as shown in Fig. 10.

Figure 9 (a) shows the rotation mode, where the rotating wave is observed in the order module 0, 1, 2. The phase difference between neighboring oscillators is approximately $2\pi/3$. Figure 9 (b) shows the partial in-phase mode, where the volumes of the two modules (1 and 2) are in phase, and the other module (0) is in antiphase with the first two modules. Figure 9 (c) shows the partial antiphase mode, where the volumes of the two modules (1 and 2) are in antiphase, and module 0 barely oscillates. Figure 9 (d) shows the intra-oscillation mode, where

---

[1] The names of the oscillatory patterns (a), (b), and (c) are based on Takamatsu's work [4].

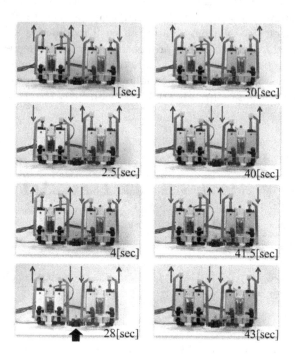

Fig. 8. Snapshots of oscillatory pattern transition in two modules. At 28 sec, the air regulator (indicated by the black arrow) is tuned to increase the fluid conductance.

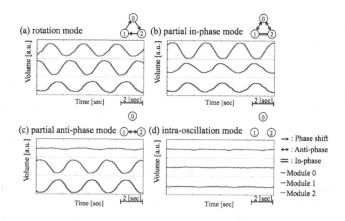

Fig. 9. Oscillatory patterns observed for three modules. Schematics of the phase relationships among the three oscillators are indicated in the upper right corner of each plot. The relationships between two modules are indicated by $\rightarrow$: $\frac{2\pi}{3}$ phase shift; $\leftrightarrow$: anti phase; and $=$: in phase.

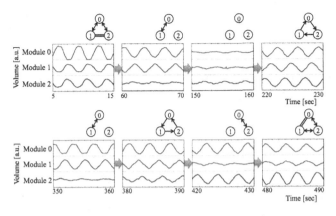

Fig. 10. Transitions between several oscillation modes. ($\omega_r = 0.305$, $\omega_l = 0.3$)

the volumes of all the modules barely oscillate. In this oscillation mode, two phase oscillators are in antiphase inside each module, which do not require a volume change oscillation. Schematic diagrams of phase relationships among the three oscillators are indicated in the upper right of each plot. The relationships between two of the modules are indicated by $\rightarrow$: $\frac{2\pi}{3}$ phase shift; $\leftrightarrow$: anti phase; and $=$: in phase.

The parameters of the robot are as follows: $\omega_r = 0.305$ rad/sec; $\omega_l = 0.3$ rad/sec; $\sigma = 0.003$; $\phi_{0,n}(t=0) = 0.0$; $\phi_{1,n}(t=0) = \phi_{2,n}(t=0) = \pi$.

4   Discussion

Consider the essential factor that induces the transition between the oscillatory patterns: one significant feature of this model is that two phase oscillators are embedded for controlling one module, which allows the module to induce an intra-oscillation mode. The intra-oscillation mode is not possible with a single phase oscillator. Note that there is a small difference between ω_r and ω_l, the intrinsic frequencies of the two phase oscillators, and we speculate that this difference is an essential factor.

In order to investigate this, we conducted additional two experiments with (i) $\omega_r = \omega_l = 0.3$ and (ii) $\omega_r = 0.33$ and $\omega_l = 0.3$. The results are shown in Figs. 11 and 12, respectively. When $\omega_r = \omega_l = 0.3$, a transition between the oscillation modes can be confirmed; however, the oscillation eventually converges to the intra-oscillation mode (see Fig. 11). The same result is observed when $\omega_r - \omega_l$ is smaller than 0.005 (see 3.3). In contrast, when $\omega_r - \omega_l$ is larger than 0.005, the oscillation switches between the rotation and intra-oscillation modes ((a) and (d); see Fig. 12).

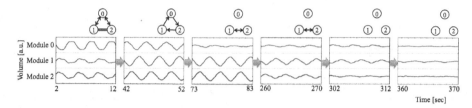

Fig. 11. Time evolution of module volumes for $\omega_r = \omega_l = 0.3$

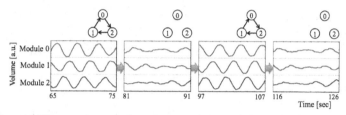

Fig. 12. Time evolution of module volumes for $\omega_r = 0.33$ and $\omega_l = 0.3$

5   Conclusions

We have presented a real physical robot that exhibits versatile oscillatory patterns and spontaneous transitions among these patterns. The two main results of this work are as follows. The first concerns the emphasis on considering the physical interaction between body parts stemming from protoplasmic streaming. The second is related to the design scheme of phase modification originating from the mechanosensory information, which is based on the discrepancy function. The experimental results clearly indicate that the synergistic effect of the phase modification and the morphological communication of the protoplasmic streaming successfully generates versatile oscillatory patterns and spontaneous transitions without relying on complex oscillators such as chaos oscillators.

Our future work will focus on investigating the essential mechanism that induces such versatile behavior and on implementing the mechanism in a robot (e.g., the robot we developed in [9]) to generate locomotion. To do so, we believe a robot should be endowed with the capability to switch behavior spontaneously according to the situation encountered without increasing the complexity of the control system.

Acknowledgments. The authors are deeply indebted to Kentaro Ito, Assistant Professor of Mathematical and Life Sciences at Hiroshima University, and Ryo Kobayashi, Professor of Mathematical and Life Sciences at Hiroshima University, for their valuable advice and considerable effort for the development of the robot. This research is partially supported by Grant-in-Aid for challenging Exploratory Research (No. 23656171).

References

1. Tero, A., Kobayashi, R., Nakagaki, T.: A Mathematical Model for Adaptive Transport Network in Path Finding by True Slime Mold. Journal of Theoretical Biology 244, 553–564 (2007), doi:10.1016/j.jtbi.2006.07.015
2. Kobayashi, R., Tero, A., Nakagaki, T.: Mathematical Model for Rhythmic Protoplasmic Movement in the True Slime Mold. Journal of Mathematical Biology 53, 273–286 (2006), doi:10.1007/s00285-006-0007-0
3. Takamatsu, A., Tanaka, R., Yamada, H., Nakagaki, T., Fujii, T., Endo, I.: Spatiotemporal Symmetry in Rings of Coupled Biological Oscillators of Physarum Plasmodial Slime Mold. Physical Review Letters 87, 0781021 (2001), doi:10.1103/PhysRevLett.87.078102
4. Takamatsu, A.: Spontaneous Switching among Multiple Spatio-temporal Patterns in Three-oscillator Systems Constructed with Oscillatory Cells of Slime Mold. Physica D: Nonlinear Phenomena 223, 180–188 (2006), doi:10.1016/j.physd.2006.09.001
5. Yoshiyama, S., Ishigami, M., Nakamura, A., Kohama, K.: Calcium Wave for Cytoplasmic Streaming of Physarum Polycephalum. Cell Biology International 34, 35–40 (2009), doi:10.1042/CBI20090158
6. Umedachi, T., Idei, R., Ishiguro, A.: A Fluid-filled Soft Robot That Exhibits Spontaneous Switching among Versatile Spatio-temporal Oscillatory Patterns Inspired by True Slime Mold. In: Proc. 2nd International Conference on Morphological Computation (ICMC 2011), pp. 54–56 (2011)
7. Umedachi, T., Takeda, K., Nakagaki, T., Kobayashi, R., Ishiguro, A.: Fully Decentralized Control of a Soft-bodied Robot Inspired by True Slime Mold. Biological Cybernetics 102, 261–269 (2010), doi:10.1007/s00422-010-0367-9
8. Umedachi, T., Takeda, K., Nakagaki, T., Kobayashi, R., Ishiguro, A.: Taming Large Degrees of Freedom -A Case Study with an Amoeboid Robot-. In: Proc. of 2010 IEEE International Conference on Robotics and Automation (ICRA), pp. 3787–3792 (2010), doi:10.1109/ROBOT.2010.5509498
9. Umedachi, T., Takeda, K., Nakagaki, T., Kobayashi, R., Ishiguro, A.: A Soft-bodied Fluid-driven Amoeboid Robot Inspired by Plasmodium of True Slime Mold. In: Proc. of 2010 IEEE/RSJ International Conference on Intelligent Robots and Systems (IROS 2010), pp. 2401–2406 (2010), doi:10.1109/IROS.2010.5651149
10. Guharay, F., Sachs, F.: Stretch-activated single ion channel currents in tissue-cultured embryonic chick skeletal muscle. The Journal of Physiology 352(1), 685–701 (1984)

CyberRat Probes: High-Resolution Biohybrid Devices for Probing the Brain*

Stefano Vassanelli[1,**], Florian Felderer[2], Mufti Mahmud[1,3],
Marta Maschietto[1], and Stefano Girardi[1]

[1] NeuroChip Lab, University of Padova, via f. Marzolo 3, 35131–Padova, Italy
[2] Max Planck Institute of Biochemistry, D-82152 Martinsried, Germany
[3] Institute of Information Technology, Jahangirnagar University,
Savar, 1342–Dhaka, Bangladesh
{stefano.vassanelli,stefano.girardi.1,marta.maschietto}@unipd.it,
mahmud@dei.unipd.it

Abstract. Neuronal probes can be defined as biohybrid entities where the probes and nerve cells establish a close physical interaction for communicating in one or both directions. During the last decade neuronal probe technology has seen an exploded development. This paper presents newly developed chip–based CyberRat probes for enhanced signal transmission from nerve cells to chip or from chip to nerve cells with an emphasis on *in–vivo* interfacing, either in terms of signal–to–noise ratio or of spatiotemporal resolution. The oxide–insulated chips featuring large–scale and high–resolution arrays of stimulation and recording elements are a promising technology for high spatiotemporal resolution biohybrid devices, as recently demonstrated by recordings obtained from hippocampal slices and brain cortex in implanted animals. Finally, we report on SigMate, an *in–house* comprehensive automated tool for processing and analysis of acquired signals by such large scale biohybrid devices.

Keywords: Neuronal probe, brain recording, brain stimulation, biohybrid devices, neuronal signal analysis.

1 Introduction

1.1 Biohybrid Devices for Neural Interfacing

In recent years the use of on–chip microelectromechanical systems (MEMS) in the biomedical field has gained increasing attention. New MEMS based scientific, diagnostic and therapeutic tools are being developed as a result of continuous improvement in micromachining and microelectronics technologies and simultaneous deepening of knowledge about cellular and molecular mechanisms in life sciences. Microchips as neuronal probes for multi–site recording of neuronal

---

* All authors contributed equally to the work.
** Corresponding author.

T.J. Prescott et al. (Eds.): Living Machines 2012, LNAI 7375, pp. 274–285, 2012.

activity were among the first to be introduced [1] and are now representing an expanding technology [2], [3]. However, the applications of these types of biohybrid devices are not limited with the ones mentioned above and there is a great potential for novel applications. The technology underwent a progressive development and is now widely adopted by neuroscientists for recording living neurons *in–vitro*. Among others we have used implantable microchips as neuronal probes for investigating brain circuits *in–vivo* while, in parallel, their potential for neuroprosthetics application has been successfully demonstrated in non–human primates [4] and assessed in clinical trials in paralyzed patients [5].

Although biohybrid devices for neuroscience applications are based on electrical signaling between neurons and microelectronics sensors, they are comprehensive of other technological approaches. These include other means of information exchange, such as, chemical or optical signals. Also, the devices' interaction with the brain may happen at different levels, either of individual cells ([2], [6], [7], [8], [9]) or ensembles ([10]) or *in–vivo* ([11], [12]), and the communication can be uni– or bi–directional, allowing scientists to explore new concepts of cognitive processing and / or capabilities of the brain.

In this paper we present high–resolution neuronal probes, named, "CyberRat Probes" for *in–vivo* interfacing with the brain.

1.2   Existing Technologies

Biohybrid devices have been developed to record and stimulate extracellularly with a number of sites distributed in space [13] for (*in–vitro*) neural tissue interfacing. The tissue is placed in an electrolyte above the surface of a solid–state chip with voltage–sensitive sites in a regular spatial arrangement. Between the tissue and the chip surface a cleft is formed, that, in the case of dissociated neurons in culture on the chip surface has been shown to be in the order of 50 nm using fluorescence interference contrast (FLIC) microscopy [2]. The sites may be made by means of metal electrode (in multi electrode arrays, MEAs [14], [15], [16]) or using a Metal–Oxide–Semiconductor Field Effect Transistor (MOSFET) where the metal of the site is replaced by the electrolyte above a transistor's dielectric (popularly known as Electrolyte–Oxide–Semiconductor Field Effect Transistor, EOSFET [17], [18]). In both approaches the recording sites are then connected to further signal–processing circuitry.

In both cases, however, realization of large high–density 2D arrays is restricted by interconnect issues: only one interconnect layer is available in the bulk material which is used to make a connection between the active sites in the center of the chips to pads at the chip borders. Thus, aiming for a significant increase in spatiotemporal resolution, (extended) Complimentary Metal–Oxide–Semiconductor (CMOS) technology and chips with related circuitry have been proposed in recent years to circumvent such interconnect problems. Moreover, such chips allow provision of signal processing circuitry in closest proximity to the related recording/stimulation sites. CMOS–based noble metal electrode arrays have been published with up to 11 k sites, and extended EOSFET arrays have been reported with up to 16 k and – very recently – 32 k sites [19], [20], [21], [22].

A number of recent developments are also aiming for *in–vivo* interfacing and have been discussed elsewhere [3]. Two classical examples of metal based multi–site electrodes for *in–vivo* interfacing are – the 'Utah probe' [23] and the 'Michigan probe' [24] (a detailed review on metal based neuronal probes can be found in [3]). Whereas extracellular recording and stimulation principles from the *in–vitro* approaches can be adopted, the chips developed in that context cannot simply be transferred as they are. Biggest concern is power: if the power is transferred wirelessly, the amount of available power is limited. If that is not the case (and the power is provided through a cable), the maximum power which can be consumed in live tissue is limited due to heat generation. Unfortunately, however, the number of sites and bandwidth of such a system increase the power consumption whereas the noise of a system shows an increase with decreasing power allowed per site [3].

The existing techniques face challenges mainly in terms of high space resolution (μm range) and bi–directional communication, i.e., simultaneous stimulation and recording with a single device. Though recent developments have seen some closed loop stimulation devices [25], [26], [27], yet, the integration of recording and stimulation sites at a higher–resolution still remains an open problem. Also, biocompatibility of the devices and robustness of recording has to improve for the stability of long–term implants. Also, due to the different neuronal network topology in different brain areas the *in–vivo* systems must always be carefully tailored depending on the related target application.

1.3 Tools for Neuronal Signal Analysis

The existing biohybrid devices for neuronal applications generate huge amount of data. Inferring meaningful conclusions from these data through appropriate processing and analyses impose a big challenge on the neuroscience and neuro-engineering communities [28], [29]. To respond to this challenge sophisticated signal processing techniques are required to support the analyses. Individual tools are available for spike detection, sorting, and EEG analysis [30], [31], [32]. But very few tools are available to process LFPs and integrate all the signal processing steps. Also, due to limited cross compatibility commercially available tools cannot be easily adapted to newly emerging requirements for data analysis and visualization.

Over the years academic and commercial software tools and/or packages are developed to process and analyze neuronal data [30], [31], [32], [33], [34], [35], [36], [37], [38], [39], [40]. They mainly deal with data visualization, spike detection and sorting, spike train analysis, and EEG analysis. Also, a couple of open platforms are under development to promote cross–laboratory software sharing through the worldwide web [41], [42]. However, the community is still in need of a comprehensive tool/package to be able to analyze spikes, LFPs, and EEG at once. Thus an umbrella tool is required for all types of neuronal signals. The

"SigMate", developed in our lab, is a comprehensive tool, designed to perform processing and analyses of spikes, LFPs, and EEG signals [43], [44], [45].

2   Interfacing with the Brain Using the CyberRat Probes

2.1   Probe Description

The CyberRat probes were fabricated from silicon–on–insulator (SOI) wafers (4 in.) with 100 μm n–type silicon (1–10 Ω cm) and 2 μm SiO$_2$ on a 400 μm thick silicon substrate (SiMat, Landsberg, Germany). Each chip consisted of two parts: a needle (2 mm long) with an array of four transistors (gate area 10 μm $\times$ 10 μm, pitch 80 μm, Figure 1, a) and a contact plate with the bond pads. For stability reasons, the chips of the prototype series were relatively massive with a thickness of 100 μm and a width of 360 μm. A rather blunt shape of the tip was fabricated in order to place the transistors close to the tip. The contact plate was 500 μm thick, 5 mm wide and 10 mm long. Its edge was displaced with respect to the needle in order to allow a visual control of the impalement [49].

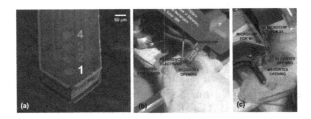

Fig. 1. Setup for recording of neuronal signals with CyberRat probe prototype. (a) Scanning electron microscope picture of the CyberRat probe prototype with the 4 FETs shown in different colors. (b) Magnification of the rat head: the opening in the skull in correspondence of the S1 cortex and the chip inserted into the cortical area are visible. The chip was carefully inserted through a slit in the meninges. A selected whisker was inserted into a tube mounted on the piezoelectric for stimulation. (c) Multiple implants where one chip is positioned in the S1 cortex and the second in the M1 cortex.

2.2   Animal Preparation

Wistar rats were maintained in the Animal Research Facility of the Department of Human Anatomy and Physiology (University of Padova, Italy) under standard environmental conditions. P30–P40 animals were anesthetized with an induction mixture of Tiletamine and Xylazine (2 mg/100 g weight and 1.4 g/100 g weight, respectively). The animal's eye and hind–limbs' reflexes, respiration, and whiskers' spontaneous movements were monitored throughout the experiment and whenever necessary, additional doses of Tiletamine (0.5 mg / 100 g weight) and Xylazine (0.5 g / 100 g weight) were provided.

Rats were positioned on a stereotaxic apparatus and fixed by teeth– and ear-bars. The body temperature was constantly monitored with a rectal probe and maintained at about 37°C using a homeotermic heating pad. Heart beat was monitored by standard ECG. Anterior–posterior opening in the skin was made in the center of the head, starting from the imaginary eye–line and ending at the neck. The connective tissue between skin and skull was then removed by means of a bone scraper. The skull was drilled to open a window in correspondence of S1 (AP −1 ÷ −4, LM +4 ÷ +8) and/or M1 (AP +1 ÷ +4, LM 0 ÷ +3) right cortices. In order to reduce brain edema, the meninges were not removed entirely from the exposed brain surface, but only cut at coordinates AP −2.5, LM +6 and AP +2.5, LM +1.5 for S1 and M1 cortices, respectively [50].

Throughout all surgical operations and recordings, the brain was bathed through a perfusion system by a standard Krebs solution (composition in mM: NaCl 120, KCl 1.99, NaHCO$_3$ 25.56, KH$_2$PO$_4$ 136.09, CaCl$_2$ 2, MgSO$_4$ 1.2, glucose 11), constantly oxygenated and warmed to 37°C. At the end of the surgery the controlateral whiskers were trimmed at about 10 mm from the rat snout.

2.3   Stimulation and Signal Recording

Neuronal signals were evoked by single whiskers mechanical stimulation performed with a piezoelectric through a connected tube. The piezoelectric was driven by a waveform generator (Agilent 33250A 80 MHz, Agilent Technologies) providing 0.5 Hz stimuli. Each whisker, starting from the posterior group, was individually inserted into the tube and the corresponding response was checked in S1 cortex IV layer (at 640 μm depth) in order to find the most responsive whisker ("principal whisker"). Depth recording profiles were obtained by moving the chip up and down in order to monitor all cortical layers. The chips were left into the brain for about 5 hours, throughout all the duration of the experiment.

Recordings from the S1 cortex were compared to standard electrophysiological measurements with borosilicate micropipettes (1 MΩ resistance), filled with Krebs solution.

The chips and the pipettes were connected to the computer by custom–built amplifiers and the neuronal signals were recorded by custom–made software developed in LabVIEW (http://www.ni.com/labview/). At each recording depth LFPs were recorded by stimulating the whisker at different angles ranging from 0° to 315° at a step of 45° [51].

2.4   Neuronal Signal Detection

In Figure 2 500 ms raw and averaged traces recorded by the microchip in the rat S1 cortex under whisker mechanical stimulation are shown. Signals, mainly in the form of local field potentials (LFPs) [59], are recorded simultaneously by the four EOSFETs of the CyberRat probe and are represented in different colors, according to Figure 1 a. The recording depths are 900, 820, 740, and 660 μm for the white, red, green, and blue colored signals, respectively.

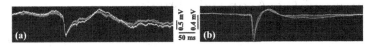

Fig. 2. Raw (a) and averaged (b) ($n = 50$) evoked signals recorded simultaneously by four EOSFETs of the CyberRat probe. The whisker stimulation onset was at 150 ms.

By means of the four recording sites in the prototypes, even small variations of LFPs amplitudes at different depths and in distinct layers can be appreciated (Figure 2).

2.5 Comparison of Recordings between CyberRat Probe and Conventional Probe

To check whether the neuronal signals recorded by EOSFETs are consistent with standard single–electrode extracellular recording, we compared the averaged LFPs detected by FET1 to a typical recording obtained with a borosilicate micropipette at the same depths in the S1 cortex (Figure 3 a, b). The comparison demonstrates that transistors record evoked potentials with standard onset, shape and amplitude characteristics, thus representing a reliable electrophysiology recording tool.

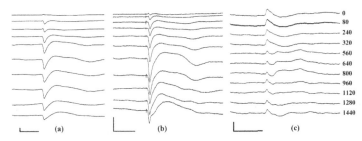

Fig. 3. Comparison of averaged ($n = 50$) evoked LFPs recorded from the rat S1 cortex by means of a micropipette (a), transistor (b), and simultaneous recordings of LFPs from S1 and M1 (c). Recording depths are measured from the cortical surface in μm as indicated on the right side of the trace. The scale bars for x–axis are 100 ms and y–axis are 0.5 mV.

2.6 Simultaneous Recordings from S1 and M1 Cortex

Multiple implants of the CyberRat probe in the S1 and M1 cortex of the rat were performed. Simultaneous recordings of LFPs evoked by whisker stimulation in S1 and M1 are shown in Figure 3 b, c. It clearly shows that whisker stimulation was evoking responses in all S1 and M1 layers. The shape and amplitude of the signals recorded in M1 were consistent with previous results shown by Ahrens and Kleinfeld [59].

3   Analysis of the Acquired Signals

As mentioned earlier, the signals acquired by the state–of–the–art neuronal probes are huge in amount and require tools for automated processing and analysis. We have developed a comprehensive tool, 'SigMate' for processing and analysis of the signals recorded using the CyberRat probes discussed in the previous section. This tool is more of a framework that incorporates popular open–source standard tools and our *in–house* tools. The following subsections will provide information about the design and existing features of the SigMate package.

3.1   SigMate Design

SigMate is designed using a multi–layered approach with three layers (Figure 4).

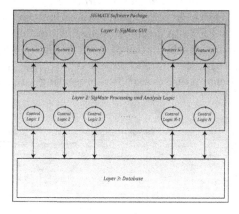

Fig. 4. Three–layered architecture of the SigMate software package

- Presentation layer (top layer): This is the topmost level of the application. The presentation layer contains the GUIs of the application. It communicates with the middle layer by requesting the user commands.
- Application layer (business logic layer, or middle layer): The logic layer is separated from the presentation layer and here an application's functionality is controlled by performing detailed processing and analysis.
- Data layer (bottom layer): This layer consists of databases and/or storages. This layer keeps data neutral and independent from applications or business logic and improves scalability and performance.

3.2   SigMate Features

SigMate is a comprehensive package as it contains features dealing with LFPs, spikes, and EEG. Figure 5 shows the initial GUI of the SigMate software package annotated for its various attributes.

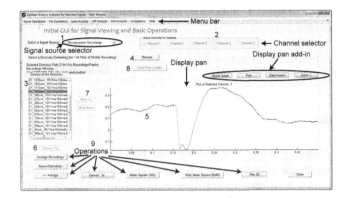

Fig. 5. Main GUI of the SigMate package. This GUI contains the menu bar that bundle the various features of the package together. There are some common attributes which are available in all the features' GUIs: 1) a dropdown box for selection of signal source, 2) a number of check boxes for channel selection, 3) a listbox to show the selected files, 4) a browse button to let the program select a folder whose contents can be seen in the listbox, 5) a display pane (with zoom, pan and data cursor add-ins) to visualize the selected signal file(s) from the listbox, 6) a remove file button to remove selective files from the listbox, 7) a pair of move up and move down buttons to set the order of the selected signal files, 8) a load data button to load the signal files present in the listbox for easy access, and 9) module specific buttons and fields.

SigMate's features at present include:

- **Signal visualization:** to visualize signals in 2D and 3D.
- **File operations:** to perform file splitting, file concatenating, and file column rearranging.
- **Stimulus artifact removal:** to remove slow artifacts induced by air-puff stimulation in case of cortical signal recording, and fast artifacts induced by intracortical microstimulations.
- **Noise characterization:** to assess the quality of the recorded signal and provide the user with noise information of the recorded signals.
- **Spike detection, sorting and spike train analysis:** we adopted a popular package "Wave_Clus" from [30].
- **Current source density calculator:** calculates current source density (CSD) of the recorded LFPs using standard CSD method, step inverse CSD method, δ–source inverse CSD method, and spline inverse CSD method.
- **Latency estimator:** estimates latencies through event and sink detection in LFPs and CSDs, respectively. Also calculates the cortical layer activation order using the calculated latencies, if a depth profile of LFPs is provided.
- **Contour based single LFP sorter:** classifies single LFPs based on their shapes to understand the possible signal variations generated by the underlying neuronal networks under same stimulation.
- **Angular tuning calculator:** calculates angular preferentiality from the recorded signals using the latency and amplitude information.

- **Signal subtractor:** it performs subtraction of control signals from the evoked potentials to better visualize the effect of the stimuli on the evoked signals.
- **EEG analysis:** we adopted a popular open-source package, "EEGLAB" from [32] for processing EEG.

4 Conclusion

Despite first evidence has been provided that biohybrid devices tailored for neuroscience applications can be employed to drive neuroprosthetic devices in humans [5], there is a long way to go before reasonable advantages can be obtained to justify a massive use in clinics of the approach. Potentially, for example, this class of biohybrid devices offer the possibility of on–chip integration of neuromorphic substitutes of brain circuits. Examples of various model based neuroprosthetic devices were reported [4], [5], [53], [54], [55], [56], [57][4-5, 62-66]. The reported applications not only included neuroprosthetic devices to restore motor functions, but also devices focusing on afferent processes, i.e., sensory processes [58][67]. Depending on the necessity and applications, however, high–resolution probes are required to better interface and understand the brain circuitry. We summarized the existing technology for neuronal probes, provided a first report on the CyberRat probes and demonstrated their experimental workability. In conclusion, brain–chip / machine interfacing represent a transdisciplinary approach allowing to investigate brain function with unprecedented resolution and acting as communication link between nervous system and neuroprostheses.

Acknowledgments. This work was carried out as a part of the European Commission funded CyberRat project under the Seventh Framework Programme (ICT-2007.8.3 Bio-ICT convergence, 216528, CyberRat) and Progetto di Eccellenza Cariparo 2008.

The authors express their gratitude to Prof. Peter Fromherz for providing us with the chips and his invaluable suggestions during the experiments & Prof. Roland Thewes for fruitful discussions and technical support.

References

1. Rutten, W.L.: Selective electrical interfaces with the nervous system. Annu. Rev. Biomed. Eng. 4, 407–452 (2002)
2. Fromherz, P.: Neuroelectronic Interfacing: Semiconductor chips with Ion Channels, Nerve cells, and Brain. In: Waser, R. (ed.) Nanoelectronics and Information Technology, pp. 781–810. Wiley–VCH, Berlin (2003)
3. Wise, K.D., et al.: Wireless implantable microsystems: high-density electronic interfaces to the nervous system. Proc. IEEE 92, 76–97 (2004)
4. Lebedev, M.A., Nicolelis, M.A.: Brain-machine interfaces: past, present and future. Trends. Neurosci. 29(9), 537–546 (2006)

5. Hochberg, L.R., et al.: Neuronal ensemble control of prosthetic devices by a human with tetraplegia. Nature 442(7099), 164–171 (2006)
6. Vassanelli, S., Fromherz, P.: Transistor probes local potassium conductances in the adhesion region of cultured rat hippocampal neurons. J. Neurosci. 19(16), 6767–6773 (1999)
7. Vassanelli, S., Fromherz, P.: Transistor records of excitable neurons from rat brain. Appl. Phys. A. 66, 459–463 (1998)
8. Hai, A., Shappir, J., Spira, M.E.: Long-term, multisite, parallel, in-cell recording and stimulation by an array of extracellular microelectrodes. J. Neurophysiol. 104, 559–568 (2010)
9. Lambacher, A., et al.: Electrical imaging of neuronal activity by multi–transistor–array (MTA) recording at 7.8 μm resolution. Appl. Phys. A. 79(7), 1607–1611 (2004)
10. Hutzler, M., et al.: High-resolution multitransistor array recording of electrical field potentials in cultured brain slices. J. Neurophysiol. 96, 1638–1645 (2006)
11. Girardi, S., Maschietto, M., Zeitler, R., Mahmud, M., Vassanelli, S.: High resolution cortical imaging using electrolyte–(metal)–oxide–semiconductor field effect transistors. In: 5th Intl. IEEE -EMBS Conf. on Neural Eng., pp. 269–272. IEEE Press, New York (2011)
12. Vassanelli, S., Mahmud, M., Girardi, S., Maschietto, M.: On the Way to Large–Scale and High–Resolution Brain–Chip Interfacing. Cogn. Comput. 4(1), 71–81 (2012)
13. MEA Meeting 2010, http://www.nmi1.de/meameeting2010/index.php
14. Berdondini, L., et al.: A microelectrode array (MEA) integrated with clustering structures for investigating in vitro neurodynamics in confined interconnected subpopulations of neurons. Sens. Actuat. B: Chem. 114, 530–541 (2006)
15. Berdondini, L., et al.: Extracellular recordings from high density microelectrode arrays coupled to dissociated cortical neuronal cultures. J. Neurosci. Meth. 177, 386–396 (2009)
16. Potter, S.M., Wagenaar, D.A., DeMarse, T.B.: Closing the loop: stimulation feedback systems for embodied MEA cultures. In: Taketani, M., Baudry, M. (eds.) Advances in Network Electrophysiology: Using Multi–Electrodes–Arrays, pp. 215–242. Springer, New York (2005)
17. Fromherz, P.: Joining ionics and electronics: semiconductor chips with ion channels, nerve cells, and brain tissue. In: 2005 IEEE International Solid–State Circuits Conference (Tech. Dig. ISSCC), pp. 76–77. IEEE Press, New York (2005)
18. Stangl, C., Fromherz, P.: Neuronal field potential in acute hippocampus slice recorded with transistor and micropipette electrode. Eur. J. Neurosci. 27, 958–964 (2008)
19. Imfeld, K., et al.: Large–scale, high–resolution data acquisition system for extracellular recording of electrophysiological activity. IEEE T. Bio.–Med. Eng. 55(8), 2064–2072 (2008)
20. Frey, U., et al.: Switch–matrix–based high–density microelectrode array in CMOS technology. IEEE J. Solid–St. Circ. 45(2), 467–482 (2010)
21. Eversmann, B., et al.: A 128 × 128 CMOS biosensor array for extracellular recording of neural activity. IEEE J. Solid–St. Circ. 38(12), 2306–2317 (2003)
22. Eversmann, B., Lambacher, A., Gerling, G., Kunze, A.: A neural tissue interfacing chip for in–vitro applications with 32 k recording / stimulation channels on an active area of 2.6 mm^2. In: 37th Solid–State Circuits Conference (ESSCIRC), pp. 211–214. IEEE Press, New York (2011)

23. Jones, K.E., Campbell, P.K., Normann, R.A.: A glass/silicon composite intracortical electrode array. Ann. Biomed. Eng. 20, 423–437 (1992)
24. Kipke, D.R., Vetter, R.J., Williams, J.C., Hetke, J.F.: Silicon–substrate intracortical microelectrode arrays for long–term recording of neuronal spike activity in cerebral cortex. IEEE T. Neur. Sys. Reh. 11(2), 151–155 (2003)
25. Lee, J., Rhew, H.G., Kipke, D.R., Flynn, M.P.: A 64 Channel Programmable Closed–Loop Neurostimulator With 8 Channel Neural Amplifier and Logarithmic ADC. IEEE J. Solid–St. Circ. 45(9), 1935–1945 (2010)
26. Azin, M., Guggenmos, D.J., Barbay, S., Nudo, R.J., Mohseni, P.: A BatteryPowered Activity–Dependent Intracortical Microstimulation IC for Brain–Machine–Brain Interface. IEEE J. Solid–St. Circ. 46(4), 731–745 (2011)
27. Venkatraman, S., et al.: In Vitro and In Vivo Evaluation of PEDOT Microelectrodes for Neural Stimulation and Recording. IEEE T. Neur. Sys. Reh. 19(3), 307–316 (2011)
28. Buzsaki, G.: Large–scale recording of neuronal ensembles. Nat. Neurosci. 7(5), 446–451 (2004)
29. Prochazka, A., Mushahwar, V.K., McCreery, D.: Neuralprostheses. J. Physiol. 533(pt. 1), 99–109 (2001)
30. Quiroga, R.Q., Nadasdy, Z., Ben–Shaul, Y.: Unsupervised spike detection and sorting with wavelets and superparamagnetic clustering. Neural Comput. 16(8), 1661–1687 (2004)
31. Kwon, K.Y., Eldawlatly, S., Oweiss, K.G.: NeuroQuest: A comprehensive analysis tool for extracellular neural ensemble recordings. J. Neurosci. Meth. 204(1), 189–201 (2012)
32. Delorme, A., Makeig, S.: EEGLAB: an open source toolbox for analysis of single–trial EEG dynamics including independent component analysis. J. Neurosci. Meth. 134(1), 9–21 (2004)
33. Bokil, H.S., Andrews, P., Kulkarni, J.E., Mehta, S., Mitra, P.P.: Chronux: A platform for analyzing neural signals. J. Neurosci. Meth. 192, 146–151 (2010)
34. Cui, J., Xu, L., Bressler, S.L., Ding, M., Liang, H.: BSMART: A Matlab/C toolbox for analysis of multichannel neural time series. Neural Net. 21(8), 1094–1104 (2008)
35. Egert, U., et al.: MEA–Tools: an open source toolbox for the analysis of multielectrode data with Matlab. J. Neurosci. Meth. 117(1), 33–42 (2002)
36. Gunay, C., et al.: Database analysis of simulated and recorded electrophysiological datasets with PANDORAs toolbox. Neuroinformatics 7(2), 93–111 (2009)
37. Huang, Y., et al.: An integrative analysis platform for multiple neural spike train data. J. Neurosci. Meth. 172(2), 303–311 (2008)
38. Magri, C., Whittingstall, K., Singh, V., Logothetis, N., Panzeri, S.: A toolbox for the fast information analysis of multiple–site LFP, EEG and spike train recordings. BMC Neurosci. 10(1), 81 (2009)
39. Vato, A., et al.: Spike manager: a new tool for spontaneous and evoked neuronal networks activity characterization. Neurocomputing 58-60, 1153–1161 (2004)
40. Versace, M., Ames, H., Lveill, J., Fortenberry, B., Gorchetchnikov, A.: KInNeSS: a modular framework for computational neuroscience. Neuroinf. 6(4), 291–309 (2008)
41. Lidierth, M.: sigTOOL:A Matlab-based environment for sharing laboratory developed software to analyze biological signals. J. Neurosci. Meth. 178, 188–196 (2009)
42. Meier, R., Egert, U., Aertsen, A., Nawrot, M.P.: FIND–A unified framework for neural data analysis. Neural Networks 21(8), 1085–1093 (2008)
43. Mahmud, M., Bertoldo, A., Girardi, S., Maschietto, M., Vassanelli, S.: SigMate: A MATLAB–based automated tool for extracellular neuronal signal processing and analysis. J. Nerusci. Meth. 207, 97–112 (2012)

44. Mahmud, M., Bertoldo, A., Girardi, S., Maschietto, M., Vassanelli, S.: SigMate: a Matlab–based neuronal signal processing tool. In: 32nd Intl. Conf. of IEEE EMBS, pp. 1352–1355. IEEE Press, New York (2010)

45. Mahmud, M., et al.: SigMate: A Comprehensive Software Package for Extracellular Neuronal Signal Processing and Analysis. In: 5th Intl. Conf. on Neural Eng., pp. 88–91. IEEE Press, New York (2011)

46. Weis, R., Muller, B., Fromherz, P.: Neuron adhesion on a silicon chip probed by an array of field–effect transistors. Phys. Rev. Lett. 76(2), 327–330 (1996)

47. Schmidtner, M., Fromherz, P.: Functional Na+ channels in cell adhesion probed by transistor recording. Biophys. J. 90, 183–189 (2006)

48. Lambacher, A., et al.: Identifying Firing Mammalian Neurons in Networks with High–Resolution Multi–Transistor Array (MTA). Appl. Phys. A. 102, 1–11 (2011)

49. Felderer, F., Fromherz, P.: Transistor needle chip for recording in brain tissue. App. Phys. A. 104, 1–6 (2011)

50. Swanson, L.W.: Brain Maps: Structure of the Rat Brain. Academic, London (2003)

51. Mahmud, M., Girardi, S., Maschietto, M., Pasqualotto, E., Vassanelli, S.: An automated method to determine angular preferentiality using LFPs recorded from rat barrel cortex by brain–chip interface under mechanical whisker stimulation. In: 33rd Intl. Conf. of IEEE EMBS, pp. 2307–2310. IEEE Press, New York (2011)

52. Maschietto, M., et al.: Local field potentials recording from the rat brain cortex with transistor needle chips (unpublished)

53. Hofstotter, C., et al.: The Cerebellum chip: an analog VLSI Implementation of a Cerebellar Model of Classical Conditioning. In: Saul, L.K., Weiss, Y., Bottou, L. (eds.) Advances in Neural Information Processing Systems, pp. 577–584. MIT Press, Cambridge (2005)

54. Hofstotter, C., Mintz, M., Verschure, P.F.M.J.: The cerebellum in action: a simulation and robotics study. Eur. J. Neurosci. 16, 1361–1376 (2002)

55. Vershure, P.F.M.J., Mintz, M.: A real–time model of the cerebellar circuitry underlying classical conditioning: A combined simulation and robotics study. Neurocomputing 38-40, 1019–1024 (2001)

56. Liu, S.C., Delbruck, T.: Neuromorphic sensory systems. Curr. Opin. Neurobiol. 20(2), 288–295 (2010)

57. Wen, B., Boahen, K.: A silicon cochlea with active coupling. IEEE T. Biomed. Circ. S. 3(6), 444–455 (2009)

58. Heming, E.A., Choo, R., Davies, J.N., Kiss, Z.H.T.: Designing a thalamic somatosensory neural prosthesis: Consistency and persistence of percepts evoked by electrical stimulation. IEEE T. Neur. Sys. Reh. 19(5), 477–482 (2011)

59. Ahrens, K.F., Kleinfeld, D.: Current flow in vibrissa motor cortex can phase-lock with exploratory rhythmic whisking in rat. J. Neurophysiol. 92, 1700–1707 (2004)

Crayfish Inspired Representation of Space via Haptic Memory in a Simulated Robotic Agent

Stephen G. Volz[1], Jennifer Basil[2,3], and Frank W. Grasso[1,3]

[1] BioMimetic and Cognitive Robotics Laboratory, Brooklyn College, Brooklyn, NY
[2] Biology Department, Brooklyn College, Brooklyn, NY
[3] City University of NY Graduate Center, NY, NY
{Stephen.gv,fwgrasso}@gmail.com, jbasil@brooklyn.cuny.edu

Abstract. Some species of crayfish can learn and remember environmental features by actively collecting spatial information with their antennae. We were able to qualitatively reproduce features of crayfish exploratory behavior by incorporating into a simulated robot a mapping strategy using cross-correlation of sensed environmental features (wall discontinuities). Our model collects environmental information along one continuous surface (in one dimension) to produce a representation of a two-dimensional space. The simulated robotic model can use this information to discriminate between familiar and novel environments and shows recognition of previously explored environnments. Our results support the hypothesis that crayfish can collect and use spatial information gathered haptically while exploring. Our model also predicts features of this spatial exploration strategy that can be subsequently tested in crayfish, allowing us to modify our model accordingly.

Keywords: Robot crayfish biomimetics tactile haptic thigmotactic spatial memory navigation.

1 Introduction

A variety of animals, including humans, can explore and learn about their environments through the use of active haptic sensing mechanisms. Information that is collected through active touch, (i.e.., the movement of the sensor on an appendage or through the body motion to direct the stream of input and select the information provided to the sense organ). This haptic information obtained through a sensory-motor loop, can be learned, and stored in memory and for use guiding decisions later in an animal's life [1-6]. Animals are known to use a variety of mechanisms to collect, learn, and subsequently use tactile information to solve spatial orientation problems [2, 5, 7] and to recognize behaviorally relevant stimuli in their environment [4, 6-8]. Freshwater crayfish are adept at collecting and using tactile information from their environment, as their sight is limited and they are active at night in the turbid waters where they live [2, 9, 10]. Crayfish gather tactile information, in part, with their long and flexible second antennae, which detect a number of features of objects they contact -- such as distance, orientation, and texture [10]. In nature, they are likely to rely

T.J. Prescott et al. (Eds.): Living Machines 2012, LNAI 7375, pp. 286–297, 2012.

heavily on their second antennae to navigate to and from shelter on foraging trips or to avoid unintended encounters with predators or conspecifics [11].

In 2000, Basil and Sandeman [2] directly tested the ability of the crayfish *Cherax destructor* to learn and remember features of an environment through the use of haptic information alone. Blindfolded crayfish were introduced once a day for 40 minutes to a simple rectangular arena, and their exploratory activity was recorded each day for 4 days. Body lengths traveled per day was the measure of their exploratory zeal. Animals in general tend to explore a familiar area less and less over time, a simple form of latent (non-rewarded) learning or habituation. This was the case for the crayfishes; they habituated to their spatial environment, decreasing exploration after one exposure, an effect that persisted for the following two days. Exploration included both active antennal sweeps at distinct spatial points and also continuous antennal trailing against the walls as they walked. Since crayfishes are thigmotactic (have a preference for contact with walls as they move), two small walls were then added to the arena. When placed into this novel configuration, the crayfishes increased exploration markedly (dishabituation). This change required that they recall the original configuration learned in the previous 4 trials, detect that it had changed in the intervening 24 hours, and then explore the new environment more to update this recalled information [2] (results reproduced in Figure 1). On the whole, these results suggest that crayfish form and use some kind of internal representation of the space they are exploring. In these studies, antennal-nerve ablations abolished habituation in crayfishes, indicating that the antennal nerve mediated the source of information for the habituation [2]. These results form the foundation of the studies reported here.

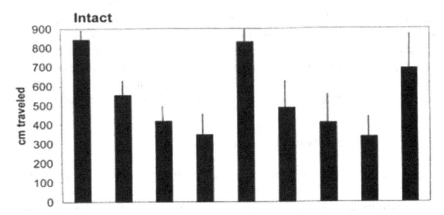

Fig. 1. Results from Basil and Sandeman's 2000 study. Note the reduction in exploration across the first four trials, and the spike in activity on the fifth, which was in a novel environment.

To explore potential mechanisms underlying this ability, we developed and tested a model of haptic exploration that could successfully replicate the behavior of the crayfish. The model considers only features of the environment that can be perceived haptically. The model is parsimonious in that we make minimal assumptions about what blindfolded animals in Basil & Sandeman's study perceived: the *lengths of the walls* punctuated by the *corners* of the testing environments. This approach is simple and, importantly, is supported by documented "step counting" mechanisms common to arthropods in nature (below). Geometry of the corners and walls of the environment can be directly perceived via exploration of the surfaces by one or both of the second antennae. The lengths of continuous wall segments can be measured by a crayfish summing the number of steps (of a known length) taken while in contact with the wall surface. This kind of "step counting," or odometry, is long known to be an integral mechanism of navigation in arthropods such as desert ants, fiddler crabs, and lobsters using a path-integration navigation strategy [12-14]. In desert ants their path-integration strategies provide highly accurate distance measures (odometry).

To better understand the integrative processes behind recognition of an environment using tactile features, we constructed a computational model that relies on a discrete time-sequence correlation algorithm to identify patterns in streams of haptic/tactile data that are known to be available to the crayfish. The model generates and tests hypotheses about possible mechanisms of tactile learning and spatial memory in crayfishes. This model takes uses its tactile sensory input to construct a representation of its environment. This stored representation can be reconstructed for comparison with the sensory input gathered from tactile exploration. In this, and perhaps uniquely arthropod case a one-dimensional time series of distances that are measured via proprioceptive odometry that are punctuated by the detection of features perceived by touch, such as obstacles or corners forms the constructed representation of the 2dimensional space the agent navigates. The aim of the study presented here was to provide evidence confirming or excluding the involvement of memory as an explanation for the behavior observed in the Basil and Sandeman study.

Our results may usefully inspire new types of navigation algorithms for use in autonomous vehicles, which capture the crayfishes' ability to solve complex spatial tasks with non-visual sensory input. This biomimetic approach allows us to build hypotheses about the internal cognitive mechanisms guiding behavior, test their utility in models, and generate predictions which may be tested in living animals. A natural byproduct of the biomimetic approach is the production and testing of biologically inspired algorithms, which may improve our current methods of robotic control.

2   Robotic Simulation of Crayfish Behavior

Our model constructed its representation each time the agent was placed in the environment and allowed to explore. Because the environment was two-dimensional and the tactile sensors provided a one-dimensional stream (time series) of sensory input memory of past sensory inputs was used a non-linear process to construct the representation. Essentially, it stored the series of distances traveled between significant events (turns) regardless of how long the interval between the events. A series of distances was collected and stored between significant events.

The representation it constructed was a series of distance traveled between turns that were imposed by the thigmotactic behavior of the agent. Autocorrelation was used to determine when the series repeated itself and a circuit of the maze was completed. The repeated sequences was then stored in memory for reference. Later introductions to the same or a different environment would involve either recognition of the present environment as an instance of a previously experienced one or its storage as a novel one. Recognition used a cross-correlation method. While the cross-correlation performed was mathematically linear, the input to it was a non-linear time-series, a set of distances traveled between events arranged in chronological order.

To test its viability we developed and implemented our theoretical model in a simulated robotic agent. The agent used thigmotaxis, an inherently haptic mechanism, to explore its environment. Thigmotaxis, as implemented in our model, is haptic and is the result of a sensory-motor loop connecting tactile sensory information, and the self generated movements of the agent's body. The agent maintains contact with the wall at a constant distance, by modifying its bearing based on its last measured distance from that wall. This feedback loop between perception and motor output results in the agent holding a relatively constant distance from the wall it is in sensor contact with, while proceeding along the surface of that wall. This exploration continues until one of two correlational processes identified the environment as novel or familiar: 1) The agent correlated features from its ongoing sensory input stream to a time-shifted copy of that stream itself. This cross-correlation mechanism simultaneously 2) correlated the current trace to stored representations of all environments it had previously "identified" by the same mechanism. In this way, the agent could "recognize" familiar environments.

The agent was run in simulations of a biomimetically-scaled model of the arena used in [2]. The size of the environment was scaled to match the length of the robot analog of antennae. The effective sensor range (a simulated infra-red sensor on a robot [15]) on the agent was scaled to proportionately match the second-antenna length of a real-world crayfish. This provided the agent with properly scaled input that paralleled that of crayfish, allowing us to make predictions about crayfishes in the real world. We then evaluated the simulated agent with its exploratory time in virtual environments scaled and ordered as in [2] and we compared the performance of the simulated agent to that of crayfishes.

3   Methods

3.1   The Model

The simulation of crayfish exploratory behavior was implemented in a custom C++ program designed to run in a Khepera II™ (K-team Mobile Robotics, Zurich, Switzerland) mobile robotic agent. This program implemented three behaviors (FIND WALL, FOLLOW WALL, and GUIDED TURN) and also a memory mechanism that monitored the progression of sensory states as the three behaviors were executed.

The agent began exploration in the middle of the arena. At the start of a trial control was taken by the FIND WALL behavior, which consisted of movement in a straight line until the agent detected an obstacle or wall with either its left or right infrared (IR) sensor. Once IR contact was made, the agent engaged in the FOLLOW

WALL behavior: keeping one robot IR sensor in contact with the wall, maintaining an IR signal near a set threshold, while also moving forward. This caused the agent to move in parallel to the contacted wall at a fixed distance, as freely moving crayfish do. The sensor that maintained contact was the same sensor that made first contact with the wall at the transition from FIND WALL to FOLLOW WALL behavior.

The transition to GUIDED TURN behavior was triggered by the detection of a significant discontinuity in the wall being followed (Figure 2). If the agent reached a closed corner, both IR sensors would be triggered. If the agent encountered an open corner, the wall-following behavior would implement a large magnitude turn. Upon encountering a corner, the agent turned in place, away from the wall, until reestablishing IR contact at the proper fixed distance with the next wall segment (Figure 2). When it reacquired contact with the wall at the appropriate distance, it returned to FOLLOW WALL behavior.

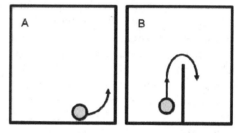

Fig. 2. Diagram the agent making a: A) 'closed' turn, when an additional wall is detected. B) 'open' turn, when the tracked wall falls away

As the agent proceeded, it recorded the intervals it had traveled using motor encoders. Motor encoders are on-board sensors typical of mobile robots that detect how much rotation has been produced by each of the agent's wheels. At each corner, ('event') the distance from the previous corner was computed and passed to the memory system for storage. In this way, the agent obtained information to construct a representation of the space it was exploring based on what was essentially the one-dimensional sequence of lengths that resulted from wall-following until facing a discontinuity. These measurements serve as an analog to odometry calculations possibly carried out in the thoracic ganglion of crayfish. This process constructs a vector each component of which corresponds to the length of a certain wall in the arena.

Exploratory behavior continued until terminated by one of two simultaneous processes. The primary process, "Identification," identifies a novel environment by recognizing that a pattern is repeated in the following way. The agent correlates the first half of the series of stored wall lengths to the second half, and compares the obtained r value to an identification threshold (.99). When an r value above threshold is obtained, identification of two completed circuits of the perimeter of the environment has been achieved. If "Identification" occurs, the program stores the pattern as an internal representation, computed as the average of the two halves of the stored pattern of that environment for later recognition. We refer to this stored pattern as a Reference Memory Trace (RMT), which is maintained between presentations of environments. A summary of this algorithm can be seen in figures 3 and 4.

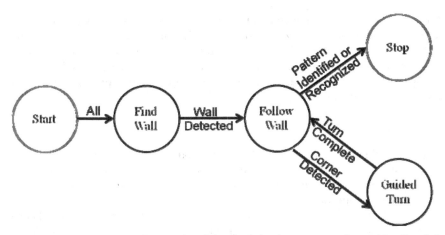

Fig. 3. A finite state acceptor diagram describing the behaviors our agent is capable of, and the transition rules connecting these behaviors

The second cause for the agent to terminate a trial is "Recognition." This condition is met when the agent correlates the last n of recent series of perceived wall lengths with each of the RMTs created during earlier trials (n = # of wall lengths in a given RMT). As in Identification, the agent performs a Recognition test each time the agent traverses a new wall length. A correlation above the threshold signals recognition of a previously encountered environment.

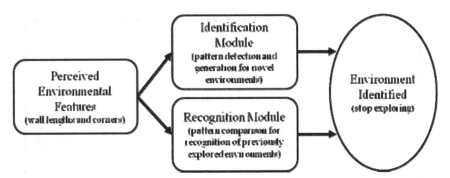

Fig. 4. Diagram of the connectivity of the memory modules in our model. Note that these processes are asynchronous, conditional on the existence and detection of features that are discovered by the agent and not set *a-priori*.

Recognition is a faster process than Identification. Because Recognition operates on a single circuit of the environment, a previously identified environment will always produce a significant r value after less exploration than the Identification process.

3.2   Materials

Simulations were carried out using Webots™ (Cyberbotics, Zuirch, Switzerland) version 4.6.1 robotic simulation software. Webots allows us to have complete control over the environment and physical forces to which the agent is exposed, but also allows us to model any robotic platform desired to implement the algorithm. For this study we chose to implement our algorithm in a simulated Khepera II mobile robotic platform, as we use Khepera IIs in our laboratory. Our model utilized two of the Khepera's front-mounted, side-facing IR sensors as analogs to the crayfish's tactile appendages, the second antennae. The simulated Khepera II was programmed with the controller described above.

3.3   Procedure

A simulated Khepera II, programmed with our model (above), experienced four exposures to each of two distinct environments, followed by a final "probe" exposure to a third configuration of the test environment. Any RMT's formed during an exposure were left intact between exposures. The series of training and testing environments consisted of rectangles of equal perimeter and dimensions. The difference between the training and testing environments was the introduction of two obstacles into the second environment: two short walls of equal length, 1/3 down the length of each of the long walls of the arena, projected into the center of the environment. These obstacles were relocated to the short walls of the arena in the third environment (Figure 5). To properly scale the size of the simulated arenas, we calculated and used the ratio of the typical second antennae length of a crayfish and the functional length of a Khepera II's IR sensor. The simulated arenas were then given dimensions that were the same number of "sensor lengths" as the arena used by Basil and Sandeman in their behavioral studies on crayfish.

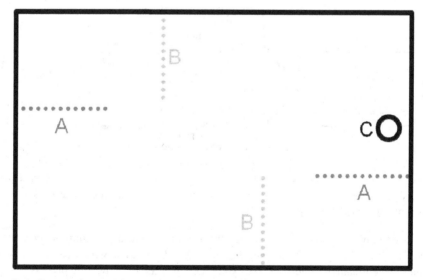

Fig. 5. A diagram summarizing the configuration of the arenas used in our simulations. A) The obstacles used during trials 5-8 B) the obstacles used during trials 9-10 C) the starting position of the agent during all trials. The agent was oriented at the start of a trail to face toward the wall on the right side of the figure.

An exposure, or trial, continued for 40 simulated minutes. The agent began each trial in the exact same location within the arena, at the middle of one of the arena's short walls, facing toward the center of the arena (Figure 5), consistent with the methods applied in the Basil and Sandman study. Upon the start of a trial, the agent would proceed to explore the environment until it either Identified or Recognized that environment. At the time of either of these events, the agent would cease its exploratory activity and remain motionless until the time for that trial had expired.

4   Results

The simulated robot explored the environment less and less over trials, paralleling the behavior of crayfish. With experience in an environment, less motor activity was required for the agent to recognize an environment it had previously explored (Figure 4, t(3)=2584 p< 0.05). Moving the agent to a new environment resulted in a marked increase in exploration. Exploration of a novel environment after prior habituation elsewhere was significantly greater than that required to identify a familiar environment (t(3)= -2934 p< 0.05)(dishabituation). The large t values reported here result from the extremely low variance across replications within a simulation environment. Overall, our results resemble those in crayfish[2]. To measure how similar our simulated robot's behavior was to that of crayfish, we normalized average exploration times by the simulated Khepera and the crayfish relative to the largest average exploration time in each study. A Pearson r was then generated using the two sets of data, (r(8)= 0.85 p < .05).

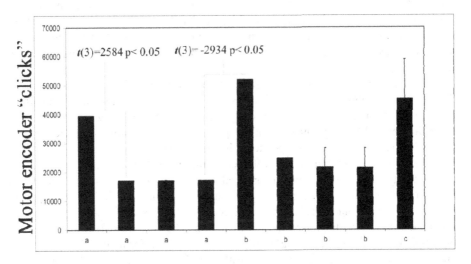

Fig. 6. Simulation results. Y-axis represents average distance traveled by the agent during each trial. X axis represents sequence of environments (environments a, b, and c). Error bars are S.E.M in the trials where variation was seen.

Surprisingly, in some replicates, total motor activity continued to decease after the second and third presentations. Thus even in one environment, there was continuous decline of motor activity across presentations and an increase in the variability of total motor activity with experience (Figure 6). This was not expected, as after the agent had formed a representation of the environment following its initial exposure, we predicted that it would then require half the amount of exploration to Recognize that environment than it did to first generate Identification. This pattern of activity revealed that identification and recognition mechanisms can be confused in symmetrical environments. The agent erroneously formed a new representation of the environment *corresponding to only half of the original environment* because the halves were symmetrical.

5   Discussion

Computational modeling is a method often used to describe cognitive and neural mechanisms that may reproduce animal behavior. Here we implemented a computational model that simulates how the crayfish, *Cherax destructor*, might habituate to novel, two-dimensional environments using its haptic sense.

We compared our computational model to the performance of *in vivo* subjects to explore hypothetical mechanisms underlying spatial learning via tactile signals collected from the environment. We used a model that we considered one of simplest possible to reproduce the behavior of crayfish described by Basil and Sandeman. The qualitative agreement between the pattern of exploration by our simulated robot and the crayfish is striking. Our simulation habituated to familiar environments, exploring less and less over time, and showed a sharp increase in activity when the environment was changed. Our model currently provides one of perhaps many possible mechanisms that might be guiding this behavior in real animals. The alternatives generated by models, however, provide predictions that can then be tested both in simulation and in real animals.

A similar correlational mechanism has been implemented previously in a Rug Warrior Pro™ robot (AAA Robotics, Cambridge MA) navigating in a real-world rectangular arena. The robot used a cross-correlation of the ongoing sequence of distances and obstacles rather than the half-sequence used in this study [15]. (Remember that while the cross-correlation operator is linear the input we used was sequential but non-linear in time). Despite these differences, the qualitative matching between the robot and our model underscores the general viability of cross correlation as a means to learn and remember environments using haptic information. Failures to replicate perfectly the exploratory behavior of crayfishes, indicates the need to augment or supplement the algorithms used in our simulation. Thus the consistent differences in the behavior of the agent compared to the crayfish are informative. One important difference arises from the symmetrical pattern in the sequence of wall lengths that defined our simulated experimental arena. From the perspective of a wall-following crayfish, the arena consists of the same sequence of wall lengths repeated twice. This pattern led our agent to occasionally identify half of the pattern of wall lengths as a complete circuit. The agent's representation of the environment would be shorter and require less exploration because it was easier to match. This does not appear to be the case in real crayfish. The current model predicts that symmetry would confuse an animal using the algorithm we have proposed, and this difference leads us to predictions that are easily testable *in vivo*. It could be that the direction of movements during exploration, in combination with wall and obstacle detection, helps to discriminate between what seem like symmetrical features otherwise. Or, perhaps crayfishes keep track of total rotation in space, coupling real-time tactile features with net angular displacement to discriminate between features that would be similar in a one-dimensional stream of tactile information alone.

Another component of the simulated agent's performance differed from the behavior of crayfishes: amount of exploration the agent required to identify previously explored environments. Here, the agent required exactly half the amount of exploration to recognize an environment than it did to initially form a representation of that environment. The agent only had to perceive the pattern of wall lengths that defined the environment a single time to produce an adequate match. Crayfishes, however, show a linear decline in exploratory activity across all subsequent visits to a previously explored environment, exploring proportionately more than our model predicts. Crayfish naturally live in an unpredictable environment. 'Over-sampling' features of their environment may be an adaptation to real-world variation at short timescales, such as changes in water depth, shifting of substrate, deposition of rocks, leaves, etc.

These simulations raise intriguing questions: do crayfishes use antennal contacts to punctuate the integration of steps they have taken (distance) or do they gate when they use their antennae based upon distance traveled? Or, possibly, a combination of both? (see [16] this meeting) Are distances traveled and number of walls encountered actually what crayfish are sensing and using? These alternatives could be explored by affine transformation of their space: testing animals in two rectangular arenas (same geometry) but of different sizes (different metrics). If crayfishes are relying upon metrics (counting steps) to determine distances at which they use their antennae, a prediction would be that those relying upon step counting to gate antennal behavior would behave differently in the two spaces: one has longer walls and would require more steps (and antennal touches) than the other to explore. Crayfishes relying upon antennal touches to gate the integration of distances traveled, for instance integrating steps already taken at each corner, would not be as affected by the metric changes in their environment. Differences in behavior in these spaces would also support the notion that the length of the walls (a distance vector) is indeed being collected and retained by the animals.

Overall, the qualitative similarity in behavior of our modeled agent and crayfishes reinforces that this is a viable class of model to study the acquisition and recognition of spatial environments based on a combination of tactile and proprioceptive (step counts) information in crayfish and perhaps other animals. Differences between the model and crayfish provide alternatives for future empirical investigation with real animals to validate or elaborate features of the model. This "feedback loop" between computational and behavioral study allows us to test competing hypotheses in very different frameworks. Crayfish exploring an environment via the haptic sense learn and navigate through space with sparse sensory input, and with limited neural "hardware" to carry out computations. Successful navigation in environments where visual information is limited is a desirable property to implement in artificial systems. By exploiting this feedback loop between living and simulated animals it is our aim to develop functional algorithms for use in autonomous navigation that will be more effective under the conditions where other navigational systems such as GPS and visually-guided frameworks are inoperative.

This research project is an incremental step towards understanding how crayfish are able to understand their 2-dimensional environment from a time-series of point-contacts. The crayfish with its navigation of dark and murky environments is an existence proof that it is possible to represent space through tactile input, in the absence of visual information. Further advances are possible in closing the gap between the performance of the simulated agents and the crayfish.

Acknowledgments. We thank the BCR lab staff for helpful discussions and technical assistance throughout the course of this project. In particular we are grateful to Kamil Klosklowski for his help with writing the C-programs.

References

1. Li, H., Listeman, L.R., Doshi, D., Cooper, R.L.: Heart rate measures in blind cave crayfish during enviromental disturbances and social interactions. Comparative Biochemistry and Physiology 127, 55–70 (2000)
2. Basil, J., Sandeman, D.: Crayfish (cherax destructor) use tactile cues to detect and learn topographical changes in their environment. Ethology 106, 247–259 (2000)
3. Harris, J.A., Harris, I.M., Diamond, M.E.: The topography of tactile learning in humans. J. Neurosci. 21(3), 1056–1061 (2001)
4. Harris, J.A., Petersen, R.S., Diamond, M.E.: Distribution of tactile learning and its neural basis. Proc. Natl. Acad. Sci. USA 96(13), 7587–7591 (1999)
5. Okada, J., Toh, Y.: Active tactile sensing for localization of objects by the cockroach antenna. J. Comp. Physiol. A Neuroethol. Sens. Neural Behav. Physiol. 192(7), 715–726 (2006)
6. Simone-Finstrom, M., Gardner, J., Spivak, M.: Tactile learning in resin foraging honeybees. Behavioral Ecology and Sociobiology 64(10), 1609–1617 (2010)
7. Li, H., Cooper, R.L.: Spatial familiarity in the blind cave crayfish, Orconectes australis packardi. Crustaceana 74(5), 417–433 (2001)
8. McMahon, A., Patullo, B.W., Macmillan, D.L.: Exploration in a T-Maze by the Crayfish Cherax destructor Suggests Bilateral Comparison of Antennal Tactile Information. The Biological Bulletin 208(3), 183–188 (2005)
9. Caine, E.A.: Comparative Ecology of Epigean and Hypogean Crayfish (Crustacea: Cambaridae) from Northwestern Florida. American Midland Naturalist 99(2), 315–329 (1978)
10. Sandeman, D.C.: Physical properties, sensory receptors and tactile reflexes of the antenna of the Australian freshwater crayfish Cherax destructor. Journal of Experimental Biology (141), 197–217 (1989)
11. Barbaresi, S., Gherardi, F.: Experimental evidence for homing in the red swamp crayfish. Procambarus Clarkii, 1145–1153 (2006)

12. Boles, L.C., Lohmann, K.J.: True navigation and magnetic maps in spiny lobsters. Nature 421(6918), 60–63 (2003)
13. Layne, J.E., Barnes, W.J., Duncan, L.M.: Mechanisms of homing in the fiddler crab Uca rapax. 2. Information sources and frame of reference for a path integration system. J. Exp. Biol. 206(pt. 24), 4425-4442 (2003)
14. Wehner, R.: Desert ant navigation: how miniature brains solve complex tasks. Journal of Comparative Physiology A: Neuroethology, Sensory, Neural, and Behavioral Physiology 189(8), 579–588 (2003)
15. Appleman, G., Basil, J.A., Grasso, F.W.: Simulation of crayfish spatial exploration with a mobile robot. Biological Bulletin 195(2), 249 (1998)
16. Grasso, F.W., Evans, M., Basil, J., Prescott, T.J.: Toward a Fusion Model of Feature and Spatial Tactile Memory in the Crayfish *Cherax destructor*. In: Prescott, T.J., et al. (eds.) Living Machines 2012. LNCS (LNAI), vol. 7375, pp. 352–354. Springer, Heidelberg (2012)

Parallel Implementation of Instinctual and Learning Neural Mechanisms in a Simulated Mobile Robot

Briana Young[1], Stefano Ghirlanda[2,3], and Frank W. Grasso[1,2]

[1] BioMimetic and Cognitive Robotics Laboratory, Brooklyn College, Brooklyn, NY, USA
[2] Department of Psychology, Brooklyn College, Brooklyn, NY, USA
[3] Centre for the Study of Cultural Evolution, Stockholm University
{fwgrasso,briana.m.young,dr.ghirlanda}@gmail.com

Abstract. The question of how biological learning and instinctive neural mechanisms interact with each other in the course of development to produce novel, adaptive behaviors was explored via a robotic simulation. Instinctive behavior in the agent was implemented in a hard-wired network which produced obstacle avoidance. Phototactic behavior was produced in two serially connected plastic layers. A self-organizing feature map was combined with a reinforcement learning layer to produce a learning network. The reinforcement came from an internally generated signal. Both the adaptive and fixed networks supplied motor control signals to the robot motors. The sizes of the self-organizing layer, reinforcement layer, and the complexity of the environment were varied and effects on robot phototactic efficiency and accuracy in the mature networks were measured. A significant interaction of the three independent variables was found, supporting the idea that organisms evolve distinct combinations of instinctive and plastic neural mechanisms which are tailored to the demands of the environment in which their species evolved.

Keywords: learning neural network instinct robots evolution.

1 Introduction

One property of living systems that is desirable in "living machines" as artifacts of human technology is the ability to self-organize behavior from experience. The fact that many, if not all, behavioral processes observed in biological organisms result from a combination of inherited and acquired mechanisms has not received as much explicit attention in cybernetic research as the extremes of fixed architectures or total plasticity. In the research reported here we explore the interaction of "instinctive," behaviors that are performed perfectly in the absence of experience, and the development of plastic behavior in an agent that has adaptive circuits.

Explanations of the nature of the relationship between instinct and learning have historically been offered from radical behaviorist, ethological and computational perspectives. Their explanations have ranged from complete dependence on learning (e.g., radical behaviorism in psychology[1, 2]) to complete independence from

T.J. Prescott et al. (Eds.): Living Machines 2012, LNAI 7375, pp. 298–308, 2012.

learning (e.g., ethology [3, 4]) to determine behavioral outcome. Ethologists did not propose that organisms do not employ or rely on learning mechanisms. Rather, the classic ethologists claimed that instinctive or immutable behavioral mechanisms are hierarchically organized so that environmental input and the internal state of the organism dictate which behavior pattern is activated. This enables the selection of a behavior that is appropriate for a given context without relying on plasticity, or learning mechanisms[3]. This framework has influenced the thinking of researchers in the field of behavior-based robotics [5-7]. Radical behaviorists, in contrast, disregarded genetically inherited behavioral predispositions. They claimed instead that all behavioral responses are stimulus-driven and that any stimulus-response pairing can be learned [2]. Although attempts have been made at an integrated perspective where both hard-wired behavior and learning are important [8-10], psychologists remain largely interested in learning, and ethologists, in behavior with a strong genetic component.

Research since the early ethologists and radical behaviorists has pointed to the idea that, rather than being either instinctive or plastic, neural mechanisms are some combination of the two that have been shaped by evolution to respond to specific environmental demands. Plastic structures are not completely plastic, but rather targeted toward certain environmental features that have persisted over multiple generations of a species [10, 11]. The components of the mechanism that reflect these consistent features that are inherited and therefore more fixed, steering the learning in a specific direction. The adaptive feature of a system, its flexibility, is not completely plastic but canalized in directions that reflect the variable features of the environment.

The counterparts in the computational literature to instinct-driven and learning-driven perspectives come in the form of fixed- and plastic-connection neural networks. A fixed-connection neural network may be used to represent a hard-wired (analogous to a genetically inherited) behavioral mechanism because nothing in the network changes to reflect experience. The motor output or behavior of a fixed connection network will only vary as a result of differing input rather than as a result of change to the internal structure of the network [12]. On the other hand, the connectionist approach employs networks that do change (adapt their structure) as a result of the inputs they processes and are analogous to plastic neural mechanisms that produce adaptive behavior in biological organisms. Underlying learning neural networks is the presumption that behavior can be learned without regard for instinctive or inherited behaviors [13, 14]. The output produced by these networks, which represents the behavior patterns produced in biological organisms, is in some way a reflection of this internal adaptive process. [15].

Some previous studies have made an effort to balance the demands of fixed, predictable network performance and the capacity to adapt behavior in the control of robots and biological systems within the context of action selection [16-18]. The study here takes a related but slightly different approach in attempting to describe the emergence of a novel behavior in a developmental context. Such behavior is produced by an agent that has a well-developed and occasionally competitive fixed behavior in full function during development.

Ethological and psychological research has suggested that instinctive and learned behaviors come into conflict with each other, and that at other times they augment or complement each other [10]. It is potentially these interactions between instinct and learning that give rise to the kind of mechanisms that have both inherited and plastic aspects and represent environment-shaped, or targeted, plasticity. Therefore, a computational approach to understanding how animals resolve such conflicts to behave adaptively is useful. To address the issue of competition/cooperation between instinct and learning, we use a neural architecture composed of fixed and plastic networks interacting to produce behavior in a single robotic agent.

Developmental studies are methodologically challenging and difficult to perform. A natural complement to studies with biological systems is studies with artificial systems, which, while simpler than their biological counterparts, offer opportunities to study the interaction of the inherited and acquired components of behavior under controlled and reproducible conditions. This report describes an abstracted version of this approach using behavior-based robotics and adaptive neural networks.

We designed a controller for our robotic agent that included an instinctive aspect (a fixed-connection neural network) and a plastic or learning aspect (a sequential arrangement of a Kohonen-style self-organizing feature map, or SOM [20, 21], and a reinforcement learning network similar to a Sutton-Barto network [19] which used internally generated reinforcement signals). We hypothesized that if we presented the agent with two types of environments, complex and non-complex, we would see differential demand for internal representation space (in the form of the SOM). Given the task of navigating toward light in a variable environment, we expected to see the agent perform more accurately and efficiently when it had a larger self-organizing layer (analogous to greater plastic neural resources in biological organisms). We expected that in a non-complex environment, the agent would be able to perform accurately and efficiently with a smaller self-organizing layer, or, with limited plastic neural resources.

2   Methods

The simulations were implemented using Webots™ 4.0.27 robotic simulation software (K-team, Zurich), and the Webots simulation environment used to simulate a Khepera II robot and a virtual world. We developed a neural net controller for the robot in C++. One aim of these procedures was to develop these controllers in simulation before we moved on to implement the controller in a physical Khepera robot. Here we report on the results from these simulations.

2.1   Agent Specification

The Webots simulator produces agents that fully simulate the physical properties of the Khepera II robot. The sensor properties possess the correct geometric properties to correspond to the sensors on a physical Khepera robot, and the motors reproduce the kinematics of a Khepera II robot. The control networks described below took inputs from eight light sensors and six infrared (IR) sensors located around the periphery of the agent's body that were oriented to sense in the horizontal plane.

Fig. 1. Diagram of the simulated Khepera. IR sensors are represented by lines; photosensor locations are indicated by white dots.

2.2   Network Design

We used three types of networks in designing a robotic controller that would model the brain of a simple biological organism that possessed both instinctive and plastic neural mechanisms. One of these networks carried out an instinctive function while two plastic networks were connected serially. The "instinctive" network implemented obstacle avoidance by steering the agent away from walls or objects detected with six infra-red proximity (IR) sensors. The function of the learning network was to evolve phototaxis by learning to steer the agent towards light based on input from eight bilaterally placed visible-light sensors (photosensors) and an error signal generated by the difference between light levels at the current and previous time step. Both the outputs of the "instinctive" and plastic networks affected the motor behavior of the agent. Their combined effect on movement was a normalized linear sum which meant that their influence on the movement of the agent could be antagonistic or synergistic.

The activity of the agent was determined by instructions from both the adaptive networks and from the instinctive network. We hypothesized that at times, the instructions from the two networks would be in agreement with each other (the instructions to steer toward the light and away from an obstacle would send similar commands to the simulated agent's wheels) and that sometimes they would contradict each other (output from the instinctive network might steer the agent in one direction while output from the learning layer might steer it in another).

The control network can be thought of as two signal processing arcs converging on two motor-output units which supplied the commands to control the left and right wheels of our differential-drive simulated Khepera robot. The "instinctive" arc used IR sensor inputs, keeping the left and right sides separate to compute activation values for the units in the motor output layer which translated directly into commands sent to the simulated wheels. This process decreased the speed of the motor on the more stimulated side by a factor of 0.25 and increased the speed of the motors on the less stimulated side by a factor of 0.75 of the scaled IR input.

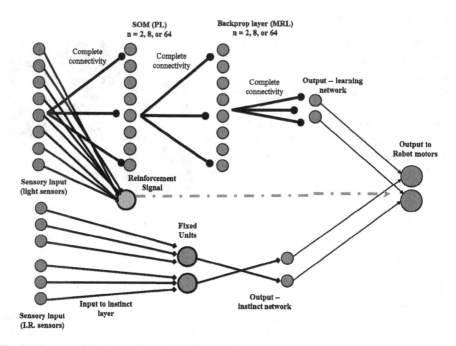

Fig. 2. Diagram of the control network. "Complete connectivity" between two layers indicates that each unit in the first layer is connected to every unit in the second layer.

The instinctive (fixed-connection) network had three layers. A sensory input layer, encoding information about the presence of contact surfaces in the environment, provided input to a "hidden" instinct layer. The instinct layer provided input to a two-unit output layer which encoded the agent's response from the instinctive network.

The adaptive portion of the control system had more parts, was dynamically more complex, and consisted of four layers. A sensory input layer, encoding the level of ambient light from the photosensors, was first. This provided input to a recurrent Kohonen-style network [20, 21] configured to form a self-organizing feature map (SOM) of the patterns of light the agent experienced as it moved through its environment. We refer to this hidden layer as the *perceptual layer* (PL). The third layer received a completely connected input from the perceptual layer and was implemented as a feed-forward reinforcement network [19] similar to a modified form of error back-propagation . We refer to this as the *motor reinforcement layer* (MRL). The MRL fed into a two-unit output layer that encoded the response from the learning network. The two sets of temporary outputs from the learning and instinctive networks were summed to produce the final motor commands that determined the activity of the agent. It was this final, summed set of outputs that was subjected to the error signal. The error signal could have one of two values. It was -1 if the sum of the light sensor readings was less than it was at the previous time step, meaning that the agent had steered away from the light and therefore made the "incorrect" choice. It was 1 if this sum was greater than at the previous time step, indicating that the agent had made the "correct" choice. This error signal was propagated back through the weights connecting the MRL to the learning output layer, and through the weights connecting the PL to the MRL. Since the error signal was restricted to one of two values, weights were always strengthened or weakened as a proportion of their value.

Before we implemented the network controller, the code developed for the perceptual layer and for the motor reinforcement layer was validated. We verified that the perceptual layer was competent to from SOM maps on its own and that the reinforcement motor layer was capable of learning input-output mappings. These tests were run by providing the SOM with a series of numerical training sets that were orthogonal to one another (meaning that they shared no common elements or were at right angles to each other in multi-dimensional state space) and observing the tuning of each unit in the SOM to a unique input set. The ability of the reinforcement layer to converge on the correct outputs was tested by supplying the network with randomly selected and paired input and output patterns. The rationale for this method was that if the network was competent to learn arbitrary associations between input and output sets, it should be capable to learn input-output pairs that followed some logical structure, such as those provided by a simulated environment.

2.3   Simulation Environment

The arena in which the agent was situated consisted of a rectangular space surrounded by four walls. The long walls of the arena were 0.14 units high by 1 unit long, and the short walls were 0.14 by about 0.79. The obstacle was a cylinder 0.03 units high with a radius of 0.08 cm. One of these Webots units of measurement is equal to about 36 cm. On trials when the obstacle was present, it was positioned flat surface-down in the center of the arena.

Fig. 3 The simulated environment with simulated Khepera, obstacle present condition

A light source was positioned in a corner of the arena (centered at Webots' Cartesian coordinates (0.3, -0.4), with an ambient intensity level of 0.25 and a radius of 100). Its starting position was the corner directly opposite from that in which the robot started each trial. The light source in these simulations was a point-source with no physical extent, the intensity of which fell off as an inverse square law. The source strength was set to an intensity that did not exceed the agent's light sensor levels at the starting position.

The obstacle was not opaque to light, so on trials in which it was included, it was detectable by the IR sensors but did not obscure or shadow the light source from the photosensors.

2.4   Simulation Trial Procedure

Before the start of a trial all adaptable network weights (PL and MRL) were re-initialized to a standard set of randomly generated values. A simulation trial was initiated by moving the robot into the standard starting position (Webots coordinates 0.3, 0.4) in a standard starting orientation (5.3 Webots radians). At the start of a trial, the controller was loaded into the agent and execution commenced. At each time step in a trial the sensors were read, their inputs were supplied to the control network, the network generated its outputs, the connection weights were updated, and motor commands from the network output were supplied to the agent motors.

The agent navigated through the arena until the trial met the criteria for either a successful or an unsuccessful trial. A trial was called successful when two or more of the agent's light sensors returned a value of 1 (the maximum possible value for a light sensor) and the light source was considered located. A trial was called unsuccessful if the agent collided with a wall and remained immobile for more than 2,000 time steps, if it remained in a continuous cycling pattern for more than 2,000 time steps, or if the total time of the trial went above 32,034 time steps. This value is the expected number of time steps it would take the agent to locate the light source if it was engaging in a random walk. Of all successful trials, it was the highest obtained time step value and represented the movement of the agent when phototaxis was least active or effective.

After the trial had attained successful or unsuccessful status, the agent was stopped and the number of time steps taken to find the light source was recorded in the case of a successful trial. A score of 32,034 was entered for an unsuccessful trial.

2.5   Experimental Design

We were interested in the degree to which learning capacity affected ability to navigate variable environments. We therefore chose to systematically vary the sizes of the two adaptive layers (PL and MRL) as well as the complexity of the agent's environment. The three independent variables in our experiment were size of the PL, size of the MRL, and the presence or absence of the obstacle (see fig. 3). The three layer sizes we used for both the PL and MRL layers were 2, 8, and 64 units, resulting in a total of 9 possible combinations of layer sizes. Environmental complexity had two experimental conditions: obstacle-present and obstacle-absent. Our dependent measures were the number of time steps required for the agent to locate the light source and the error rate, or number of unsuccessful trials per condition. A total of 23 trials in each of 18 conditions were run.

3   Results

When no obstacle was present, an agent configured with a PL sized 2 and an MRL sized 64 units used, on average, the fewest time steps to locate the light (see Fig. 4a). Thus in an environment free from obstacles, smaller PL sizes and large MRL sizes led to the acquisition of more efficient phototaxis.

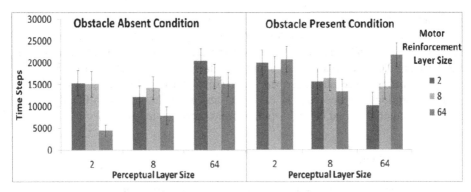

Fig. 4a & 4b. Mean time steps required for the mature agent to locate the light source for each combination of perceptual and reinforcement layer sizes (2, 8 or 64), a. for obstacle absent and b. obstacle present conditions.

When the obstacle was present, an agent configured with a PL sized 64 and a MRL sized 2 used the fewest mean time steps to locate the light (Fig 4b). For environments with obstacles, large PL produced more efficient phototaxis. In this condition, smaller MRL's are sufficient to store adequate motor output patterns.

We found that the patterns in error rate (number of unsuccessful trials per) data were similar to the patterns in efficiency data in that the conditions that resulted in fewer time steps to locate the light also produced more successful trials (Figs. 4 & 5).

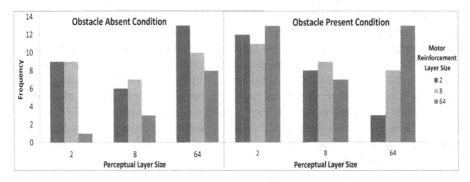

Fig. 5a & 5b. Error rate (number of unsuccessful trials) for each combination of hidden and reinforcement layer sizes (2, 8 or 64), a. obstacle absent b. obstacle present conditions.

The study was designed for analysis with a 3-factor ANOVA to interpret and seek effects of each network type on the time the agent took to reach the light source. However, Levene's test of equality of error variance showed significant violation of the assumption of homogeneity of variance ($F(8, 198)=7.243$, $p < 0.01$ for obstacle-present, $F(8, 198)=14.574$, $p < 0.01$ for obstacle-absent, which prevented us from carrying out the planned ANOVA. Inspection of the data revealed that the source of the violation is the relatively high rate of failed trials, which were scored as a time of 32,034 (see above). This produced a large deviation from the times obtained when the agent successfully located the light source. We therefore analyzed time-step and error rate data with a bootstrapped randomization test comparing the obstacle-absent and obstacle-present conditions (R software version 2.13.1 with the boot library version 1.3.2, using 10,000 re-samplings of simulation data). This analysis confirmed that the qualitative pattern of results described above (small MRL and large PL in the object-present condition, and vice-versa in the object-absent condition) is significant for both time steps required to find the light ($p<0.01$) and success rate ($p<0.01$).

4   Discussion

Our results confirm that the complexity of the environment (obstacle vs. no obstacle) affects which network architecture develops better phototaxic performance, and agree with our *a priori* hypothesis that complex environments require a large PL and simple environments can be navigated by agents with a small PL. We were surprised by the inverse relationship between PL and MRL size in the best-performing conditions. Such a three-way interaction between PL size, MRL size, and environmental complexity is of great interest as it indicates an interaction between instinctual and learned components of behavior, mediated by the environment. While these results might be explained by the differences in the geometry of the two environments in relation to a fixed physical body, they point to the utility of plasticity in matching developed performance to the requirements of the environment. This is a result that has been indicated in theoretical studies of foraging [14].

In the obstacle absent condition, a network with a small PL (2 units) and a large MRL (64 units) was most efficient, suggesting that 2 perceptual units were enough to represent the relevant features of such a simple environment. That the MRL needed to be quite large further suggests that this layer had to make finer discriminations.

When an obstacle was added to the simulated arena, the most efficient network had a PL of 64 units and MRL of 2 units. This agrees with the intuition that a more complex environment requires larger perceptual resources, but it is not immediately clear why a small MRL is best in this condition, Two possibilities, not mutually exclusive, are 1) a large PL layer needs more time to self-organize, leaving less time to adapt MRL connections and favoring a small MRL layer, and 2) a PL that is better able to disambiguate environmental states supports efficient learning even by a small MRL layer.

Note that our adaptive network cannot be characterized as simply "learning phototaxis," because its task was modified by the presence of the fixed-connection network. The adaptive network, that is, did not have to provide motor commands that would be appropriate *per se* to navigate toward the light, but motor commands that would do so when summed with the motor commands that were concurrently provided by the fixed-connection network. A study of how the behavior of the adaptive network was modified by the presence of the fixed-connection network is being planned. Here we would like to stress that the interplay between adaptive and fixed mechanisms demonstrated above is likely to be very common in biological organisms, because learning systems typically do not replace phylogenetically older fixed input-output mechanisms, but rather complement them, operating within an already functioning behavioral system.

Matching memory capacity to environmental requirements is not trivial. Simply making the system large is not a practical solution since, as we know, larger brains are energetically costly to maintain. The variation in the performance of the adaptive networks demonstrated here are representative, in an elementary way, of the information processing trade-offs biological, multi-layer nervous systems have traversed in their phyologenetic histories. Adaptive networks must allow for the simultaneous developmental trajectories of all the adaptive and fixed networks present in a given brain. The constraints hold for self-organizing networks in artificial robot systems. In the design of such networks, reasonable allowance must be made for both the variability of the environment and the progressive alterations of functional properties of all the sub-systems present in an agent and their interactions.

Acknowledgments. We thank the BCR lab members for helpful discussions and technical assistance throughout the course of this project. In particular David Brown and David Grinberg for their help with writing the C-programs.

References

1. Skinner, B.: Why I Am Not a Cognitive Psychologist. Behaviorism 5(2), 1–10 (1977)
2. Watson, J.B.: What is Behaviorism? In: Behaviorism, pp. 6–19. W. W. Norton, New York (1924)
3. Tinbergen, N.: The hierarchical organization of nervous mechanisms underlying instinctive behaviour. Physiological Mechanisms in Animal Behavior (1950)
4. Tinbergen, N.: On aims and methods of ethology, pp. 114–137 (1963)
5. Arkin, R.C.: Behavior-based robotics, p. 491. The MIT Press (1998)
6. Brooks, R.A.: Cambrian Intelligence the early history of the new aI, p. 199. A bradford book. The MIT press (1999)
7. Brooks, R.A.: Flesh and machines how robots will change us, vol. 260. Pantheon books (2002)
8. Hogan, J.A.: Developmental psychobiology and behavioral ecology. In: Blass, E.M. (ed.) Handbook of Behavioral Neurobiology, pp. 63–106. Plenum Publishing, Philadelphia (1988)
9. Hinde, R.A. (ed.): Animal Behavior, a synthesis of ethology and comparative psychology, 2nd edn. McGraw-Hill Book Company, New York (1970)
10. Shettleworth, S.J.: Cognition, Evolution, and Behavior. Oxford University Press (2010)
11. Dunlap, A.S., Stephens, D.W.: Components of change in the evolution of learning and unlearned preference. Proceedings of the Royals Society B, London (2009)
12. Anastasio, T.J.: Tutorial on neural systems modeling, 1st edn. Sinauer Associates Inc., Sunderland (2010)
13. Touretzky, D.S., Saksida, L.M.: Skinnerbots. In: Maes, M.M.P., Meyer, J.A., Pollack, J., Wilson, S.W. (eds.) Animals to Animats 4: Proceedings of the Fourth International Conference on Simulation of Adaptive Behavior, pp. 285–294. MIT Press, Cambridge (1996)
14. Touretzky, D.S., Saksida, L.M.: Operant conditioning in skinnerbots. Adaptive Behavior 5(3/4), 219–247 (1997)
15. McCelland, J.L., Rumelhart, D.: Parallel distributed processing: explorations in the microstructure of cognition. In: Psychological and Biological Models, vol. 2, p. 611. A bradford book–the MIT press, London, Cambridge (1986)

16. Gurney, K., Redgrave, P., Prescott, T.J.: A computational model of action selection in the basal ganglia II. Analysis and simulation of behaviour. Biological Cybernetics 84, 411–423 (2001)
17. Gurney, K., Redgrave, P., Prescott, T.J.: A computational model of action selection in the basal ganglia I. A new functional anatomy. Biological Cybernetics 84, 411–423 (2001)
18. Lengyel, M., Dayan, P.: Hippocampal contributions to control: The third way. In: Platt, J.C., et al. (eds.) Neural Information Processing Systems Foundation, NIPS 2007, Vancouver, Canada (2007)
19. Barto, A.G., Sutton, R.S., Anderson, C.W.: Neuronlike adaptive elements can solve difficult learning control problems. IEEE Transactions on Systems, Man and Cybernetics 13, 834–846 (1983)
20. Kohonen, T.: Self-organized formation of topologically correct feature maps. Biological Cybernetics 43, 59–69 (1982)
21. Kohonen, T.: Self-organizing maps. Springer, Berlin (1997)

Distributed Control of Complex Arm Movements

Reaching Around Obstacles and Scratching Itches

David Zipser

Professor Emeritus,Department of Cognitive Science, UCSD and Visiting Scholar,
UC Berkeley
1408 Milvia St. Berkeley CA 94709
dzipser@ucsd.edu

Abstract. This paper presents a computational theory for generating the complicated arm movements needed for tasks such as reaching while avoiding obstacles,
or scratching an itch on one arm with the other hand. The required movements are
computed using many control units with virtual locations over the entire surface
of the arm and hand. These units, called *brytes*, are like little brains, each with its
own input and output and its own idea about how its virtual location should move.
The paper explains how a previously developed gradient method for dealing with
ill-posed multi-joint movements [1] can be applied to large numbers of spatially
distributed controllers. Simulations illustrate when the arm movements are successful and when and why they fail. Many of these failures can be avoided by a
simple method that adds intermediate reaching goals. The theory is consistent
with a number of existing experimental observations.

1   Introduction

Explaining how we make the complicated movements needed to reach an object while
avoiding intervening obstacles has proved to be a challenge for motor control theory.
Yet these movements are made frequently, as when we pick fruit or service a car.
Even when there is a clear path from the current hand position to the goal, the actual
movement may be perturbed by the need for other parts of the arm to avoid obstacles.
The task becomes more challenging when both the reach goal and the obstacles are
moving as when scratching an itch. The problems that need to be solved to do these
tasks are massively ill-posed, because there are both more degrees of freedom in a
real arm than in the space it operates in, and an infinite number of paths to a goal.

Here I show that a strategy using distributed control units, which I have called
brytes [2], together with some relatively simple central guidance can be used to do
these complex tasks.

1.1   The Bryte Way

Imagine that you are shrunk to a tiny size and attached someplace on an arm, say the
back of the wrist. You are provided with a joystick that allows you to move freely in
space by indirectly controlling the arm's joint angles. You go where the joystick

T.J. Prescott et al. (Eds.): Living Machines 2012, LNAI 7375, pp. 309–320, 2012.
© Springer-Verlag Berlin Heidelberg 2012

points while remaining fixed to the same place on the arm. If you see someplace you want to go, you point the joystick in that direction and the arm takes you there. If, on the other hand, you want to avoid something, you point the joystick away form it. Of course, you are far too smart for this job. What is needed is a much simpler computational unit with just those features of your brain required for the task. This unit can't be too simple since it has to have sensory input to inform it about its environment and an output to indicate movement direction. It also needs some internal computation ability. In short the unit that replaces you has to be a brain, but a very basic one, which I call a *bryte*. To accomplish complicated tasks brytes are distributed all over the arm. These brytes do not have to be physically located on the arm so long as they get input in a reference frame that has a *virtual* location on the arm. The place to put the brytes is obviously the brain.

To make this paradigm work, information about the location of objects in space has to be transformed from the sensory frames where it is obtained, i.e. vision or touch, to frames at the virtual locations of the brytes. How and where the brain does these transformations has been extensively investigated, so I will assume that spatial location information in any reference frame can be transformed to any other. The evidence for this *fungibility of reference frames* in the context of arm movement is quite strong and will be discussed in more detail later.

Once a bryte knows the location of objects of interest it can compute its response as a 3D vector of appropriate magnitude and direction. The details of these computations are given below, but simply stated, goal-seeking brytes output a vector of constant length pointing to the goal, while obstacle-avoiding brytes produce vectors pointing away from obstacles. The length of the avoider-vectors decreased rapidly with distance from obstacles.

Finally, the movement instructions of all the brytes have to be combined and converted from 3D values into joint angle rotations. This requires dealing with the ill posed nature of arm movements.

The redundancy of arm movements is generally dealt with using optimization in which a scalar variable depending on all the quantities that vary during arm movement is minimized. This is referred to as optimal control [3]. One widely used optimal control paradigm construes the problem as one in classical mechanics and tries to find the path along which some physical quantity is minimized. In simple cases, this reduces to a set of solvable partial differential equations. In more complicated cases, where there is both joint and path redundancy, extensive numerical computations are generally required. One important advantage of this approach is that the solution provides not only a postural path, but also the joint torques needed to get a real physical arm to the goal.

Another approach to optimum control is to treat the arm as a geometrical rather than a physical object and solve only the kinematic problem. This is simpler and can be solved by gradient methods that do not require a lot of computation. While it does not provide the required torques it does give a stream of joint angles change values which, when applied to the geometry of an arm, will bring it to the goal. If this approach is used by the brain there must be a "down-stream" controller to generate the torques using the joint angle changes supplied by the gradient method. One reason to think this is what happens is that the geometry of the postural paths obtained by the gradient method resemble experimentally observed arm movements. The gradient method is ideal for generating arm movements with brytes because it can be rapidly computed online and the outputs of the brytes can be combined by simple addition because the gradient is a linear operator.

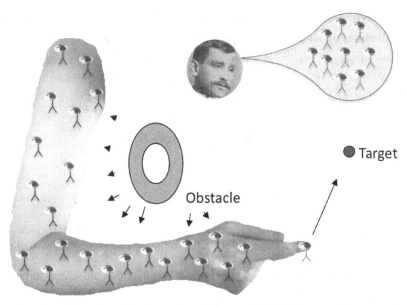

Fig. 1. The virtual location of the brytes, i.e. big eyes with little legs, are shown on the arm. The actual location of the brytes is in the brain. All brytes except the one on the finger tip are avoiders. The finger tip bryte is a seeker. (Photo is author's grandfather c. 1905)

1.2   Using Brytes

Consider the problem of a hand moving to reach a goal while all other parts of the arm try to avoid any number of arbitrarily placed obstacles. Only two types of brytes—*obstacle-avoiders* and a *goal-seeker*--are needed to control this task. Roughly speaking, the avoiders move away from nearby obstacles and ignore far away ones, while the goal-seeker moves toward the goal of the reach, Figure 1.

The information about the scene needed for the task comes from the visual and tactile sensory systems. However, sensory processing is not the subject here, so I avoid this aspect of the problem by using an 'internal model' in which scenes are represented using sets of parameterized surfaces. For example, arms are made up of several foreshortened and flattened cylinders and spheres. This representation makes it straightforward to provide the brytes with direct access to information about each point on the surface of all the objects in a scene. Although all brytes have access to the same scene, each bryte views it from its own virtual location. This means that each bryte has to transform the scene information into its own egocentric reference frame.

Seeker Brytes. Goal-seeking brytes compute a 3D movement vector of unit length, $\mathbf{m}^s$, where $\mathbf{s}$ extends from the bryte's virtual location to the location of the goal (equation 1 and Figure 2).

$$\mathbf{m}^s = \mathbf{s}/\|\mathbf{s}\| \tag{1}$$

Avoider Brytes. Avoider brytes have a more complicated problem because they have to take all points on all the obstacles into account. The avoider movement vector, $\mathbf{m}^a$, is the sum of vectors, $\mathbf{m}^{a(i)}$, with directions, $\mathbf{a}(i)$, from many points distributed over the obstacles to the virtual origin of the bryte. The index i ranges over all points on all obstacles. The magnitudes of the $\mathbf{m}^{a(i)}$ depend on the distance to the obstacle point and the angular displacement of the point (equation 2 and Figure 2).

$$\mathbf{m}^{a(i)} = \mathbf{a}(i)\,\mathrm{f}\left(\|\mathbf{a}(i)\|\right)\cos\theta(i)$$

$$\text{If: } \cos\theta(i) > 0 \tag{2}$$

$$\text{Else: } \mathbf{m}^{a(i)} = 0$$

Where:

f is a weighting function that decreases as $\mathbf{a}(i)$ increases, typically a Gaussian.

$\theta(i)$ is the angle between the normal vector to the arm surface at the virtual origin of the bryte and the vector $\mathbf{a}(i)$.

The $\cos\theta(i)$ term discriminates against obstacle points that are lateral to the bryte because a collision with them is less likely. It is also used to eliminate obstacle points behind the bryte whose cosines are negative, Figure 2.

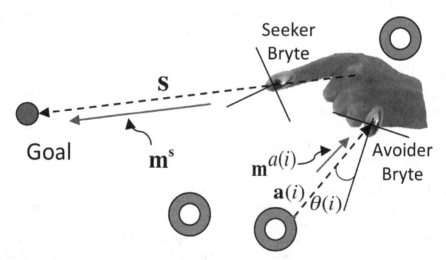

Fig. 2. The virtual reference frames of a seeker bryte and a single avoider bryte are shown. The symbols and their use are described in the text. Only one point on one obstacle is shown, but each avoider bryte has to access many such points on all the obstacles. The movement vector of the avoider bryte shown will get contributions from the two obstacles below the hand but not from the one above because the $\cos\theta(i)$ is negative. This obstacle will, however, contribute to the movement vector of brytes on the other side if the hand.

Combining Bryte Outputs. In order to generate arm movement, all the 3D bryte movement vectors have to be combined into a single vector in joint angle space. It has been shown previously that this vector can be implemented as the gradient of the function that maps points in joint angle space to points in 3D space [1], see methods in supplementary materials for details. The gradient points in the direction in joint angle space that will most rapidly reduce the difference between the current and goal postures. Since the gradient is a linear function, the gradients of each 3D movement vector can be added to get the gradient of the required joint angle vector. This computation is slow because it involves multiplying each movement vector by a huge Jacobian matrix. Fortunately, linearity works both ways. This means that all the movement vectors for a single arm segment, i.e. forearm, palm, etc., can be added together before multiplying by the Jacobian for their arm segment. The resulting gradient vectors are then added. This greatly speeds computation because there are only a few arm segments instead of the large number of brytes. While this method is valid mathematically, it not obvious that it will solve the problem and thus must be tested by simulation.

2   Results

(Supplementary material containing methods, the .avi movies animating the figures and MatLab GUIs containing all the code used for the simulations is available at: http://crcns.org/data-sets/movements/zipser-1.)

2.1   Simulation of Obstacle Avoidance

The theory was evaluated by simulating an arm with 7 joint angles reaching for a goal while trying to avoid up to 6 obstacles. The simulations compute the path on line at reasonable speeds. There is no pre-computation of the path and the simulations run rapidly enough so that many different configurations of arm starting location and obstacle position can be evaluated conveniently. A graphical user interface, or 'GUI', was used to display the simulation. The GUI allows parameters such as the starting arm posture, goal location, and the number and location of the obstacles to be manipulated. GUIs containing all the simulation code are available in the supplementary material. Simulations demonstrated that the arm could move to the goal and around obstacles in many but not all configurations. A variety of obstacle-avoiding movements are illustrated with still picture figures here and with .avi movies in the supplementary material. In the simulations of the obstacle avoidance task, 172 avoider brytes where distributed on the arm and there was one seeker bryte on the fingertip.

Fig. 3 above shows a comparison of movement without obstacles to a path avoiding one obstacle. When an obstacle is present the arm approaches the goal along the same initial path as without the obstacle, but deviates around the obstacle as the hand nears it. The reason that the hand has to get near the obstacle before it starts to avoid it is that the avoider brytes only respond to nearby objects.

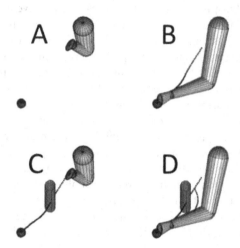

Fig. 3. The arm at the start and end of movements.Pictures taken from the GUI simulation. A: The starting posture for all movements in this figure and also Figures 4 and 5. B: The end posture of a movement in the absence of obstacles. C: The starting posture as in A, but with an obstacle on the path taken in B. D: The end posture of a movement that has avoided the obstacle. The fingertip path taken around the obstacle is shown.

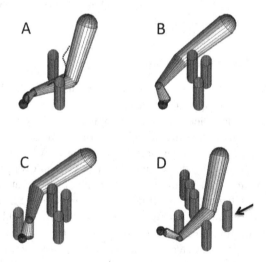

Fig. 4. The end postures of movements avoiding various configurations of obstacles. The details are discussed in the text.

In Figure 4A the arm goes between two obstacles. In Figure 4B the hand cannot fit through the gap between the obstacles, but it finds a successful path around the far side and reaches the goal. The arm initially heads for the gap between obstacles because it is on a direct line to the goal. When it nears the first obstacle the arm deviates to one side and then continues this sideways deviation till it passes the second obstacle and has a clear path to the goal. These complex movements are not planed, but arise from the resolution of the seeker and avoider movement vectors.

Placing an additional obstacle near the goal forces the arm to make a complicated movement that takes advantage of its extra degrees of freedom, Figure 4C. This demonstrates why the arm has redundant degrees of freedom, and also shows the power of the gradient method for utilizing redundancy. Figure 4D shows a situation in which an obstacle has been placed near the elbow and upper arm. This obstacle is not on the hand path to the goal, but the elbow and upper arm have to avoid it. The avoider brytes on the upper arm make this possible. This example is an illustration of the power of distributed, bryte based on-line control.

Unfortunately, the simple local movement policy used in the above examples can be stymied by obstacles that can easily be avoided. An example is shown in Figure 5A. The two obstacles are a little bit closer together than in the earlier examples and also more symmetrically placed on the direct path to the goal. The result is that when the hand gets near this pair of obstacles the goal-seeker and object-avoider movement vectors cancel and movement stops. This is an example of a local minimum to which the local movement policy is prone. This is not a trivial problem and can only be solved by using movement policies that exploit "global" information. Policies of this kind will be explored below.

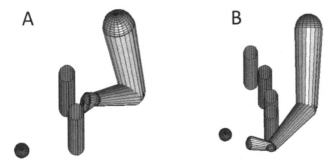

Fig. 5. A: The end posture of a movement that failed to reach the goal because of a local minimum. B: The end posture of a movement that failed because of the physical impossibility of reaching the goal.

Moving the obstacles in the previous example just a bit closer together and adding two more obstacles lets the arm avoid a local minimum and find a path around the obstacles; however, the hand still does not reach the target, Figure 5B. In this case it is not a local minimum that stops movement, but the physical impossibility of reaching the goal, at least from the side it was approached. This is a problem which cannot be solved by any clever policy since it arises from the physical constraints of the situation. It illustrates that some goals simply cannot be reached. A smart system might be able to detect such situations in advance and try another route, but real life experience shows that it is often hard to tell in advance. As a consequence, we sometimes try reaches that fail.

Intermediate Goals. The local minimum problem is not experienced in real life. One reason for this is that we *do not* move directly toward the final reaching goal and deviate only when near an obstacle as in Figure 3B. Rather, typical reaching paths start by heading in a direction that avoids the obstacle. It is *as if* the hand initially moved to an intermediate target on a safe route. Once the hand passes the obstacle, it changes course and heads toward the final goal.

I implemented an intermediate goal movement policy to see if it could solve the local minima problem observed in the simulations. Implementing this strategy does not require any change in the local movement policy. Rather, it simply adds global information. The details of the search policy used are given in the supplemental material.

Simulation showed that the intermediate goal policy can avoid the local minimum that caught the hand, Fig. 6A. The local minimum is avoided by examining the two potential safe targets between the obstacles. This reveals that the path to one goes through an unsafe region, while the other is found to have an obstacle just beyond it. Having eliminated the interior path, the shorter of the two exterior paths is taken.

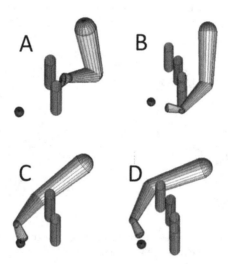

Fig. 6. A: Using only the local policy the arm is stopped by a local minimum. C: Using both the local and intermediate goal policies a path is found that avoids the trap and reaches the goal. **B:** The arm is physically unable to reach the goal using the local policy. D: Using both policies the arm takes a shorter path and gets closer to the goal but still can't reach it.

Intermediate goals cannot overcome physical limitations although, by taking a shorter path it can get closer to the goal, Figure 6B and D.

2.2   Simulating Scratching an Itch

Bryte based distributed control can be used in tasks more challenging than the one described above. The next example uses the same general paradigm to simulate a two-arm task in which the hand on one arm scratches an itch anywhere on the other arm. In this case both arms move to bring the hand to the itch while the moving arms avoid hitting each other.

The strategy used is the basically same as in the first example except that now each arm is an obstacle for the other and each arm has a goal location on it. The arm with the itch moves the itch site toward the fingertip on the scratching arm and the scratching arm moves its fingertip to the site of the itch on the other arm. The same two kinds of brytes are used, i.e. avoiders and seekers, however, there are some differences. In the obstacle example the finger's task was simply to reach the location of the goal. In the scratch task the finger also has to be oriented perpendicular to the surface of the arm at the itch site. The reason for this is to encourage the arms to rotate to expose the itch site to the scratching finger. It also gives a bit more realism. Orientation matching requires an additional term to be included in the cost function to minimize the angle between the surface normals at the itch site and at the tip of the scratching finger.

The scratch task was simulated using the local policy. With this policy each of the 120 avoider brytes has to continually check 120 points distributed around the surface of the opposite, moving arm. While this increases the number of computations required, simulations of the scratch task run reasonably fast. One reason for speed is that the arm-relative locations and orientations of the brytesare constants that can be pre-computed.Their current space based locations and orientations are obtained using the changing joint angle values.

A problem arises when the scratching fingertip nears the itch site because the avoider movement vector cancels out the action of the seeker bryte thereby preventing the finger from reaching the itch site. To overcome this, all avoiding is turned off when the fingertip gets near the itch site, allowing the finger to reach the itch site.

Examples of scratching itches starting from various arm postures and itch locations are shown in Figure 7A-D. The start (top) and end postures (bottom) illustrated here do not do justice to the complex realistic coordinated movements of the arms that occur in the simulations. These movements can be viewed in the movies provided in the supplemental material.

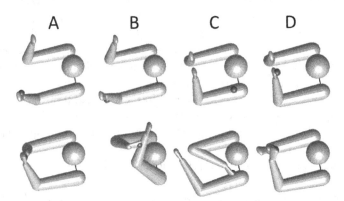

Fig. 7. Examples of starting and ending postures of both arms moving to scratch an itch. In A, B and C the movements are successful while in D a local minimum is encountered. (The local minimum appears as a small local limit cycle in the simulations due to the use of a finite step size).

In Fig. 7A the itch is located on the side of the palm. The finger moves on a fairly direct path to the itch site while the itch hand rotates slightly to orient the itch. In Fig. 7B the itch is located near the mid point of the forearm on the outer side. Both arms move to bring the scratching finger to the itch. Their final position is near the center of the space in front of the body. In Fig. 7C the itch is located on the lower side of the upper arm near the shoulder. The itch arm rotates and moves up to make the itch more accessible while the scratch arm lowers and bends sharply to reach the itch site.

The simulation accomplishes the task in a realistic appearing way. However, it was not possible to quantitate the realism of the model's behavior because there are no experimental data to compare it with. Subjectively, the movements appear reasonably natural and the final postures appear quite plausible.

Unfortunately, because this is a local model, it can get caught in local minima due to cancellation of the seeker and avoider bryte vectors. An example is shown in Figure 7D. A small change in starting posture eliminates the local minimum. An analytical approach to finding safe paths is given the methods section of the supplemental materials.It was not implemented, but is likely to eliminate most of this kind of local minima.

3   Discussion

The role played by brytes in this work could just as well be described using conventional jargon. This raises the question of whether brytes are more than an entertaining didactic device. I have conjectured that brytes, i.e. simple brains, are good entities for constructing brain models, and also actual entities adopted by evolution to facilitate the emergence of bigger, better brains. Preliminary steps elucidating this conjecture are recorded in a series of lectures at UC Berkeley's Redwood Center for Theoretical Neuroscience [2].

The computations done by the avoider and seeker brytes used here are consistent with the experimental observations of neurophysiology. The remarkable properties of visual tactile cells[4] are ideally suited to the computational needs of avoider brytes. These neurons have visual receptive fields overlaying tactile fields on the surface of the hand, arm, face and presumably other body parts. These fields move with the body surface they are attached to. Vision is required to initiate their visual response to an object, but not for its maintenance, and the visual fields are not affected by eye movements. This behavior requires a continual transformation of their reference frames—an impressive example of the dynamic fungibility of coordinate systems required by bryte based distributed control. It has been speculated, quite reasonably, that these cells play a role in object avoidance. This is consistent with the local nature of their visual receptive fields and their location in a high level motor area.

Seeker brytes, on the other hand, need information enabling then to locate the goal relative to their current virtual location. Experiments with reaching to targets at various locations have identified reach-activated neurons that respond to target distance in hand based reference frames[5]. These neurons are found in parietal reach areas and have just the properties required by seeker brytes. Taken together, these experimental observations give support to the use of brytes as computational units for modeling brain function.

Brytes as used here receive their input from internal representations of a scene and not directly from visual images. This is a significant conjecture with important implications for how the brain functions. It implies that maintaining internal representations and processing visual images are separate aspects of cognition. They interact of course, but can be studied and modeled separately. The relevant parameters often come directly from sensory input, but they can also come from memory or 'imagination'. Sensory systems in addition to vision, such as touch, can also contribute input to scene representation. Touch is particularly relevant to obstacle avoidance in tight situations or when vision is obscured. The fact that some of the neurons involved have both visual and tactile receptive fields suggests that both sources of information can be equivalent for generating scene representations and thus movements.

For obstacle avoidance the objects in the internal representation do not have to look exactly like those in the visual scene since the details of objects are of little significance. It is only necessary to know enough about an obstacle to avoid it. So, for example, to avoid a cactus, a parameterized spheroid enclosing the cactus together with its thorns would suffice. Avoidance does no require a detailed representation of the thorns. This does not mean that the brain cannot represent fine detail; rather it implies that the part of the brain concerned with dynamically avoiding obstacles my use a cruder representation than required for analysis of detail, an insight that might be helpful in identifying the neural basis of scene representation.

Coordinated arm movements require that the controller generating the movements of each arm know the location and orientation of the normal to every point of the other arm is at all times. This raises the question of how the hemispherically separated motor controllers get the information about the other arm. Each hemisphere has a robust motor and sensory representation of the arm it controls, but little is known about how it gets information about the other arm. This information could come through the *corpus callosum* but, since the location of each point on the ipsolateral arm has to be known by each point on the contralateral arm, this would require a lot of callosal fibers. In the simulations described here this problem is simplified by pre-computing a data structure that contains the relative location of many points on each arm segment relative to a fixed base location and orientation of that segment. Updating these locations requires knowing only the current location and orientation of the segment, information that can be computed form the joint angles. If a similar strategy is used in the brain, only joint angle information needs to be passed from side to side. The pre-computed data structures, being static, might be encoded in synaptic weights, making them difficult to identify explicitly. Their existence could only be inferred by experimental analysis of relevant behaviors and the absence of the information in callosal fibers.

References

1. Torres, E.B., Zipser, D.: Reaching to Grasp with a Multi-jointed Arm (I): A Computational Model. Journal of Neurophysiology 88, 1–13 (2002)
2. Zipser, D.: Brytes or How to make big brains out of lots of small brains, parts 1, 2, 3. Redwood Center for Theoretical Neuroscience (2009, 2010),
   http://archive.org/search.php?query=david+zipser
3. Todorov, E.: Optimality principles in sensorimotor control. Nature Neuroscience 7, 907–915 (2006)
4. Graziano, et al.: Coding of visual space by premotor neurons. Science 11, 1054–1057 (1994)
5. Ferraina, S., Brunamonti, E., Giusti, M.A., Costa, S., Genovesio, A., Caminiti, R.: Reaching in Depth: Hand Position Dominates over Binocular Eye Position in the Rostral SuperioParietal Lobule. The Journal of Neuroscience 29(37), 11461 (2009)

Cerebellar Memory Transfer and Partial Savings during Motor Learning: A Robotic Study

Riccardo Zucca[1,*] and Paul F.M.J. Verschure[1,2]

[1] Laboratory of Synthetic Perceptive, Emotive and Cognitive Systems (SPECS),
DTIC, Universitat Pompeu Fabra (UPF), Barcelona, Spain
[2] Catalan Institute of Advanced Studies (ICREA), Barcelona, Spain
{riccardo.zucca,paul.verschure}@upf.edu

Abstract. Faster re–learning of an already acquired motor skill is a common phenomenon observed in classical conditioning of motor responses. The cerebellum is critically involved in the acquisition and retention of those responses. Nevertheless, it remains unclear whether the memory of the association is stored in the cerebellar cortex or in the nuclei. In a previous study we have demonstrated that a neuronal model reflecting basic properties of the cerebellum can acquire well–timed conditioned responses by mediation of plasticity in the cerebellar cortex when controlling a robot in an obstacle avoidance task. Here we extended the model to investigate a possible mechanism of memory transfer and consolidation based on plasticity at the level of the deep nucleus. Experimental results show how the collaboration of these two sites of plasticity can drive the robot to timely adapt and maintain a long term memory in solving the obstacle avoidance task.

Keywords: Cerebellum, motor learning, classical conditioning, savings, memory.

1 Introduction

A common observation when we learn a new motor skill is that re–learning the same activity later in time appears to be faster. In the context of Pavlovian classical conditioning of motor responses this phenomenon is known as savings and it has been subject of different studies [1,2]. Classical conditioning of discrete motor responses, such as the eye–blink reflex, involves the contiguous presentations of a neutral conditioned stimulus (CS; e.g. a tone or a light, which normally does not elicit a response) with a reflex evoking unconditioned stimulus (US; e.g. an air–puff towards the cornea that causes the closure of the eyelid) [3]. The repeated paired presentations of these two stimuli, CS and US, finally result in the expression of a conditioned response (CR) that predicts and anticipates the occurrence of the US (e.g., the eyelid closure in response to the tone). Non–reinforced presentations of a previously reinforced CS induce a progressive elimination of the CR that is known as extinction. However, even if no more CR is produced, a residual of the original learning is conserved as demonstrated by the faster reacquisition of the CR when the CS is again paired with the US [1].

T.J. Prescott et al. (Eds.): Living Machines 2012, LNAI 7375, pp. 321–332, 2012.
© Springer-Verlag Berlin Heidelberg 2012

A large number of lesion, inactivation and electro–physiological studies generally agree that classical conditioning of motor responses is strictly dependent on the cerebellum (see [4,5] for a comprehensive review). However, the relative contributions of the cerebellar cortex and the deep nucleus in the acquisition and retention of the CR still remain quite elusive. A change in synaptic efficacy seems to be involved in both sites and to be responsible in controlling different aspects of the CR.

One hypothesis [6,7] is that two mechanisms of plasticity – at cortical and nuclear level – jointly act to regulate different sub–circuits (fast and slow) involved in the acquisition and retention of the CR. In a sort of 'cascade' process a faster sub–circuit – including the Purkinje cells, the deep nucleus and the inferior olive – 1) is responsible for learning a properly timed CR, as indicated by the fact that lesions of the cerebellar cortex produce responses which are disrupted in time [8,9,10], and, 2) it signals to the second circuit that the two stimuli are related. A slower sub–circuit – involving the Pontine nuclei, the deep nucleus and the Purkinje cells – driven by the faster one, regulates then the expression of the CR and stores a long–term memory of the association.

In the context of the Distributed Adaptive Control (DAC) framework [11], a computational neuronal model – VM model from now on – based on the anatomy of the cerebellum has been previously developed to control a behaving robot [12,13]. In those studies, the authors investigated how plasticity in the cerebellar cortex can be sufficient to acquire and adaptively control the timing of the CR. Nevertheless, the model was limited to the role of the cerebellar cortex and deliberately did not include other components of the cerebellar circuit that are supposed to be related to classical conditioning (i.e. the role of the deep nucleus). Here we propose, in the light of different findings on bi–directional synaptic plasticity in the deep nucleus [10,14,15,16], an extension of the VM model with the goal of investigating whether synaptic plasticity between the mossy fibre collaterals originating from the Pontine nuclei and the deep nucleus can support partial transfer and consolidation of the memory of the CS–US association. The plausibility and robustness of the model has been tested on a mobile robot solving an obstacle avoidance task within a circular arena.

2   Materials and Methods

2.1   Cerebellar Neuronal Model

The cerebellar model used in this study extends the original VM neural model [12,13] and it has been implemented within the IQR neuronal network simulator [17] running on a Linux machine. The original model follows the classical Marr–Albus idea that learning in the cerebellum is dependent on the depression of the parallel fibres synapses in the cerebellar cortex [18,19] and it includes several assumptions. The representation of time intervals between conditioned and unconditioned stimulus events is assumed to depend on intrinsic properties of the parallel fibre synapses (spectral timing hypothesis). The model also integrates the idea that cerebellar cortex, deep nucleus and the inferior olive form a negative feedback loop which controls and stabilizes cerebellar learning [20].

A schematic representation of the model components and connections and its anatomical counterparts is illustrated in Fig. 1.

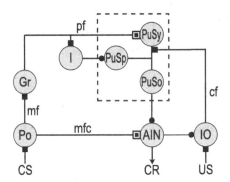

Fig. 1. The cerebellar circuitry. CS– and US–inputs converge on a Purkinje cell (which is composed of three compartments: *PuSy*, *PuSp*, and *PuSo* referring respectively to the dendritic region, the spontaneous activity and the soma of the Purkinje cell). The inferior olive (*IO*), Purkinje cell and deep nucleus (*AIN*) form a negative feedback loop. The CS–indirect pathway conveys the CS from the Pontine nuclei (*Po*) through the mossy fibres (*mf*), the granular layer (*Gr*) and the parallel fibres (*pf*) to the Purkinje cell. A CS–direct pathway conveys the CS signal from *Po* to the *AIN* through the collaterals of the mossy fibres (*mfc*). The US–signal is conveyed to the Purkinje cell through the climbing fibres (*cf*) generating from *IO*. The inhibitory inter–neurons (*I*) receive input from *pf* and inhibit *PuSp*. Excitatory connections are represented by filled squares, inhibitory connections by filled circles and sites of plasticity by surrounded squares.

The main modification to the original model is based on the assumption that long–term potentiation (LTP) and long–term depression (LTD) can be also induced in the synapses between the mossy fibres collaterals and the deep nucleus by release of Purkinje cell inhibition [6,10,14,15,16].

2.2   Model Equations

As in the original model, the network elements are based on a generic type of integrate–and–fire neurons. The dynamics of the membrane potential of neuron i at time $t + 1$, $V(t + 1)$, is given by

$$V_i(t + 1) = \beta \cdot V_i(t) + E_i(t) + I_i(t) \tag{1}$$

where $\beta \in (0, 1)$ is the persistence of the membrane potential, defining the speed of decay towards the resting state, $E_i(t)$ represents the summed excitatory and $I_i(t)$ the summed inhibitory input. $E_i(t)$ and $I_i(t)$ are defined as

$$E_i(t) = \gamma^{\mathrm{E}} \sum_{j=1}^{N} A_j(t) w_{ij}(t) \tag{2}$$

$$I_i(t) = -\gamma^I \sum_{j=1}^{N} A_j(t)w_{ij}(t) \tag{3}$$

where γ^E is the excitatory and $-\gamma^I$ the inhibitory gain of the input. N is the number of afferent projections, $A_j(t)$ is the activity of the presynaptic neuron $j \in N$, at time t, and w_{ij} the efficacy of the connection between the presynaptic neuron j and the postsynaptic neuron i. The activity of an integrate–and–fire neuron i at time t is given by

$$A_i(t) = H(V_i(t) - \theta^A) \tag{4}$$

where θ^A is the firing threshold and H is the Heavyside function

$$H(x) = \begin{cases} 1 & \text{if } x > 0 \\ 0 & \text{otherwise .} \end{cases} \tag{5}$$

If V_i exceeds the firing threshold θ^A, the neuron emits a spike. The duration of a spike is 1 simulation time–step (ts) and is followed by a refractory period of 1 ts. Pontine nuclei, granule cells and inferior olive are modelled with such integrate–and–fire neurons.

Purkinje Cell. The Purkinje cell is defined by three different compartments: *PuSp* accounts for the tonic, spontaneous activity of the Purkinje cell, *PuSo* represents the soma and *PuSy* represents the dendritic region where synapses with parallel fibres are formed.

PuSp is spontaneously active as long as it is not inhibited by the inhibitory inter–neurons (I). *PuSo* receives excitation from *PuSp*, *PuSy* and *IO*. *PuSy* represents the postsynaptic responses in Purkinje cell dendrites to parallel fibres stimulation. *PuSy* does not emit spikes but shows continuous dynamics according to

$$A_i(t) = H(V_i(t) - \theta^A)V_i(t) . \tag{6}$$

In order for the Purkinje cell to form an association between the CS and the US, the model assumes that a prolonged trace of the CS (an eligibility trace) is present in the dendrites (*PuSy*). An exponentially decaying trace with a fixed time constant defines the duration of this eligibility trace for synaptic changes in Purkinje dendrites.

Deep Nucleus. *AIN* neurons are constantly inhibited by Purkinje cell activity unless a pause in *PuSo* activity occurs. Following disinhibition, the *AIN* neuron shows a characteristic feature called rebound excitation [21]. The output of the *AIN* is then responsible for the activation of the CR pathway downstream of the deep nucleus and for the inhibition of the *IO*. A variant of a generic integrate–and–fire neuron was used to model rebound excitation of the *AIN*. The membrane potential of neuron i at time $t + 1$, $V_i(t + 1)$, is given by

$$V_i(t) + 1 = \beta \cdot V_i(t) + [H(V_i(t) - \theta^{\mathrm{R}})H(\theta^{\mathrm{R}} - V_i(t - 1)) * \mu] + I_i(t) , \qquad (7)$$

where θ^{R} is the rebound threshold and μ is the rebound potential. The potential of *AIN* is kept below θ^{R} by tonic input from *PuSo*. When disinhibited, *AIN* membrane potential can repolarize and, if the rebound threshold is met, the membrane potential is set to a fixed rebound potential. *AIN* is then active until it is above its spiking threshold.

Simulation of Plasticity. Two sets of connections undergo bi–directional plasticity: (1) *pf–Pu* synapse, defined as the connection strength between *PuSy* and *PuSo*; (2) *mf–AIN* synapse, defined as the connection between *Po* and *AIN*, can both undergo long–term potentiation (LTP) and long–term depression (LTD). In order to learn, the *pf–Pu* synaptic efficacy has to be altered during the conditioning process. LTD can occur only in the presence of an active CS driven stimulus trace ($A_{\mathrm{PuSy}} > 0$) coincident with *cf* activation [22]. At each time–step the induction of LTD and LTP depends on

$$w_{ij}(t + 1) \begin{cases} \epsilon w_{ij}(t) & \text{if } A_{\mathrm{PuSy}} > 0 \ \& \ A_{\mathrm{cf}} > 0 \quad \text{(LTD)} \\ w_{ij}(t) + \eta(w_{ij}^{\max} - w_{ij}(t)) & \text{if } A_{\mathrm{PuSy}} > 0 \quad \text{(LTP)} \\ w_{ij}(t) & \text{otherwise} . \end{cases} \qquad (8)$$

The learning rate constants (ϵ and η) are set to allow one strong LTD event as the result of simultaneous *cf* and *pf* activation, and several weak LTP events following a *pf*–input. As a result, *pf* stimulation alone leads to a weak net increase of *pf–Pu* synapse, while *pf* stimulation followed by *cf* stimulation leads to a large net decrease.

A second mechanism of plasticity is implemented at the *mf–AIN* synapse. Contrarily to the well–established induction of LTP and LTD at *pf–Pu* synapse, evidence for a potentiation of the mossy fibres collaterals to deep nucleus synapses has been sparse. LTP and LTD effects of this synapses in vitro [23,15,16] as well as increases in the intrinsic excitability of the AIN [21] have been shown. Theoretical studies [24] also suggest that LTP in the nuclei should be more effective when driven by Purkinje cells. Hence, in our model the induction of LTP depends on the expression of a rebound current (δ_{R}) in the *AIN* after release from *PuSo* inhibition and the coincident activation of *mf*s, while LTD depends on the *mf*s activation in the presence of inhibition. The weights at this synapses are updated according to the following rule

$$w_{ij}(t + 1) \begin{cases} w_{ij}(t) + \alpha(w_{ij}^{\max} - w_{ij}(t)) & \text{if } A_{\mathrm{mf}}(t) > 0 \text{ and } \delta_j^{\mathrm{R}} \text{ (LTP)} \\ w_{ij}(t) - \beta(w_{ij}^{\max} - w_{ij}(t)) & \text{if } A_{\mathrm{mf}}(t) > 0 \quad \text{(LTD)} \\ w_{ij}(t) & \text{otherwise} . \end{cases} \qquad (9)$$

The learning rate constants at *mf–AIN* synapse are chosen in order to obtain a slower potentiation and depression than at the *pf–Pu* synapse.

2.3   Simulated Conditioning Experiments

Our simulations implement an approximation of Pavlovian conditioning protocols [3] and their impact on the cerebellar circuit. CS events triggered in the direct pathway ($Po{\rightarrow}mf$–collaterals${\rightarrow}AIN$) are represented by a continuous activation of the mossy fibres lasting for the duration of the CS. In the indirect pathway ($Po{\rightarrow}Gr{\rightarrow}pf{\rightarrow}$Purkinje cell) CS events are represented by a short activation of *pf*, corresponding to 1 ts, at the onset of the CS. US events are represented by a short activation (1 ts) of the *cf* co–terminating with the CS. The behavioural CR is simulated by a group of neurons receiving the output of the *AIN* group.

The performance of the circuit is reported in terms of frequency of effective CRs. A response is considered an effective CR if, after the presentation of the CS, the CR pathway is active before the occurrence of the US and the *AIN* is able to block the US–related response in the *IO*. To avoid the occurrence of unnatural late CRs as observed in the previous study due to the long persistence of the eligibility trace, a reset mechanism is introduced to inhibit the CS trace whenever the US reaches the cerebellar cortex.

2.4   Robot Associative Learning Experiments

In order to assess the ability of our model to generalize to real–world conditions and a variable set of inter–stimulus intervals (ISIs), the performance of the model has been evaluated interfacing the circuit to a simulated E–puck robot performing an unsupervised obstacle avoidance task in a circular arena (Fig. 2). During the experiments the occurrences of the CSs and the USs were not controlled by the experimenter, instead the stimuli occurred simply as a result of the interaction of the device with the environment.

The robot is equipped with a set of URs triggered by stimulating the proximity sensors (Fig. 2 *right*). The activation of the four frontal sensors due to a collision with the wall of the arena (US) triggers a counter–clockwise turn. A camera (16x16 pixels) constitutes the distal sensor and a group of cells in a different process responds to the detection of a red painted area around the border of the arena signalling the CS. Visual CSs and collision USs are conveyed to the respective pathways of the cerebellar circuit resulting in an activation of $A = 1$ for 1 ts. The activation of the CR pathway triggers a specific motor CR, i.e. a deceleration and a clockwise turn.

During the experiment the robot is placed in a circular arena exploring its environment at a constant speed. Thus, in the UR–driven behaviour, collisions with the wall of the arena occur regularly. The red patch CS is detected at some distance when the robot approaches the wall and is followed by a collision with the wall. The collision stimulates the proximity sensors thus triggering a US. Consequently, as in a normal conditioning experiment, the CS is correlated

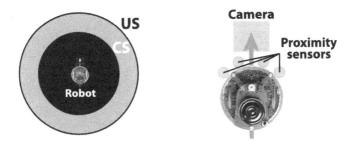

Fig. 2. Robot simulation environment. *Left*: the robot is placed in a circular arena where a red patch, acting as a CS stimulus, allows to predict the occurrence of the wall (US stimulus). *Right*: top view of the E–puck sensor model.

with the US, predicting it. The ISIs of these stimuli are not constant during this conditions due to the variations in the approaching angle to the wall. The aim of the experiments with the robot is twofold: on the one hand, to determine if the circuit can support behavioural associative learning, reflected by a reduced number of collisions due to adaptively timed CRs. On the other hand, to determine if the robot can maintain a memory of the association when re–exposed to the same stimuli after the CRs have been extinguished.

3   Results

3.1   Results of the Simulated Cerebellar Circuit

Acquisition, extinction and reacquisition sessions consisting of 100 trials (paired CS–US, CS alone, and paired CS–US presentations) with a fixed ISI of 300 ts (1 ms timestep) were performed. We have previously shown that the model can acquire the timing of the CR over a range of different ISIs [13]. As illustrated in Fig. 3a the circuit is able to reproduce the same kind of performance usually observed during behavioural experiments. CRs are gradually acquired until they stabilize to an asymptotic performance fluctuating around 80–90 % of CRs after the third block of paired presentations. CRs acquisition is due to the collaboration of the two sites of plasticity implemented in the model. The paired presentation of CSs and USs gradually depresses the efficacy of the pf–Pu synapses (black line in Fig. 3d) and in turns it leads to a pause in $PuSo$ (not shown). Potentiation at the nuclear site is only possible after the inhibition from $PuSo$ starts to be revoked. The appearance of the first CR is then critically dependent on the level of membrane potential excitability of the AIN that can subsequently repolarize.

Following acquisition, training the circuit without the reinforcing US leads to extinction of the acquired CRs (Fig. 3b). This is primarily due to the reversal effect of LTP in the pf–Pu synapse (Fig. 3e) which gradually restores the activity in the Purkinje cell during the CS period. A partial transfer of the memory will

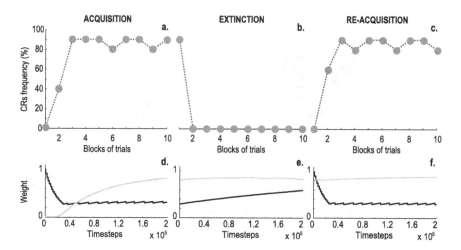

Fig. 3. Performance of the circuit during simulated acquisition, extinction and reacquisition. *Top*: Learning curves for experiments with an inter–stimulus interval of 300 ts. The percentage of CRs is plotted over ten blocks of ten trials. *Bottom row*: Synaptic weights changes induced by plasticity at the *pf–Pu* synapse (black trace) and *mf–AIN* (grey trace) synapse over the entire sessions.

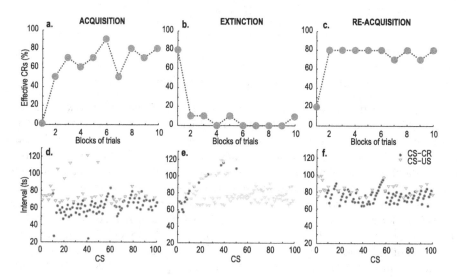

Fig. 4. Learning performance of the robot during the obstacle avoidance task. *Top*: Effective CRs per blocks of 10 CSs. *Bottom*: changes in the inter–stimulus interval (CS–US, triangles) and CR latency (CS–CR, circles) for the 300 CSs occurred during an experiment.

then still occur during the first part of the extinction training until the Purkinje cell activity is not completely recovered. Net LTD effects (extinction) in the nuclear synapse are then only visible when the nuclei are strongly inhibited by the recovered Purkinje cell activity. The retention of a residual nuclear plasticity is dependent on the small LTD ratio chosen to reverse the effects of the long term potentiation. This residual plasticity in the *mf–AIN* synapse is what actually leads to a faster expression of the CRs in the reacquisition session, as shown in Fig. 3e. The results of this simulation illustrate that, due to its increased excitability, less plasticity in the cortex is necessary to induce a rebound in the *AIN* and consequently express a robust CR. The magnitude of observable savings is therefore dependent on the amount of residual plasticity in the nuclei. Longer extinction training can completely reverse the plasticity so that reacquisition will require the same amount of conditioning trials as in the naïve circuit to express the first CR. This is in agreement with the findings of a graded reduction in the rate of reacquisition as a function of the number of extinction trials in conditioned rabbits [24,25].

3.2   Robotic Experiments Results

Cerebellar mediated learning induced significant changes in the robot behaviour during the collision avoidance task. Learning performance during acquisition training directly follows the results observed in the previous study [13]. Initially, the behaviour is solely determined by the URs, the robot moves forward until it collides with the wall and a turn is made. While training progresses, the robot gradually develops an association between the red patch (CS) and the collision (US) that, finally, is reflected in an anticipatory turn (CR).

The overall performance of the system during acquisition is illustrated by the learning curve in Fig. 4a. The CRs frequency gradually increases until it stabilises to a level of 70 % CRs after the fourth block of trials. The changes in ISI and CR latencies over the course of a complete session help explaining the learning performance of the model (Fig. 4b). During the first part of the experiment the CS is always followed by a collision (US) with latencies between 70–90 ts. The first CR occurs with a longer latency that can not avoid the collision with the wall and consequently more LTD at the *pf-PU* is induced. These long latencies CRs are due to the low excitability of the AIN that just started to undergo potentiation and a strong rebound can not be induced. As a result of LTD induction, in the following trials the circuit adapts the CRs latencies until most of the USs are prevented. When the CSs are no more followed by a US, a gradual increase in the CR latency is observed due to the sole effect of LTP. By the end of the session the balanced induction of LTD and LTP stabilizes the CRs latencies at a value of 15–20 ts shorter than the ISI. In a couple of cases very short latencies responses were elicited because the CS was intercepted while the robot was still performing a turn. These short ISIs didn't allow the *AIN* membrane potential to return at rest, therefore inducing a new rebound. In some other cases the robot occasionally missed the CS and a UR was then observed.

The next experiment was designed to test the extinction of the previously acquired CRs. In order to avoid the induction of LTD at the *pf–Pu* synapse, the US pathway of the circuit was disconnected. The extinction performance of the robot is illustrated by the learning curve in Fig. 4b. The CRs frequency gradually drops during the first two blocks of CSs until no more responses are observed after the fifth block. The consecutive presentations of CSs not followed by the US increase the CRs latencies (Fig. 4e) due to the effect of LTP at the *pf–Pu* synapse, that is no more counterbalanced by the LTD. Some residual CRs are still expressed for longer ISIs and are more resistive to extinction (i.e. trial 38 and 52). This is explained by the fact that for longer ISIs a pause can still be expressed by *PuSo* allowing the *AIN* to rebound.

Finally, a re–acquisition experiment was performed to investigate whether a new paired CS–US training session leads to a faster expression of the first CR. As illustrated by the learning curve in Fig. 4c a CR is already expressed during the first block of CS–US presentations while the performance of the system stabilizes during the second block of trials. As observed in the simulation results, the faster expression of the first CR is due to the residual plasticity in the *mf–AIN* synapse that allows for a faster repolarization of the *AIN* membrane potential. Consequently, a more robust rebound can be observed whenever *PuSo* releases *AIN* from inhibition. When compared to the acquisition session, the performance also appears more stable both in terms of USs avoided and latencies to the CR (Fig. 4f). Since during the acquisition session the *mf–AIN* synapse is still undergoing potentiation and a robust rebound can not be elicited until late during the training, weaker CRs are elicited that can not fully inhibit the *IO*. More LTD events are then induced at the *pf–Pu* synapse and a shortening of the latency is observed.

4   Discussion

Here we investigated the question whether the AIN provides the substrate of the cerebellar engram. In order to test this hypothesis we have presented a computational model of the cerebellar circuit that we explored using both simulated conditioning trials and robot experiments. We observed that the model is able to acquire and maintain a partial memory of the associative CR through the cooperation of the cerebellar cortex and the cerebellar nuclei. The results show how nuclear plasticity can induce an increase in the general excitability of the nuclei resulting in a higher facilitation of rebound when released from Purkinje cell inhibition. Nevertheless, the minimum numbers of trials necessary to produce a CR are still dependent on the induction of plasticity in the cortex. A limit of the model is that learning parameters in the two sites of plasticity need to be tuned in order to obtain a stable behaviour and prevent a drift of the synaptic weights. Moreover, given no direct evidence of a slower learning rate in the nuclear plasticity, we based our assumption on the evidence that, following extinction, short latency responses have been observed in rabbits after disconnection of the cerebellar cortex [24]. Savings have been shown to be a very strong phenomena and normally very few paired trials are sufficient to induce again a CR [2]. Our model showed that part of the memory can be copied to the nuclei, however this does not exclude that other mechanisms dependent on alternative parts of the cerebellar circuit could complement the ones we investigated so far. Different forms of plasticity have been discovered in almost all the cerebellar synapses [26] and they could be the loci of storage of long–term memories as well. In a different modelling study Kenyon [27], for example, proposed that memory transfer could take place in the cerebellar cortex due to bi–directional plasticity at the synapses between parallel fibres and the inhibitory interneurons, but this direct dependence has not been confirmed.

In relation to other models of the cerebellum, the present approach lays between purely functional models [28] and more detailed bottom-up simulations [7]. It embeds assumptions already implemented in those models – like the dual plasticity mechanism – but with the aim of keeping the model minimal in order to allow simulations with real–world devices.

In future work we aim to investigate other alternative hypothesis of memory transfer to be tested with robots in more realistic tasks.

Acknowledgments. This work has been supported by the Renachip (FP7–ICT–2007.8.3 FE BIO–ICT Convergence) and eSMC (FP7–ICT–270212) projects.

References

1. Napier, R., Macrae, M., Kehoe, E.: Rapid reacquisition in conditioning of the rabbit's nictitating membrane response. Journal of Experimental Psychology 18(2), 182–192 (1992)
2. Mackintosh, N.: The psychology of animal learning. Academic Press, London (1974)
3. Gormezano, I., Schneiderman, N., Deaux, E., Fuentes, I.: Nictitating membrane: classical conditioning and extinction in the albino rabbit. Science 138(3536), 33 (1962)
4. Hesslow, G., Yeo, C.: The functional anatomy of skeletal conditioning. In: A Neuroscientist's Guide to Classical Conditioning, pp. 88–146. Springer, New York (2002)
5. Thompson, R., Steinmetz, J.: The role of the cerebellum in classical conditioning of discrete behavioral responses. Neuroscience 162(3), 732–755 (2009)
6. Miles, F., Lisberger, S.: Plasticity in the vestibulo ocular reflex: a new hypothesis. Annual Review of Neuroscience 4(1), 273–299 (1981)
7. Mauk, M., Donegan, N.: A model of Pavlovian eyelid conditioning based on the synaptic organization of the cerebellum. Learning & Memory 4(1), 130–158 (1997)
8. Perrett, S., Ruiz, B., Mauk, M.: Cerebellar cortex lesions disrupt learning-dependent timing of conditioned eyelid responses. The Journal of Neuroscience 13(4), 1708–1718 (1993)
9. Bao, S., Chen, L., Kim, J., Thompson, R.: Cerebellar cortical inhibition and classical eyeblink conditioning. Proceedings of the National Academy of Sciences of the United States of America 99(3), 1592–1597 (2002)
10. Ohyama, T., Nores, W., Mauk, M.: Stimulus generalization of conditioned eyelid responses produced without cerebellar cortex: implications for plasticity in the cerebellar nuclei. Learning & Memory (Cold Spring Harbor, N.Y.) 10(5), 346–354 (2003)

11. Verschure, P., Kröse, B., Pfeifer, R.: Distributed Adaptive Control: The self-organization of structured behavior. Robotics and Autonomous Systems 9, 181–196, ID: 114; ID: 3551 (1992)
12. Verschure, P., Mintz, M.: A real-time model of the cerebellar circuitry underlying classical conditioning: A combined simulation and robotics study. Neurocomputing 38-40, 1019–1024 (2001)
13. Hofstötter, C., Mintz, M., Verschure, P.: The cerebellum in action: a simulation and robotics study. European Journal of Neuroscience 16(7), 1361–1376 (2002)
14. Ohyama, T., Nores, W., Medina, J., Riusech, F.A., Mauk, M.: Learning-induced plasticity in deep cerebellar nucleus. The Journal of Neuroscience 26(49), 12656–12663 (2006)
15. Pugh, J., Raman, I.: Mechanisms of potentiation of mossy fiber EPSCs in the cerebellar nuclei by coincident synaptic excitation and inhibition. The Journal of Neuroscience 28(42), 10549–10560 (2008)
16. Zhang, W., Linden, D.: Long-term depression at the mossy fiber-deep cerebellar nucleus synapse. The Journal of Neuroscience 26(26), 6935–6944 (2006)
17. Bernardet, U., Verschure, P.: iqr: A tool for the construction of multi-level simulations of brain and behaviour. Neuroinformatics 8, 113–134 (2010), doi:10.1007/s12021-010-9069-7
18. Marr, D.: A theory of cerebellar cortex. J. Neurophysiology 202, 437–470 (1969)
19. Albus, J.: A theory of cerebellar function. Math. Biosci. 10, 25–61 (1971)
20. Hesslow, G., Ivarsson, M.: Inhibition of the inferior olive during classical conditioned response in the decerebrate ferret. Exp. Brain. Res. 110, 36–46 (1996)
21. Aizenman, C., Linden, D.: Rapid, synaptically driven increases in the intrinsic excitability of cerebellar deep nuclear neurons. Nature Neuroscience 3(2), 109–111 (2000)
22. Ito, M.: Cerebellar long-term depression: characterization, signal transduction, and functional roles. Physiological Reviews 81(3), 1143–1195 (2001)
23. Racine, R., Wilson, D.A., Gingell, R., Sunderland, D.: Long-term potentiation in the interpositus and vestibular nuclei in the rat. Experimental Brain Research 63(1), 158–162 (1986)
24. Medina, J., Garcia, K., Mauk, M.: A mechanism for savings in the cerebellum. The Journal of Neuroscience: the Official Journal of the Society for Neuroscience 21(11), 4081–4089 (2001)
25. Weidemann, G., Kehoe, E.: Savings in classical conditioning in the rabbit as a function of extended extinction. Learning & Behavior 31, 49–68 (2003)
26. Hansel, C., Linden, D., DAngelo, E.: Beyond parallel fiber LTD: the diversity of synaptic and non-synaptic plasticity in the cerebellum. Nature Neuroscience 4, 467–475 (2001)
27. Kenyon, G.: A model of long-term memory storage in the cerebellar cortex: a possible role for plasticity at parallel fiber synapses onto stellate/basket interneurons. PNAS 94, 14200–14205 (1998)
28. Porrill, J., Dean, P.: Cerebellar motor learning: when is cortical plasticity not enough? PLoS Computational Biology 3, 1935–1950 (2007)

A Biomimetic Approach to an Autonomous Unmanned Air Vehicle

Fotios Balampanis[1] and Paul F.M.J. Verschure[1,2]

[1] Laboratory of Synthetic, Perceptive, Emotive and Cognitive Systems – SPECS,
Universitat Pompeu Fabra, Roc Boronat 138, 08018, Barcelona, Spain
fbalaban@cs.teiath.gr
[2] ICREA Institució Catalana de Recerca i Estudis Avançats, Passeig Lluís Companys 23,
E-08010 Barcelona, Spain
paul.vershure@upf.edu

Abstract. In an effort to create an autonomous flying robot, we sought inspiration in the evolution of insects. Previous research on their visual system has led to the creation of navigational models exploiting insect optomotor principles; these are applied in the present study, by using a fast and robust quadrotor. The results show that while a biomimetic approach for an unmanned air vehicle (UAV) is plausible the reliance on vision alone shows instabilities. These can be overcome by using gyroscopes, the technological equivalent of the haltere system found in flying insects.

Keywords: UAV, Vision, Biomimetic Robot, Quadrotor, Haltere.

1   Introduction

Small aerial vehicles could serve various tasks like environmental monitoring or wildlife tracking. Several biomimetic approaches have been using models of the visual system of insects to control these devices [1]. Such an approach also allows us to better understand evolutionary solutions to the problem of flight control. Here we investigate the use of an insect based optomotor based strategy for the autonomous control of a quadrotor. Our results show that a biomimetic approach is feasible but that vision alone is not enough; additional information from gyroscopes are required to obtain stability, resembling the mechanosensory haltere system of flying insects. The latter provide an essential source of information since they can provide short latency direct feed-forward motor commands stabilizing the flight trajectory [2].

2   Materials and Methods

In this study, we make use of models of the insect visual system [1][3] implemented using the neuronal simulator iqr (iqr.sourceforge.net). The system provides an elementary processing of the visual stimuli, extracting information about the speed and direction of the robot (figure 1). For our experiments we used the commercial quadrotor Ar.Drone© (Parrot S.A., Paris, France), which has a robust stability algorithm by using its gyroscopes. Finally, we used the XIM [4] as an arena.

T.J. Prescott et al. (Eds.): Living Machines 2012, LNAI 7375, pp. 333–334, 2012.

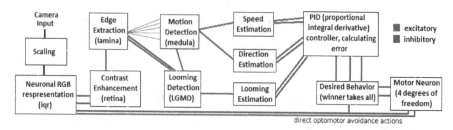

Fig. 1. Proposed optomotor image processing model as adapted from [3]

In the bottom camera experiments, the robot was navigating randomly in the arena, having the goals of avoiding red borders and landing / foraging 3 randomly distributed green patches. In the frontal camera set, the robot was spinning in a left-wise turn, at 3 different speeds, so to validate the speed and direction estimation systems.

3   Results and Discussion

In the bottom camera experiments, the robot had a landing success rate (landing in a 20cm radius around the patch) of 85% (mean landing efficiency of 3,735 sec.), a foraging success index of 70% (mean of 2,1 patches per trial)   and the avoidance maneuver error index was 0,122 and 0,140 respectively for the two scenarios. In the frontal camera experiments, the speed estimation neuron gave an overall correlated speed reading regarding the actual speed. The direction estimation group indicated the correct direction in a 82,48% of the flight time.

In this study we proposed a solution for an autonomous UAV. This quadrotor, with an appropriate behavioral layer, can manage to autonomously achieve the tasks of taking off, landing and retrieve goal patches. The results are encouraging for creating an autonomous biomimetic artificial flying insect, using a single visual sensor for navigation and stabilization sensors for robust flight trajectories.

References

1. Bermudez i Badia, S., Bernardet, U., Verschure, P.F.M.J.: Non-Linear Neuronal Responses as an Emergent Property of Afferent Networks: A Case Study of the Locust Lobula Giant Movement Detector. PLoSComputational Biology 6(3), e1000701 (2010), doi:10.1371/journal.pcbi.1000701
2. Bender, J.A., Dickinson, M.H.: A comparison of visual and haltere-mediated feedback in the control of body saccades in Drosophila melanogaster. J.Exp. Biol. 209, 4597–4606 (2006), doi:10.1242/jeb.02583
3. Bermúdez i Badia, S.: The Principles of Insect Navigation Applied to Flying and Roving Robots: from Vision to Olfaction. ETH ZURICH, Zurich (2006)
4. Bernadet, U., Bermúdez i Badia, S., Verschure, P.F.M.J.: The eXperience Induction Machine and its Role in the Research on Presence. Presence, 329–333, 388 (2007)

Towards a Framework for Tactile Perception in Social Robotics

Hector Barron-Gonzalez, Nathan F. Lepora, Uriel Martinez-Hernandez, Mat Evans, and Tony J. Prescott

Department of Psychology, University of Sheffield, Western Bank, S10 2TP, UK
{hector.barron,n.lepora,uriel.martinez,m.evans,
t.j.prescott}@sheffield.ac.uk

The first step for building reliable humanoid systems is to provide them with perceptual mechanisms that have human attributes, such as the skill development, social interaction, environmental embodiment and sensorial integration. Despite tactile perception being one of the most important elements for human interaction with the world, its implementation within artificial systems has been tardy, principally because it requires a complete integration with the motor systems and an environmental coupling to extract comprehensible information [1]. Thus, this work aims to generate a platform based on haptic information, allowing humanoids to perceive and represent surrounding objects using concepts fully grounded in the sensorial data.

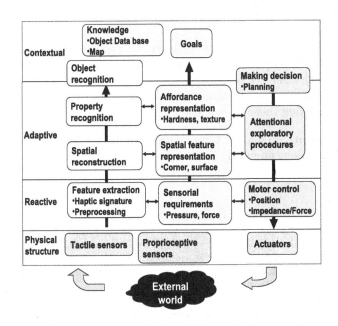

Fig. 1. Functional blocks of the tactile framework

The proposed architecture, inspired in the Distributed Architecture Control [2], has a modular design defined by three levels of processing that represent

T.J. Prescott et al. (Eds.): Living Machines 2012, LNAI 7375, pp. 335–336, 2012.

degrees of cognition (see Fig. 1). Such an architecture is not only targeted to build up the perceptual basis for social robotics, but also to explore neuropsychological models for understanding the development of human haptic skills.

The lowest layer is composed of proprioceptive sensors, tactile sensors and actuators. The respective biological elements in the human being are muscles and the mechanoreceptors located in tendons and skin. A higher level of the framework provides enactive mechanisms to extract significant haptic information and control the hand. Engineering advances related to the function of the cerebellum could provide a substantial contribution to haptic manipulation, for example by representing forward and inverse models of the interaction of the manipulator with its environment for optimal control [3].

The second level provides a type of processing that is more representational, such as the physical characteristics of the objects being identified. The implementation of this layer requires that the haptic signatures are grouped in spatiotemporal patterns, generating a hierarchical representation of the object [4]. For instance, probabilistic approaches have been efficiently used to reconstruct the spatial configuration while performing object manipulation [5]. On the other hand, qualitative information is linked to the action that objects afford. Thus, we propose that the appropriate actions should be learnt as new grounded tactile concepts are formed. Such exploratory procedures, driven by attention, provide a mechanism for active perception, as implemented in Bayesian approaches to cortical modelling [6]. Making decisions, recognizing objects and long-term memory are the executive functions for the haptic perception that comprises the highest level of the framework.

We conclude that this framework, designed with an engineering methodology, can in principle be mapped onto a cortical implementation. Furthermore, we propose that the framework be implemented using development cycle that allows the gradual improvement of haptic capabilities in androids, until potentially reaching or exceeding those of humans.

Acknowledgments. This work was supported by FP7 grants EFAA (ICT-248986) and the Mexican government under CONACYT.

References

1. Lederman, S.J., Klatzky, R.L.: Haptic perception: A tutorial. Attention, Perception and Psychophysics 71(7), 1439–1459 (2009)
2. Verschure, P.F.M.J., Voegtlin, T., Douglas, R.J.: Environmentally mediated synergy between perception and behaviour in mobile robots. Nature 425, 620–624 (2003)
3. Wolpert, D.M., Kawato, M.: Multiple paired forward and inverse models for motor control. Neural Networks 11(7-8), 1317–1329 (1998)
4. Hawkins, J., George, D., Niemasik, J.: Sequence memory for prediction, inference and behaviour. Philosophical Transactions of the Royal Society B: Biological Sciences 364(1521), 1203–1209 (2009)
5. Thrun, S., Burgard, W., Fox, D.: Probabilistic robotics. MIT Press (2006)
6. Friston, K.: The free-energy principle. A unified brain theory. Nat. Rev. Neurosci. 11(2), 127–138 (2010)

A Locomotion Strategy
for an Octopus-Bioinspired Robot

Marcello Calisti, Michele Giorelli, and Cecilia Laschi

The BioRobotics Institute, Scuola Superiore Sant'Anna, Pisa, Italy
m.calisti@sssup.it

Abstract. In this paper a locomotion strategy for a six-limb robot inspired by the octopus is shown. A tight relationship between the muscular system and the nervous systems exists in the octopus. At a high level of abstraction, the same relationship between the mechanical structure and the control of the robot is presented here. The control board sends up to six signals to the limbs, which mechanically perform a stereotypical rhythmical movement. The results show how by coordinating only two limbs an effective locomotion is achieved.

Keywords: robot locomotion, soft robotics, bioinspiration.

Currently, biology and robotics are working together to their mutual advantage. By copying the fewest anatomical structures or by reproducing the key principles, roboticists can mimic biological systems across the different levels of detail. Recently, much work has highlighted the interesting features of the octopus, and robotic artifacts mimicking the octopus arm [1] and octopus locomotion [2] have been presented. Biological studies demonstrate how arm extension and fetching are stereotyped movements, and therefore can be defined as basic motion patterns. The executive control of these basic patterns are embedded within the neuromuscular system itself, which suggests that brain commands are issued only for scaling, adjusting and combining movements to achieve the desired end results [3]. In this extended abstract, an octopus-bioinspired robotic platform able to move in all directions is used to test a coordination strategy to achieve a crawling locomotion. In the mechanical structure of the robot limb, a single stereotyped movement is embedded. By copying the complex connections of the neuromuscular system of the octopus arm, the movements performed by the octopus during crawling are easily reproduced within this mechanical structure [4], even though the full richness of the octopus arm behaviour is lost.

The robot is depicted in Fig. 1a. It is composed of a central base with 6 limbs radially distributed, each actuated by DC motors. One dsPIC microcontroller with 6 h-bridges drives the 6 motors in both directions, while 6 current sensors ensure current feedback. Driving the motor in a clockwise direction, a pulling action is achieved, while in the anti-clockwise direction a pushing action is performed. The current during several pulling actions, with the limb either in contact or detached from the ground, was recorded. Higher peaks were reached when the limb touched the ground, and an empirical current threshold,

T.J. Prescott et al. (Eds.): Living Machines 2012, LNAI 7375, pp. 337–338, 2012.
© Springer-Verlag Berlin Heidelberg 2012

T_{max}, was defined as in Fig. 1b. A second threshold was empirically defined as $t_{min} = 0.65T_{max}$. A simple trigger-coordination was implemented: when a frontal limb performing a pulling action has a current that rises above T_{max}, the opposite limb is activated to perform a pushing action; conversely, when current in the frontal limb comes down to t_{min}, the rear limb is deactivated.

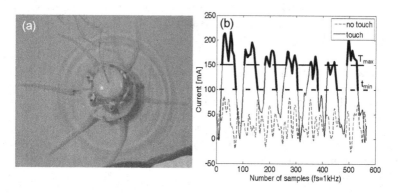

Fig. 1. In picture (a) a snapshot of the robot is shown. In picture (b) the tracks of the currents are shown.

Only one pair of arms is used at the same time. The trigger-coordination strategy introduced here leads to two opposite results: a motion status when the robot performs a step forward; or a stall status when the robot fails to move. Since the pushing limb is controlled relative to the pulling limb, the starting and ending position cycle by cycle are not known. Thus, during some cycles the pushing limb is in the right position to cooperate with the pulling limb; on the other hand, in few cycles it touches the ground too early or too late and hinders the pulling action of the frontal limb. Qualitatively, this strategy leads to more motion statuses than stall statuses. Conversely, if both arms are activated without coordinating them at all, the same two results are obtained, but with an inverse ratio: the robot performs more stall statuses than motion statuses.

References

1. Laschi, C., Mazzolai, B., Cianchetti, M., Margheri, L., Follador, M., Dario, P.: A Soft Robot Arm Inspired by the Octopus. Advanced Robotics 26 (2012)
2. Calisti, M., Giorelli, M., Levy, G., Mazzolai, B., Hochner, B., Laschi, C., Dario, P.: An octopus-bioinspired solution to movement and manipulation for soft robots. Bioinsp. Biomim. 6, 036002 (2011)
3. Sumbre, G., Fiorito, G., Flash, T., Hochner, B.: Neurobiology Motor control of flexible octopus arms. Nature 433, 595–596 (2005)
4. Calisti, M., Arienti, A., Renda, F., Levy, G., Hochner, B., Mazzolai, B., Dario, P., Laschi, C.: Design and development of a soft robot with crawling and grasping capabilities. In: IEEE International Conference on Robotics and Automation, St. Paul, Minnesota USA (to appear, 2012)

Design and Modeling of a New Biomimetic Robot Frog with the Ability of Jumping Altitude Regulation

Sadjad Eshgi, Vahid Azimirad, and Hamid Hajimohammadi

The Center of Excellence for Mechatronics, School of Engineering Emerging Technologies,
University of Tabriz, Tabriz, Iran
azimirad@tabrizu.ac.ir

Abstract. This paper introduces a new designed biomimetic robot frog which has the capability of balancing and controlling the height and length of jumping process. Hence a payload is considered inside the robot to keep its balance in high altitude. The positions of motors which generate the force of jumping operation are evaluated. The obtained results showed significant effect of the new design on jumping, balancing and altitude regulation.

Keywords: Robot frog, Balancing, Biomimetic robot.

1   Introduction

Most of the jumping animals use a complex system in the front and back legs with ability to store energy and release it at the special moment [1]. In the proposed mechanism the robot can instantly jump to a specific location or even going up the stairs. New robot frog is designed which have capability of balancing in jump process as well as jumping with calculated altitude and length.

2   The Modeling and Simulation

The robot at the moment before jumping holds his feet in front of the vertical sides of the body [2]. It helps the robot to jump longer and higher but it should open the front legs immediately after launch to increase the jumping height and moving the center of mass gravity. The next step is for the landing of the frog. The legs need to be open enough for the robot to be able to retain its balance and have a safe landing. For this action the legs which have been opened in previous step, have to be back at 30 degrees with respect to the previous position. As it is shown in Fig. 1a one of the important issues in the design of the robot is to have the capability of jumping with specified height and length. It is done by two small linear motors installed on the lever of the shafts on both sides and both of the servo motors use this mechanism, so we are able to move the pins. One of the specifications of this mechanism is non-linear motion. The robot can move arbitrarily and not necessarily in a linear direction and- reach the target point.

T.J. Prescott et al. (Eds.): Living Machines 2012, LNAI 7375, pp. 339–340, 2012.

As it is shown in Fig. 1b the robot at the moment before jumping holds his feet in front of the vertical sides of the body. It helps the robot to jump longer and higher but it should open the front legs immediately after launch to increase the jumping height and moving the center of mass.

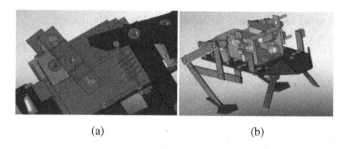

(a)   (b)

Fig. 1. Mechanism - of changing stored energy in the springs (a). The cad models in the moments before landing (b).

The dynamic equations are calculated. Then by using react force the robot can jump and its gravity center position and body mass should be known [2-3]. The coordinates of its center of mass motion can be obtained based on the angles of the two links [1]. Then by one can obtain the reaction force exerted on the earth by robot by geometrical calculation of forces. The maximum length is achieved when the robot reaction force is 45 degrees with respect the ground and the highest altitude jumping of the robot is obtained when reaction force is 90 degrees with respect the ground.

3   Conclusion

Modeling and design of a new robot frog with capability of balancing and jumping with different height and length is presented. By computing the energy of system the height and length of jump is achieved and by adding a counterweight, safety and balancing of system was increased. The simulation results showed the effectiveness of the work.

References

1. Fan, J.Z.: Biological jumping mechanism analysis and modeling for frog robot. Journal of Bionic Engineering 5, 181–188 (2008)
2. Tsukagoshi, H., Mori, Y., Sasaki, M., Tanaka, T., Kitagawa, A.: Development of Jumping & Rolling Inspector to Improve the Debris-traverse Ability. Journal of Robotics and Mechatronics 15-5, 482–490 (2003)
3. Redd, N.S., Ray, R., Shome, S.N.: Modeling and Simulation of a Jumping Frog Robot. In: Proceedings of the IEEE International Conference on Mechatronics and Automation, Beijing, China, August 7-10 (2011)

Sensation of a "Noisy" Whisker Vibration in Rats

Arash Fassihi, Vahid Esmaeili, Athena Akrami, Fabrizio Manzino,
and Mathew E. Diamond

Scuola Internazionale Superiore di Studi Avazanti (SISSA)
{fassihi,vesmaeil,akrami,manzino,diamond}@sissa.it

Abstract. To provide a biological framework to be later applied in robotics, we have devised a delayed comparison task in which subjects discriminate between pairs of vibration delivered either to their whiskers, in rats, or fingertips, in human, with a delay inserted between the two stimuli. The task is to compare two successive stimuli, with different position standard deviations defined by σ_1 and σ_2. By varying the stimulus duration we have observed that rats' performance improves for longer stimuli, suggesting that for stimuli with a probabilistic structure, evidence can be accumulated over time. On the other hand a change in stimulus duration biased human subjects. This experiment constrains models for the integration of tactile information in robotics.

1 Introduction

Tactile sense has been difficult to engineer in machines. To deepen the knowledge of tactile processing with the long term aim of developing robots that emulate living biological organisms we have designed a vibration discrimination task to study how rats and humans extract tactile vibration signals and accumulate such signals over time.

2 Methods

Stimuli are composed of a random series of positions ("noise") taken from Gaussian distributions with 0 mean and standard deviation σ, ranging from 0.25 to 3 mm. Sample position distributions characterized by $\sigma1$ (Base) and $\sigma2$ (Comparison) are shown in figure 1. Trial starts when the rat places its snout in a nosepoke. The whiskers are in contact with a plate which is driven by a vibration generator. Base and Comparison stimuli are delivered to whiskers with an intervening delay. A "Go" cue is then sounded and rat must compare Base and Comparison stimuli, determining whether $\sigma2 > \sigma1$ or $\sigma2 < \sigma1$, and turn to the left or right spout. Each trial is rewarded only for a correct choice of the animal. Human participants performe the same discrimination task by putting their left index fingertip in contact with tip of a vibrating rod and responding by pressing keys on a keyboard. They receive feedback on a monitor after each response.

T.J. Prescott et al. (Eds.): Living Machines 2012, LNAI 7375, pp. 341–342, 2012.

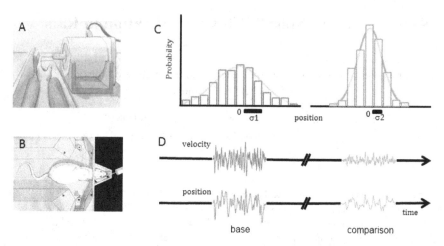

Fig. 1. The apparatus for human (A) and rats (B). Sample trial position distribution overlaid with ideal probability density function (C). Velocity and position, over time, for the same sample trial (D)

3   Results

Both rats and humans perform well in this task. By fixing σ1 at 0.9mm and varying σ2, we measured discrimination threshold (0.99mm and 1.06mm for human and rat respectively).

By varying the duration of the stimulus we have found that the sensory system accumulates information for at least 400 ms in rats. In contrast in human, subjects fail to improve their performance even if stimulus duration increases.

The next step will be to determine the stimulus features that most affect the subjects' choice and to develop algorithms for extracting the same signal from a prosthetic hand with synthetic skin, courtesy of Pisa Biorobotics Laboratory. We believe our comparative study of sensory discrimination in rat and human provides us with a comprehensive framework to nurturing good biomimetic strategies.

Integrating Molecular Computation and Material Production in an Artificial Subcellular Matrix

Harold Fellermann, Maik Hadorn, Eva Bönzli, and Steen Rasmussen

Center for Fundamental Living Technology, University of Southern Denmark,
Campusvej 55, 5230 Odense M, Denmark
harold@sdu.dk
http://flint.sdu.dk/

Keywords: biochemical information processing and production, supermolecular self-assembly, DNA computing, membrane computing, stochastic pi-calculus.

Living systems are unique in that they integrate molecular recognition and information processing with material production on the molecular scale. Predominant locus of this integration is the cellular matrix, where a multitude of biochemical reactions proceed simultaneously in highly compartmentalized reaction compartments that interact and get delivered through vesicle trafficking.

The European Commission funded project MatchIT (Matrix for Chemical IT) aims at creating an artificial cellular matrix that seamlessly integrates information processing and material production in much the same way as its biological counterpart [1]: the project employs addressable chemical containers (chemtainers) interfaced with electronic computers via mechano-electronic microfluidics.

As chemtainers we utilize artificial vesicles, oil droplets, water droplets in ionic liquids, and DNA nano cages. Chemtainers are decorated with short DNA tags that serve as programmable addresses and allow for chemtainer–chemtainer and chemtainer–matrix interaction through reversible hybridization [2,3], see Fig. 1. left, as well as DNA computing operations such as relabeling and simple Boolean

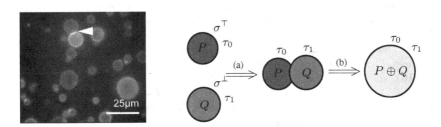

Fig. 1. left) Micrograph of vesicle-vesicle adhesion promoted by complementary DNA tags. **right)** Schematic of an operation within the MatchIT calculus that hybridizes and fuses two complementary tagged chemtainers.

T.J. Prescott et al. (Eds.): Living Machines 2012, LNAI 7375, pp. 343–344, 2012.

computations [4]. Chemtainers can carry molecular cargo and promote chemical reactions upon specific or unspecific fusion [5].

MatchIT embeds chemtainers within a microfluidic environment that enables electronic control of chemtainer creation, loading, and manipulation. In addition, special microfluidic channels allow for chemtainer docking, electrophoretic separation and reencapsulation of cargo. Paired with realtime feedback, this allows for control of chemical reaction cascades at the molecular level.

Here, we present a special purpose calculus that we have developed in order to cope with the inherent complexity of MatchIT objects and operations in a straightforward manner. Our calculus closely follows the line of *brane calculi* developed for expressing nested membrane systems [6]. It can both be used for modelling chemtainers and interactions, as well as ultimately programming the microfluidic device. Elements of this calculus are (possibly nested) chemtainer systems, their cargo, and free floating as well as chemtainer associated DNA strands (address tags). Operations of the calculus allow for chemical reactions of cargo, chemtainer and tag manipulation. Additional operations can encode for MEMS specific functionalities such as chemtainer creation and content separation by electrophoresis. The example of tag specific chemtainer fusion is shown in Figure 1. right. Equipping operations with rate constants allows for quantitative analysis of system transitions [7], e.g. by means of a Gillespie simulation where the MatchIT calculus is used to dynamically identify the set of possible transitions for the current system state.

We are particularly interested in the transitive and reflexive closure of the calculus in order to identify the set of MatchIT producible objects. Furthermore, we identify means to automatically derive operation sequences (programs) necessary to produce a given object, which forms the core of a MatchIT compiler.

References

1. cf, http://fp7-matchit.eu/
2. Hadorn, M., Hotz, P.E.: DNA-mediated self-assembly of artificial vesicles. PLoS One 5(3), e9886 (2010)
3. Hadorn, M., Bönzli, E., Hotz, P.E., Hanczyc, M.: Programmable and reversible DNA-directed self-assembly of emulsion droplets (submitted, 2012)
4. Benenson, Y., Gil, B., Ben-Dor, U., Adar, R., Shapiro, E.: An autonomous molecular computer for logical control of gene expression. Nature 429, 423–429 (2004)
5. Caschera, F., Sunami, T., Matsuura, T., Suzuki, H., Hanczyc, M.: Programmed vesicle fusion triggers gene expression. Langmuir 27(21), 13082–13090 (2011)
6. Cardelli, L.: Brane Calculi. In: Danos, V., Schachter, V. (eds.) CMSB 2004. LNCS (LNBI), vol. 3082, pp. 257–278. Springer, Heidelberg (2005)
7. Cardelli, L., Mardare, R.: The measurable space of stochastic processes. In: 7th International Conference on the Quantitative Evaluation of Systems, QEST 2010, pp. 171–180. IEEE Publishing (2010)

WARMOR: Whegs Adaptation and Reconfiguration of MOdular Robot with Tunable Compliance

Max Fremerey[1], Goran S. Djordjevic[2], and Hartmut Witte[1]

[1] Ilmenau University of Technology, Department of Biomechatronics, Ilmenau, Germany
{maximilian-otto.fremerey,hartmut.witte}@tu-ilmenau.de
[2] University of Niš, Faculty of Electronic Engineering, Niš, Serbia
goran.s.djordjevic@elfak.ni.ac.rs

Abstract. This paper introduces the idea of WARMOR: Whegs™ adaptation and reconfiguration of modular robot with tunable compliance. WARMOR robot revisits the purpose of using Whegs for robust propelling the robot over unknown and scattered terrain by approaching two topics necessary for improved performance: modularity for reconfiguration and tunable compliance for adaptation. The paper describes the robot, especially the design of wheg-derivatives and controller.

Keywords: Whegs™, tunable compliance, modular robotics, autonomous systems.

1 Introduction

In addition to existing Wheg™-driven robots (e.g. R-Hex series [1]) the WARMOR project aims on compliant wheel/leg design and modularity. The hexapod R-Hex robot was able to adapt to environment due to: inherent nonlinear flexures, a form of preflexes to dynamically stabilize the gait, and frequency of Wheg™ rotation. However, the adaptability of the original concept was fairly limited by the fact that preflexes were unchanged regardless of terrain and additional weight. Additionally they were not designed to support reconfigurable robots. Besides adaptive locomotion strategy for optimal running, it is estimated that the most critical improvement would be to have some kind of tuneable flexibility in the wheel/leg combination. Thus our research focuses on a real-time tuneable and adaptive compliance to smoothen locomotion, leading to optimized gait patterns and better energy efficiency. Tuneable compliance and overall locomotion parameters are combined into a lumped locomotion model. The modularity concept in body design strives to a reconfigurable robot body based on elemental blocks of actuation, and different kind of linkages. Such an elemental block is illustrated in figure 1b.

That way we aim to achieve smoother and faster running, better targeting, and more efficient obstacle negotiating over a family of WARMOR robots using the same principles of preflex tuning, modularity and adaptation.

T.J. Prescott et al. (Eds.): Living Machines 2012, LNAI 7375, pp. 345–346, 2012.

2   Design of WARMOR 1

The design of WARMOR 1 follows the guideline for mechatronic development VDI 2206. Material used is POM (polyoxymethylene). The robot's overall dimension is 140 mm · 120 mm · 75 mm. Four modified BMS-303 servo drives generate the torque required for locomotion. The legged wheels have four spokes arranged rectangular in a circle. Material used is glass fiber reinforced plastics, the thickness of each spoke amounts 0.3 mm. Current diameter of the legged wheels is 75 mm. For investigation of tunable compliance, wheel elements of WARMOR 1 are manually exchangeable. Currently, WARMOR 1 is open-loop remote controlled.

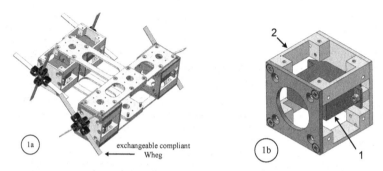

Fig. 1. 1a: CAD drawing of WARMOR 1, 1b: Elemental block of actuation, 1= servo drive, 2= framework (dimensions: 26 mm · 26 mm · 26 mm)

3   Experiments

First experiments demonstrate the function of WARMOR 1. Next step is the identification of different locomotion modes proposed by the scientific community [1,2].

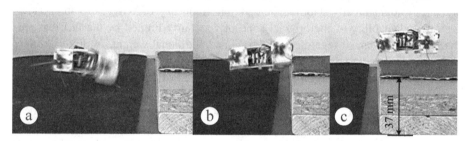

Fig. 2. First functional test of WARMOR 1, overcoming a step (height: 37 mm)

References

[1] Saranli, U., Buehler, M., Koditschek, D.E.: RHex: A Simple and Highly Mobile Hexapod Robot. The International Journal of Robotics Research 20, 616–631 (2001)
[2] Full, R.J., Koditschek, D.E.: Templates and anchors: neuromechanical hypotheses of legged locomotion on land. Journal of Experimental Biology 202, 3325–3332 (1999)

Inverse and Direct Model of a Continuum Manipulator Inspired by the Octopus Arm[*]

Michele Giorelli[**], Federico Renda[**], Andrea Arienti, Marcello Calisti,
Matteo Cianchetti, Gabriele Ferri, and Cecilia Laschi

The BioRobotics Institute, Scuola Superiore Sant'Anna, Pisa, Italy
{michele.giorelli,federico.renda,andrea.arienti,
marcello.calisti,matteo.cianchetti,gabriele.ferri,
cecilia.laschi}@sssup.it

Abstract. The extraordinary manipulation capabilities of the octopus arm generate a big interest in the robotics community. Many researchers have put a big effort in the design, modelling and control of continuum manipulators inspired by the octopus arm. New mathematical tools could be introduced in the robotics community and new sources of inspiration are needed in order to simplify the use of the continuum manipulator. A geometrical exact approach for direct model and a Jacobian method for the inverse model are implemented for a continuum manipulator driven by cables. The first experimental results show an average tip error of less than 6% of the manipulator length, and a good numerical performance to solve the inverse kinematics model.

Keywords: Tendon-driven manipulator, continuum manipulator, soft robotics, continuum mechanics, inverse kinematics.

The octopus arm has a particular structure called muscular hydrostat. This structure lacks a skeletal support, but it is entirely composed of muscle fibres, which are incompressible. Therefore, the volume of the octopus arm remains constant during all typical movements (elongation, shortening, bending and torsion), given by the muscle fibres oriented in three directions: parallel to the longitudinal axis (*longitudinal muscles*), perpendicular to the longitudinal axis (*transverse muscle*), and oblique around the long axis (*oblique muscles*).

In order to build an octopus-inspired manipulator we take in account three main characteristics: the conical shape, the soft isovolumetric structure and the arrangement of muscle fibres. In particular, after different tests, we found that silicone is the best material in term of compliance, density (close to the octopus arm), and performance. To mimic the shape and the longitudinal muscles of the octopus arm we build a conical shape manipulator with n nylon cables attached at different points of the structure. Pulling the cables generates several shape configurations in three dimensional space (fig. 1).

In order to find the direct model of the continuum manipulator driven by the cables, the exact geometrical approach has been used [1]. The shape is described by

---

[*] This work was supported in part by the European Commission in the ICT-FET OCTOPUS Integrating Project, under contract no. 231608.
[**] Both authors contributed equally to this work.

four beam deformations: the torsion, the two curvatures and the longitudinal strain of

Fig. 1. A 3-D configuration of a cable-driven robot applying the exact geometrical approach

the robot arm, respectively indicated with τ, ξ, k, and q. The direct model maps the cable tensions $\mathbf{t} \in \Re^n$ into the deformations, leading to the following equations:

$$\frac{d}{ds}\tau = -\tau A(y,\mathbf{t}) - k\xi B(y,\mathbf{t}) + kC(y,\mathbf{t}) + \xi D(y,\mathbf{t})$$

$$\frac{d}{ds}\xi = -\xi E(y,\mathbf{t}) - \tau k F(y,\mathbf{t}) + \tau G(y,\mathbf{t}) + H(y,\mathbf{t})$$

$$\frac{d}{ds}k = -kE(y,\mathbf{t}) + \tau\xi F(y,\mathbf{t}) + \tau G(y,\mathbf{t}) + H(y,\mathbf{t})$$

$$q = I(k,\xi,y,\mathbf{t})$$

(1)

where y is the distance of the cables from the manipulator midline.

An Inverse Kinetics Model (IKM) has been found applying the Jacobian method in a simple case [2]. We consider only two coplanar cables attached to the tip. The two cables can move the arm in the 2-D space. The IKM is performed in two stages. In the first one, we have found a map between the tip position $\mathbf{x} \in \Re^2$ and the cable tensions $\mathbf{t} \in \Re^2$: $\mathbf{x} = \mathbf{f}(\mathbf{t})$. In the second one, we have found the velocity model $\dot{\mathbf{x}} = \mathbf{J}(\mathbf{t})\dot{\mathbf{t}}$, where $\mathbf{J} = \dfrac{\partial \mathbf{f}}{\partial \mathbf{t}} \in \Re^{2x2}$ is the Jacobian matrix. The classical iterative algorithm employed for solving the inverse kinetics problem is defined as follows:

$$\mathbf{t}_{k+1} = \mathbf{t}_k + \mathbf{\eta}^T \mathbf{J}^{-1}(\mathbf{t}_k)(\mathbf{x}^d - \mathbf{x})$$

(2)

where $\mathbf{\eta} \in \Re^2$ are the constants that control the convergence rate of (2), and $\mathbf{x}^d \in \Re^2$ is the desired tip position.

A preliminary experimental validations of the direct and inverse models show an average tip error of less than 6% of the total robot length. Furthermore, the inverse kinetics algorithm is computed in less than fifteen iterations, showing a good degree of convergence.

References

1. Renda, F., et al.: A 3D Steady State Model of a Tendon-Driven Continuum Soft Manipulator Inspired by Octopus Arm. Bioinspiration & Biomimetics 7 (2012)
2. Giorelli, M., et al.: A Two Dimensional Inverse Kinetics Model of a Cable Driven Manipulator Inspired by the Octopus Arm. In: Proc. of IEEE ICRA 2012, St. Paul, USA (2012)

A Biomimetic, Swimming Soft Robot Inspired by the *Octopus Vulgaris*

Francesco Giorgio Serchi, Andrea Arienti, and Cecilia Laschi

The Biorobotics Institute, Research Centre on Sea Technologies and Marine Robotics
Viale Italia, 57126 Livorno, Italy
{f.serchi,andrea.arienti,cecilia.laschi}@sssup.it
http://www.bioroboticsinstitute.it

Abstract. This paper describes a first prototype of a cephalopod-like biomimetic aquatic robot. The robot replicates the ability of cephalopods to travel in the aquatic environment by means of pulsed jet propulsion. A number of authors have already experimented with pulsed jet thrusting devices in the form of traditional piston-cylinder chambers and oscillating diaphragms. However, in this work the focus is placed in designing a faithful biomimesis of the structural and functional components of the *Octopus vulgaris*, hence the robot is shaped as an exact copy of an octopus and is composed, to a major extent, of soft materials. In addition, the propelling mechanism is driven by a compression/expansion cycle analogous to that found in cephalopods. This work offers a hands-on experience of the swimming biomechanics of chephalopods and an insight into a yet unexplored new mode of aquatic propulsion.

Keywords: Soft robots, biomimetic propulsion, biomechanics.

1 Introduction

Lately, cephalopod-inspired pulsed-jet propulsion has been suggested as an alternative to the existing traditional and bioinspired ways of aquatic locomotion [1]. Cephalopods travel in water via a repetition of discontinuous jet pulses [2]. It is demonstrated [1] that this pulsed-jet propulsion offers significant benefits when compared against a continuous jet, like the one produced by a traditional propeller. A few pulsed-jet propelled underwater vehicles have recently been presented (i.e.[4,3,5]), however the present work focuses on the development of a completely new type of UUV which faithfully draws inspiration from the swimming technique adopted by cephalopods. The robot presented herein (Fig. 1) is unique in that it tightly matches the shape of a real octopus, it is made of flexible material and it employs an efficient propulsion mechanism which is very closely reminiscent of that one of cephalopods.

2 Biomimetic Soft Vortex Thruster

A polyurethane mould of an actual octopus was produced so that an exact copy of an original specimen could be obtained by filling the polyurethane mould

T.J. Prescott et al. (Eds.): Living Machines 2012, LNAI 7375, pp. 349–351, 2012.

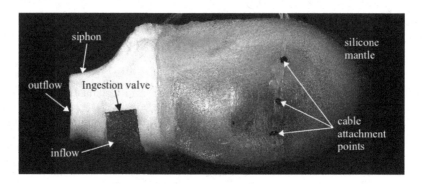

Fig. 1. Side view of the octopus-inspired prototype

with silicone. The silicone octopus mantle achieved in this way (see Fig. 1) was endowed with a cable driven actuation which drives the inward compression of the cavity and hence the expulsion of the internal fluid. Once the cables are released, the expansion of the mantle cavity is driven by the stresses generated within the silicone walls. This inflation of the chamber generates a sufficient negative pressure for ambient water to be sucked from outside through an orifice and, in this way, refill the mantle cavity. The repetition of these two phases allows the robot to effectively travel in water according to a propulsion routine which closely resembles the one of a swimming cephalopod.

After assemblage, the robot comprises of the silicone mantle, a nozzle, an inflow valve (Fig. 1) and the internal actuator and is supplied by two ion-lithium batteries which are immersed in the thicker layer of silicone of the upper portion of the mantle.

3   Results

One set of preliminary tests has been performed in water in order to roughly assess the swimming performances of this first prototype. The tentative analysis of this prototype seems to suggest that the implementation of a soft, collapsible syphon could aid in increasing the jetting performances of the robot. In addition, a more efficient pulsation cycle should entail low frequencies and an impulsive mantle contraction.

Acknowledgements. This work was supported by the CFDOctoProp project European Reintegration Grant and the European Commission in the frame of the ICT-FET OCTOPUS Integrating Project.

References

1. Krueger, P.S., Gharib, M.: The significance of vortex ring formation to the impulse and thrust of starting jet. Physics of Fluid 15, 1271–1281 (2003)

2. Gosline, J.M., DeMont, M.E.: Jet-propelled swimming squids. Scientific American 252, 96–103 (1985)
3. Moslemi, A.A., Krueger, P.S.: Propulsive efficiency of a biomorphic pulsed-jet underwater vehicle. Bioinspiration and Biomimetics 5, 1–14 (2010)
4. Krieg, M., Mohseni, K.: Thrust characterization of a bio-inspired vortex ring generator for locomotion of underwater robots. IEEE Journal of Ocean Engineering 33 (2008)
5. Ruiz, L.A., Whittlesey, R.W., Dabiri, J.O.: Vortex-enhanced propulsion. Journal of Fluid Mechanics 668, 5–32 (2011)

Toward a Fusion Model of Feature and Spatial Tactile Memory in the Crayfish *Cherax Destructor*

Frank W. Grasso[1], Mat Evans[2], Jennifer Basil[3], and Tony J. Prescott[2]

[1] BioMimetic and Cognitive Robotics Laboratory, Department of Psychology,
Brooklyn College, NY, USA
[2] ATLAS Laboratory, Department of Psychology, University of Sheffield, Sheffield, UK
[3] LIBE Laboratory, Department of Biology, Brooklyn College, NY, USA

Keywords: BioMimetics, Crustacean, Crayfish, Memory, Spatial Exploration.

1 Introduction

Previous studies of the crayfish have demonstrated an ability to remember enclosure spaces and that this memory is informed by tactile information supplied by the animal's moving antennae. Simulation and robotic studies form a suitable method for exploring central mechanisms and excluding non-viable alternatives.

Behavior. Crayfish use their haptic sense (active touch) to navigate in enclosed spaces in the absence of visual input. Indeed as a nocturnal species that evolved to live in murky or sunless environments they are likely to possess a unique haptic-spatial sense. A characteristic of the haptic sense is that it requires active movement of the touch-sensitive appendages into mechanical contact with a surface to collect information that distinguishes shapes or determines the layout of an environment. Crayfish actively use their antennae to explore novel objects [1,2] and to remember, in some manner, the configuration of the surfaces in their environment [3]. As crayfishes are adept at navigating complex landscapes by relying on their haptic sense, our goal is to implement crayfish-like algorithms on the Craybot robotic platform to solve similar problems in real-world settings.

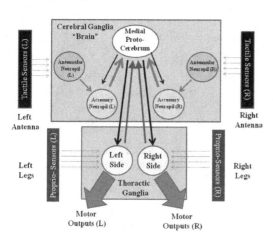

Fig. 1. The neural components of the crayfish navigation system. All are bilateral except the MPC. Two memory systems (one likely thoracic, the other cerebral) inform both independent and combined ongoing behaviors.

T.J. Prescott et al. (Eds.): Living Machines 2012, LNAI 7375, pp. 352–354, 2012.
© Springer-Verlag Berlin Heidelberg 2012

Neurobiology. The crayfish brain, while small ($< 10^7$ CNS neurons) compared to mammals, contains a sophisticated hierarchical structure with several processing levels. We hypothesize that the Medial Protocerebrum (MPC), an unpaired, central brain structure that receives highly processed perceptual information from all the senses and the thoracic ganglia is a site for spatial memory processing as well as higher order integration [4]. The two key streams of information for spatial mapping in crayfish, tactile input from the antennae and proprioceptive information from the legs ('step-counting' odometry) are known to converge in the MPC. (Fig 1). Our CrayTouchBot model models broad aspects of this neurobiological system.

Biorobotics. In previous cybernetic research [5, 6] we demonstrated the plausibility of an interrupt-based cross-correlation (IBCC) mechanism (where an event-punctuated series of traveled distances is compared to stored time-series) to account for crayfish spatial memory abilities. Specifically, to recognize a previously visited location in the context of a series of turns on a continuous wall that forms a closed circuit. Those studies were conducted in simulations with robots using infra-red proximity sensors to model the function of antennae and on-board wheel encoders to model the path-integration mechanism assumed by the model.

CrayTouchBot, *biomimetic robot platform for testing theories of crayfish spatial exploration*

New Sensors, New Information. Here we report on efforts to extend those biorobotics results from a BioMimetic perspective with a mechanical touch sensor developed to model the rat whiskers. These studies take advantage of the new information supplied by this sensor to inform a second memory model that captures additional aspects of crayfish behavior. A memory for the contours of contacted surfaces provides context and validation for positional hypotheses in the IBCC. Patterns of contact from a pair of antennae swept through standard arcs will be identical if the body (crayfish or agent) upon which the antennae are mounted has returned to a spatial location in the same orientation. Numerical correlation between the current sweeps or 'antennation' sensation patterns (ASP's) and stored previously experienced ASPs permit the classification of the current location as familiar or novel.

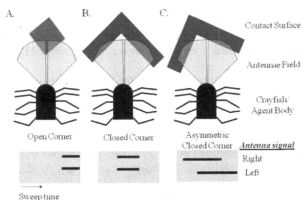

Fig. 2. Different surface contours lead to contact patterns that can be stored following stereotyped antennae sweeps. Relative orientation matters.

Combining Memory Systems. The uniqueness of an ASP-location pair cannot be guaranteed because features may repeat in a given environment and because deviations of body position and or orientation on subsequent visits to a location where a stored 'antennation' pattern may lead to a degradation of the 'antennation pattern'. Therefore the CrayTouchBot control system coordinates information in the ASP and IBCC memory systems to provide mutual information. The positional confidence of the IBCC biases the ASP recall process and vice versa to arrive at combined recall readouts that use both forms of stored information and recognition. These in turn inform the ongoing behavioral processes of CrayTouchBot during exploration to guide its steering decisions.

2   Significance

This modeling approach makes it possible to explore general hypotheses at the levels of observable behavior and known neurobiology. Crayfish exploring mazes are known to use thigmotaxis (wall-following) with antennal sweeps while they learn and remember the maze's spatial configuration [3]. The combination of the ASP and IBCC memory systems is a novel computational approach that is constrained by the neuroanatomy and neurobiology outlined above. The combination of left and right side originating information in the MPC finds expression in the CrayTouchBot in the mutual biasing of each memory system's content. The thigmotactic character of the IBCC reduces the ambiguities of orientation and position that limit the generalizability of ASP content and might be used to scaffold more advanced memory systems. Studies are underway to determine whether the behavior of the CrayTouchBot endowed with these interacting memory systems provides explanation of the dual aspects of crayfish exploratoration: thigmotactic movement punctuated by episodes of antennation at points containing tactilely inhomogeneous structures. Such intelligent systems will be useful in applications where GPS and vision are not available for spatial guidance.

References

1. Sandeman, D.: Crayfish antennae as tactile organs: their mobility and the responses of their tactile proprioceptors to displacement. Journal of Comparative Physiology 157, 363–373 (1985)
2. Zeil, J., Sandeman, R., Sandeman, D.: Tactile lobalization: the function of active antennal movements in the crayfish cherax destructor. Journal of Comparative Physiology 157, 607–617 (1985)
3. Basil, J., Sandeman, D.: Crayfish (cherax destructor) use tactile cues to detect and learn topographical changes in their environment. Ethology 106, 247–259 (2000)
4. Sandeman, D.: Structural and functional levels in the organization of decapod crustacean brains. In: Wiese, K., et al. (eds.) Frontiers in Crustacean Neurobiology, pp. 223–239. Birkhauser verlag, Basel (1990)
5. Appleman, G., Basil, J.A., Grasso, F.W.: Simulation of crayfish spatial exploration with a mobile robot. Biological Bulletin 195(2), 249 (1998)
6. Volz, S.G., Basil, J., Grasso, F.: Crayfish inspired representation of space via haptic memory in a simulated robotic agent. In: Prescott, T.J., et al. (eds.) Living Machines 2012. LNCS (LNAI), vol. 7375, pp. 286–297. Springer, Heidelberg (2012)

Development of Sensorized Arm Skin for an Octopus Inspired Robot – Part I: Soft Skin Artifacts

Jinping Hou[1], Richard H.C. Bonser[2], and George Jeronimidis[1]

[1] Centre for Biomimetics, University of Reading, UK
[2] School of Engineering and Design, Brunel University, UK

Abstract. Octopus skin was characterized to set design criteria for the artificial skin of an octopus-inspired robot. Young's moduli, failure strain and ultimate stress were obtained via uniaxial tensile tests. The fracture toughness of the skin was measured by using scissors cutting tests. Silicone rubber is waterproof and has a failure strain higher than 500%, but its fracture toughness is much lower than that of the real octopus skin. To overcome this problem, a knitted nylon fabric was chosen as reinforcement. A fabrication process was developed to optimize the stiffness and improve the quality of the skin artifact. Test results showed that the skin artifact is more flexible and tougher than the real skin.

Keywords: Octopus skin, biomimetic, skin artifact.

1 Introduction

Like animal skins, the man-made skin for an octopus-inspired robot needs to provide protection and sensing functions to the muscular units of the robotic arms. In addition, octopus suckers play a very important role in the animal's behaviour. The objective of this work is to develop an artificial skin composite material that is flexible, tough, waterproof and with embedded sensors and suckers. This work is presented her in a series of three papers. The first covers the development of the biomimetic skin itself. The second paper will present the development of flexible tactile sensor. The development of biomimetic suckers is presented in the last paper in the series.

2 Characterization of Octopus Skin

ESEM images show that the octopus skin is reinforced by randomly distributed collagen fibers. Skins of two Octopus vulgaris arms were tested using an Instron 5564 testing machine. For uniaxial tensile tests, the skin samples were clamped at the two ends and load was applied at a speed of 4mm/min. Fracture toughness of the skin was measured by scissors cutting tests.

Test results showed that material properties along and across the length of the arms are not significantly different. No difference was found between properties of skins from different arms. It was found that the Young's modulus and ultimate stress decrease along the longitudinal direction of the arm from base to tip. The average

T.J. Prescott et al. (Eds.): Living Machines 2012, LNAI 7375, pp. 355–356, 2012.
© Springer-Verlag Berlin Heidelberg 2012

Young's modulus, failure strain, ultimate stress and fracture toughness of the octopus skin were 3.2 MPa, 37.8%, 0.68 MPa and 0.23 kJm^{-2}.

3   Fabrication of Bomimetic Skin

Silicone rubber is widely used and was chosen for make the artificial skin mainly because of its high flexibility, high elastic strain and ease of moulding. Ecoflex 0030 made by Smooth-On has a Young's modulus of 0.05 MPa at 100% strain and a fracture toughness of 0.14 kJ/m^2. The low modulus may not be a problem so long as the designed shape is maintained under gravitation force when filled with water. But low fracture toughness mean less energy is needed during failure or damage. To overcome this problem, one layer of knitted nylon fabric, cut from 10 denier women's tights, was used to reinforce the silicone rubber. Knitted nylon fabric has fracture toughness and can be extended to two times its original length. The composite was fabricated by sliding the knitted nylon onto a cylinder of 90 mm in diameter and impregnate it with silicone rubber gel.

4   Properties of Skin Artifact and Arm Skin Prototype

The biomimetic skin samples were tested using the same methods mentioned earlier and the measured Young's modulus, elastic strain and fracture toughness are 0.08 MPa, 200% and 0.3 kJ/m2, respectively. The ultimate stress was not obtained as all tests terminated due to specimen slippage instead of failure. Even though the Young's modulus of the fabricated skin is much lower than that of the real octopus skin, it is stiff enough to maintain the designed conical shape when filled with water. Actually wrinkles were observed in the octopus skin to increase its compliance. It can be concluded that the artificial skin is more flexible, stronger and tougher than the real octopus skin. Figure 1 shows a skin prototype for a conical-shaped octopus arm when filled with water. The thickness of the skin is about 0.5 mm.

Fig. 1. An arm skin prototype for robotic octopus' arms

Acknowledgements. This work was supported by the European Commission in the ICT-FET OCTOPUS Integrated Project, under contract #231608.

Development of Sensorized Arm Skin for an Octopus Inspired Robot – Part II: Tactile Sensors

Jinping Hou[1], Richard H.C. Bonser[2], and George Jeronimidis[1]

[1] Centre for Biomimetics, University of Reading, UK
[2] School of Engineering and Design, Brunel University, UK

Abstract. Presented in this paper are two types of contact sensors developed for an octopus inspired robot. The first type was fabricated using QTC pills. The second type is a soft sensor composed of a thin silicone rubber plate sandwiched between two electrolycra sheets. There is a hole at the centre of the silicone plate. Under contact force, both types of sensor transform from insulators to conductors. Sensitivity tests have been carried out. The soft sensor was implemented into the skin prototype and detected contacts with surfaces, whilst remaining relatively insensitive to lateral or longitudinal extension. Its size and sensitivity can be customized according to the requirement of specific applications.

Keywords: tactile sensor, QTC, extendable soft sensor.

1   Introduction

Sensors are important for robots as well as for animals. Different types of tactile sensors have already developed for various applications. The arms of the octopus inspired robot need to sense contact before grabbing. Thus contact sensors are required. New sensors with high extensibility needed to be developed as the arms of the robotic octopus should be able to extend more than 60% in length, like the real octopus. Two types of sensors were fabricated during this work and have been presented here.

2   QTC Sensor

The first sensor developed was using quantum turning composites (QTC) pill purchased from Middlesex University Teaching Resources. Electrolycra strips glued on both faces of the QTC pill were acting as electrodes and connected to wires. Characterizations of the sensors were conducted using an Instron 5564 Testing Machine. The resistance of the QTC pill reduces almost linearly with increasing compressive load applied. The original dimensions of the QTC pill were 3mm x 3mm. The sensitivity of this size of sensor in terms of applied load was not sufficiently high for the robotic octopus arm. The sensitivity can be increased by reducing the area of the sensors and it was found that sensors with dimensions of 1.5mm x 1.5mm were sensitive enough

T.J. Prescott et al. (Eds.): Living Machines 2012, LNAI 7375, pp. 357–358, 2012.
© Springer-Verlag Berlin Heidelberg 2012

for the arm skin with an active threshold around 0.1N. But the smaller size of the sensors means that a relatively larger number of sensors would be needed to work effectively, which, in turn, means more space needed for wires. Thus, another type of sensor was developed.

3   Soft Sensor

The soft sensor is composed of a plate made of Ecoflex 0030 silicone rubber manufactured by Smooth-On with a circular hole at the centre sandwiched between two conductive electrolycra sheets. The silicone plate can be stretched to more than five times its original length and the length of electrolycra sheets can be doubled. Thus the sensor can extend in the in-plane directions by about 100%. Under contact force, the two conductive sheets come into contact and the sensor transforms from insulator to conductor. The sensitivity of the sensor changes with the thickness of the silicone plate and the size of the hole. The thickness of the sensors was optimized to reduce the space taken while maintaining stability. Sensitivity of the sensors as a function of the diameter of the holes was obtained from testing results. The minimum force needed to activate a soft sensor was 0.05N. Stability of the sensors was good under cyclic load.

4   Integrating Sensors with the Arm Skin

The sensors were glued using Ecoflex 05 to specified positions on the inner skin layer. After wires had been attached to the sensors at the ends of the electrolycra strips, the sensors were sealed using a sheet of biomimetic skin to make them watertight. To minimize the restrictions that the wires might exert to the flexibility of the arms, they were wound around the arm skin in the shape of coil springs. The arm muscular unit will slide into this inner skin lay.

Fig. 1. QTC sensor (left) and soft sensor (graduation = 1 mm)

Acknowledgement. This work was supported by the European Commission in the ICT-FET OCTOPUS Integrating Project, under contract #231608.

Development of Sensorized Arm Skin for an Octopus Inspired Robot – Part III: Biomimetic Suckers

Jinping Hou[1], Richard H.C. Bonser[2], and George Jeronimidis[1]

[1] Centre for Biomimetic, University of Reading, UK
[2] School of Engineering and Design, Brunel University, UK

Abstract. Biomechanical properties of squid suckers have been studied. The stiffness of the sucker rings was measured. Sucking force at sea level and maximum possible sucking force were obtained by using specially-designed tensile tests. Two biomimetic passive suckers were developed based on the studies of the squid sucker. The first one was a direct copy of the squid sucker while the second one is a much more simplified version, a non-stalked sucker like the octopus suckers. The second design works very effectively and has been implemented in the prototype skin of the octopus arm.

Keywords: Octopus skin, biomimetic, skin artifact.

1   Introduction

Both octopus and squid suckers generate sucking forces by attaching to a surface and increasing internal volume. The increased volume of the octopus suckers is caused by active contraction of muscles, which are numerous and located in the sucker itself. The squid sucker, on the other hand, has a structure like cylinder and piston system and, more importantly, works passively. It was decided to develop biomimetic suckers for the octopus inspired robot based on the squid suckers.

2   Biomechanical Properties of Squid Suckers

Rings found inside the suckers of squid (*Loligo vulgaris*) have a "T" shaped cross section. They provide radial support to the sucker under deformation. The stiffness of the ring was measured by bending tests and was found to be in the range 0.1-0.4 N/mm, independent of size. The maximum sucking force was found to be limited by stalk strength which was obtained from tensile test of the stalks. Sucking force at ambient pressure was measured by pulling tests after attaching the sucker to the bottom surface of a cold cast acrylic container filled with water. Both the ultimate and normal sucking force increased with sucker size.

3   First Design of Biomimetic Sucker

The first design prototype was a direct mimic of squid suckers made of 15 denier nylon sheet reinforced silicone rubber connected to the arm skin via nylon strips

T.J. Prescott et al. (Eds.): Living Machines 2012, LNAI 7375, pp. 359–360, 2012.

acting as stalks. O-rings were used to give radial support to the artificial suckers. A prototype of 10 mm in diameter could generate a sucking force of up to 4 N. A relatively large initial compressive force was needed to get the sucker to work and the flexibility of the stalk made the orientation of the sucker uncontrollable.

4   Second Design of Biomimetic Sucker

The second design was far simpler and easier to fabricate yet still produced acceptable levels of suction. The fabrication process of this type of sucker is extremely simple. By filling mixed silicone gel into a cylindrical hole up to 75% full, the high viscosity of the gel causes the level of the gel at the centre of the hole significantly lower than that around the edge of the hole, because of capillary forces. Thus, a concave meniscus is formed. Results from measurements show that the depth of the meniscus increases slightly with the diameter of the hole.

To test the level of sucking force generated by this type of sucker, a cyclic load was applied using an Instron 5564 testing machine. The artificial sucker was glued to a smooth surface fixed to the crosshead of the testing machine. The sucker was first driven into contact with the bottom surface a water-filled container until certain level of compressive force was reached and then pulled up. The maximum pulling force is the sucking force generated. Results on 8mm diameter suckers show that the sucking force starts to build up at an initial compression of 0.5N and then stabilizes to a constant value when the initial compression reaches around 2 N Testing results show that the level of sucking force of this type of sucker artifact is equivalent to those generated by the squid's suckers.

Fig. 1. Photos of a squid sucker (left), first prototype (middle) and second prototype (right)

Acknowledgements. This work was supported by the European Commission in the ICT-FET OCTOPUS Integrated Project, under contract #231608.

Decentralized Control Scheme That Enables Scaffold-Based Peristaltic Locomotion

Akio Ishiguro[1,2,*], Kazuyuki Yaegashi[1], Takeshi Kano[1], and Ryo Kobayashi[2,3]

[1] Research Institute of Electrical Communication, Tohoku University
2-1-1 Katahira, Aoba-ku, Sendai 980-8577, Japan
[2] Japan Science and Technology Agency, CREST
7 Goban-cho, Chiyoda-ku, Tokyo 102-0075, Japan
[3] Department of Mathematical and Life Sciences, Hiroshima University
1-3-1 Kagamiyama, Higashihiroshima 739-8526, Japan
{ishiguro,yaegashi,tkano}@riec.tohoku.ac.jp
ryo@math.sci.hiroshima-u.ac.jp

Abstract. We propose an autonomous decentralized control scheme of an earthworm-like robot for its scaffold-based peristaltic locomotion, on the basis of an analysis using a continuum model. We verify its validity through simulations.

Keywords: Autonomous decentralized control, Earthworm, Peristaltic locomotion.

Limbless crawlers exhibit adaptive and efficient locomotion in real time under unstructured real-world constraints. This locomotion is realized by coordinating the movements of many homogeneous body segments [1]. Implementing this remarkable mechanism into robots will pave the way for realizing robots that perform well in undefined environments. Here we take a scaffold-based earthworm-like locomotion as a practical example, and propose an autonomous decentralized control scheme of an earthworm-like robot.

An image of the robot is shown in Fig. 1. It consists of multiple homogeneous body segments that are concatenated one dimensionally. An actuator is mounted on each segment so that it can contract and elongate actively. Each segment is designed such that the frictional coefficient between the body and the ground increases as it contracts. The sensor devices for detecting the segment length and local shear stress on the bottom of the body are implemented.

The autonomous decentralized control scheme of the robot is constructed in the following way: First, we theoretically analyze the earthworm locomotion using a continuum model, and we derive an optimal force distribution for efficient locomotion. This analysis yields the result that the robot moves most efficiently when it generates a force proportional to the spatial derivative of the ratio of expansion and contraction. Second, a local reflexive mechanism that enables exploitation of scaffolds is added to the obtained results; it is designed such that several body segments cranial to the segment that detects shear stress elongate.

---

* Corresponding author.

T.J. Prescott et al. (Eds.): Living Machines 2012, LNAI 7375, pp. 361–362, 2012.

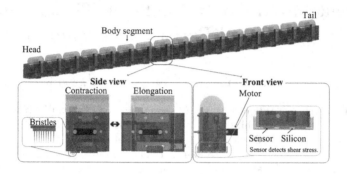

Fig. 1. Image of the earthworm-like robot

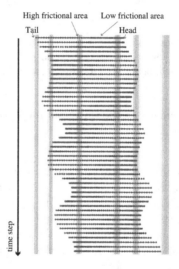

Fig. 2. Simulation results. Elongated and contracted segments are represented by red and blue colors, respectively.

We conducted simulations to investigate the validity of the proposed control scheme. As shown in Fig. 2, the simulated robot with the proposed control scheme moves effectively by exploiting high frictional areas as scaffolds.

Reference

1. Stephenson, J.: The Oligochaeta. Clarendon Press, Oxford (1930)

Autonomous Decentralized Control Mechanism in Resilient Ophiuroid Locomotion

Takeshi Kano[1], Shota Suzuki[1], and Akio Ishiguro[1,2]

[1] Research Institute of Electrical Communication, Tohoku University
2-1-1 Katahira, Aoba-ku, Sendai 980-8577, Japan
[2] Japan Science and Technology Agency, CREST, 7 Goban-cho, Chiyoda-ku,
Tokyo 102-0075, Japan
{tkano,ssuzuki,ishiguro}@riec.tohoku.ac.jp
http://www.cmplx.riec.tohoku.ac.jp/

Abstract. Ophiuroids are a suitable model for understanding the mechanism of resilient animal locomotion, because they can move by self-organizing their arm movements even when the arms are arbitrarily cut off. We observed the locomotion of an ophiuroid that has only one arm, and found that it can move by exploiting subtle terrain irregularities. We modeled this behavior using a simple local reflexive mechanism.

Keywords: Ophiuroid, Autonomous decentralized control.

Ophiuroids exhibit resilient and adaptive locomotion in real-time under unstructured real-world constraints. An ophiuroid has a radially symmetric body consisting of a central disk and five arms that radially diverge from the disk [1]. Although there is no nerve center in the nervous system, it moves on unstructured terrains by self-organizing its arm movements even when the arms are arbitrarily cut off [1]. Clarifying the autonomous decentralized control mechanism underlying this ability will pave the way for realizing highly resilient robots that function in undefined environments. In this study, as a first step, we conduct a behavioral experiment using an ophiuroid that has only one arm, and propose a simple model that describes this experimental result.

We removed four of five arms of an ophiuroid (*Ophiarachna incrassata*), and observed its locomotion on a substrate with several rectangular objects (21 × 28 × 3 mm). Figure 1 shows photographs of the locomotion. First, the arm lifts off the ground and surges forward (0.2 s). When the arm contacts the edge of the sheet, it bends at the proximal side of the contact point and pushes itself against the object edges to propel forward effectively (0.4-1.2 s).

We propose a model on the basis of the above result. Due to space limitations, we briefly describe an outline of the model. Although real ophiuroids exhibit three-dimensional motion, we model it two dimensionally, for simplicity (Fig. 2). Instead, a variable z_j that virtually determines the height of the jth arm joint is introduced, and its time evolution is treated independently to the motion on the xy-plane. The terrain irregularities are modeled by using a potential $\Phi(\mathbf{r})$

T.J. Prescott et al. (Eds.): Living Machines 2012, LNAI 7375, pp. 363–364, 2012.

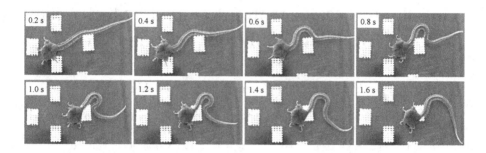

Fig. 1. Locomotion of *Ophiarachna incrassata* with one arm on irregular terrain

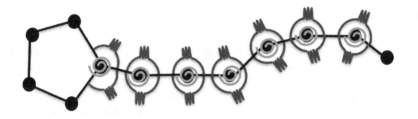

Fig. 2. Proposed model

that represents the altitude of the ground. The friction is modeled by viscous friction.

Antagonistic muscles and a torsional spring are embedded on each arm joint. The bending moments both on the xy-plane and along z-axis are generated at the joints where the muscles contract. The muscles on the proximal arm end contract rhythmically, whereas the other muscles do not contract unless the arm detects a local pressure. When the arm detects a pressure at a certain point, the muscles on its contralateral-distal and ipsilateral-proximal sides contract. This local reflexive mechanism leads to push the arm against the sheet edges effectively. The simulation results will be shown in the presentation.

Acknowledgments. The authors would like to thank Prof. Ryo Kobayashi of the Graduate School of Hiroshima University and Wataru Watanabe for helpful suggestions.

Reference

1. Arshavskii, I.Y., Kashin, M.S., Litvinova, M.N., Orlovskii, N.G., Fel'dman, G.A.: Types of locomotion in ophiurans. Neurophysiol. 8, 398–404 (1976)

A Multi-agent Platform for Biomimetic Fish

Tim Landgraf, Rami Akkad, Hai Nguyen, Romain O. Clément,
Jens Krause, and Raúl Rojas

Freie Universität Berlin, FB Mathematik u. Informatik
Aminallee 7, 14195 Berlin, Germany
tim.landgraf@fu-berlin.de
http://biorobotics.mi.fu-berlin.de

Abstract. Through interactions with live animals biomimetic robots can be used to analyze social behaviors. We have developed a robotic fish enabling us to examine complex interactions in fish shoals. The system uses small wheeled robots under a water tank. The robots are coupled to a fish replica inside the tank using neodymium magnets. The fish integrate a battery pack and two infrared LEDs that are used to track the replicas in the tank. Here, we describe the procedure to build a fish replica, review the implementation details of our hardware and software and compare it to a previous plotter-based system.

Keywords: biomimetic robots, biomimetics, swarm intelligence, social behavior, tracking.

1   Introduction

The use of biomimetic robots that help understanding complex biological systems has several advantages over conventional methodologies in behavioral biology. Foremost the study of social behaviors can benefit from biomimetic robots. Once the robot is accepted as a conspecific, the experimenter is in full control over the interaction with the animals, which augments conventional, static setups drastically and gives access to an entire new set of manipulations. Intruiging example of the recent past include robotic cockroaches to explore group shelter seeking [2], robotic bees for investigating the honeybee dance communication system [3], robotic bowerbirds for the analysis of courtship behavior [4] and a robotic fish to study group decision making [1]. The proposed system builds up on the results of the latter, a plotter-like positioning system under a water tank that can move a fish replica in the tank via strong magnets. Using this system a new set of intriguing experiments is made possible. We have built a new prototype that utilizes wheeled robots instead advancing the system to a multi-agent platform. Using more than one fish one could, for instance, study which morphologies or behaviors make up a better leader.

2   General Setup

For the development we used a smaller testing environment (Figure 1), a glass tank with dimensions 50 x 30 x 30 cm filled with 25 cm of water. For the experiments the

T.J. Prescott et al. (Eds.): Living Machines 2012, LNAI 7375, pp. 365–366, 2012.

tank can be as large as 2 x 2 m. A camera is positioned approximately 1,30 meters above the tank. The wheeled robot steering the replica fish is positioned beneath the water tank. The replica fish is coupled magnetically to the wheeled robot. The camera is connected to a personal computer which runs the main application. This software includes the visual tracking system, a communication layer used to steer the robot and the end-user GUI.

Fig. 1. The experimental setup: The wheeled platform holds a magnet up to the bottom side of a water tank (left image). The replica can move freely in the water, coupled only through magnetic forces. Right: The fish replica has built-in infrared LEDs that are powered by a small battery inside a plastic container that serves also to stabilize the body. Two neodymium magnets with horizontal polarity keep the fish aligned to the robotic base beneath the tank.

References

1. Faria, J.J., Dyer, J.R.G., Clément, R.O., Couzin, I.D., Holt, N., Ward, A.J.W., Waters, D., Krause, J.: A novel method for investigating the collective behaviour of fish. Springer (2010)
2. Halloy, J., et al.: Social integration of robots into groups of cockroaches to control self-organised choices. Science 318, 1155–1158 (2007)
3. Landgraf, T., Oertel, M., Rhiel, D., Rojas, R.: A biomimetic honeybee robot for the analysis of the honeybee dance communication system. In: Proceedings of the 2010 IEEE/RSJ International Conference on Intelligent Robots and Systems, Taipeh, Taiwan, October 18-22, pp. 3397–3403 (2010)
4. Patricelli, G.L., et al.: Male displays adjusted to female's response macho courtship by the satin bowerbird is tempered to avoid frightening the female. Nature 415, 279–280 (2002)

The State-of-the-Art in Biomimetics

Nathan F. Lepora[1], Paul F.M.J. Verschure[2], and Tony J. Prescott[1]

[1] ABRG, Department of Psychology, University of Sheffield, UK
{n.lepora,t.j.prescott}@sheffield.ac.uk
[2] SPECS, Department of Technology, Universitat Pompeu Fabra, Barcelona, Spain
paul.verschure@upf.edu

Biomimetics is the development of novel technologies through the distillation of principles from the study of biological systems. It can, in principle, extend to all fields of biological research from physiology and molecular biology to ecology, and from zoology to botany. Another key focus is on complete behaving systems in the form of biomimetic robots. Historically, the term was first used by Otto Schmitt during the 1950s, when he described a biological approach to engineering that he termed biomimetics.

Over the last decade there has been an explosion of important discoveries within the many research areas comprising biomimetics. The scientific and technological impacts expected to emerge from these advances will have future benefits for our health and quality-of-life, due to advances in information and computation technologies, robotics, brain-machine interfacing and nanotechnology applied to life sciences. Given this potential of biomimetics, many international funding initiatives are underway to drive the field forward. However, within the European research area there are currently no coherent mechanisms and processes that can guide the development of the field and the realization of its potential. A recent initiative towards this goal is the Convergent Science Network (CSN) of biomimetic and biohybrid systems (http://csnetwork.eu), which aims to facilitate surveying and roadmapping exercises combined with carrying out actions such as organizing workshops, schools and conferences to accelerate the development of these areas of research and their future technological impact.

This extended abstract describes a survey of the state-of-the-art of biomimetics that is in preparation by members of the Convergent Science Network. Rather than writing another review of biomimetics, of which there are many excellent accounts already, our aim is to report an objective statistical survey of the area using an information analysis of the topics researched. This analysis can then provide a summary to researchers and scientific policy makers on the current state of the fields comprising biomimetic research, how these fields evolved to their present state, and where they appear to be heading in the future.

Our general strategy is to construct a comprehensive database of publications on biomimetic research, using general resources available from internet sites for journals and conferences, such as IEEE Xplore and the Thomson-Reuters Web of Science. The terms 'biomimetic' and a range of synonyms, such as 'bio-inspired', 'bionic' and 'biomimicry', are used as search terms, from which we analyze this

T.J. Prescott et al. (Eds.): Living Machines 2012, LNAI 7375, pp. 367–368, 2012.

Fig. 1. Popular terms in biomimetics displayed using a word cloud

database with information analysis techniques to infer the general breakdown of the field. Example criteria include papers year, journal or conference published in and subject area judged by common words in the title.

The results of this analysis show that biomimetics now spans a diverse range of applications, as seen from the most frequent concepts from paper titles visualized using a word cloud (Fig. 1). Leading concepts include robot (occurrence frequency 19%), followed by biomimetic (16%) and then control (10%), which indicates that much of contemporary biomimetic research is focussed on applications in robotics. The word cloud also shows a wide variety of research interests. Leading examples includes terms taken from biomedical research, such as tissue (3%), cell (4%) and bone (2%), reflecting the large impact of biomimetics upon this research area. Also evident are the abilities of biological organisms that inspiration is being taken from, such as bio-inspired (2%), vision (0.5%), walking (0.5%) and underwater (2%). In addition, concepts from control engineering and artificial intelligence are also represented, including networks (4%), adaptive (2%), algorithm (2%) and optimization (2%).

Another result is that in the first decade of this century there has been an explosive growth in biomimetic research, with the number of published papers each annum doubling every two to three years. From a relatively small field in the mid-1990s of just ten or so papers per year, biomimetics has expanded exponentially thereafter to reach critical mass of several hundred papers per annum by 2003-2005. More than 1000 papers are now being published in bio-inspired engineering and technology every year. Furthermore, this growth does not appear to be saturating, so this expansion of the area of biomimetics can be expected to continue into the near future.

These findings indicate that biomimetics is entering an important period in which it becomes a dominant paradigm for robotics and other technologies. The implications of this paradigm shift are expected to result in significant new discoveries in many areas of the physical and biological sciences, and enable revolutionary new technologies with significant societal and economic impact.

Acknowledgments. This work was supported by the EU coordination action 'Convergent Science Network (CSN)' (ICT-248986).

Action Development and Integration in an Humanoid iCub Robot

Tobias Leugger[1,2] and Stefano Nolfi[1]

[1] Institute of Cognitive Sciences and Technologies, CNR, Roma, Italy
stefano.nolfi@istc.cnr.it
[2] EcolePolytechnique Federale de Lausanne, Lausanne, Switzerland
tobias.leugger@epfl.ch

Abstract. One major challenge in adaptive/developmental robotics is constituted by the need to identify design principles that allow robots to acquire and display different behavioral skills by consistently and scalably integrating new behaviors into their existing behavioral repertoire. In this paper we briefly present a novel method that can address this objective, the theoretical background behind the proposed methodology, and the preliminary results obtained in a series of experiments in which a humanoid iCub robot develops progressively more complex object manipulation skills through an incremental language mediated training process.

1 Theoretical Background and Proposed Methodology

A first theoretical assumption behind our approach is that behavioral and cognitive processes in embodied agents should be conceived as dynamical processes with a multi-level and multi-scale organization (Nolfi, 2006). This means that behavior (and cognitive skills) are: (i) dynamical processes originating from the continuous interaction between the robot and the physical and social environment, and (ii) display a multi-level and multi-scale organization in which the combination and interaction between lower-level behaviors, lasting for limited time duration, give rise to higher-level behaviors, extending over longer time scale and in which higher-level behaviors later affect lower-level behaviors and the robot/environmental interaction from which they arise. This assumption implies that a behavioral unit does not necessarily correspond to a dedicated control unit or modules of the robot's neural controller. Moreover, it implies that the development of additional and higher-level behavior can occur through the recombination and re-use of pre-existing motor skills even when these skills do not correspond to separated physical entities but rather to processes that ultimately emerge from the robot/environmental interactions.

The second theoretical assumption postulates that social interaction, with particular reference to linguistic mediated social interactions and self-talk (i.e. the possibility to talk to yourself) can promote the development and the integration of multiple action skills (for a related view see Zhang & Weng, 2007; Mirolli & Parisi, 2011).

T.J. Prescott et al. (Eds.): Living Machines 2012, LNAI 7375, pp. 369–370, 2012.

2   Model and Experimental Scenario

In line with the assumptions described above we investigated whether a humanoid iCub robot provided with an non-modular neural network controller can: (1) develop a set of elementary actions, such as REACH-OBJECT, REACH-LOCATION, GRASP, OPEN (i.e. reach a colored object placed on a table, a given target location in the robot's peripersonal space, close the fingers around an object, open the fingers) and then (2) develop higher-level integrated actions such as MOVE-OBJECT (i.e. reach a colored object, grasp it, move and then release the object in a target location).

The developmental process has been carried out in two phases. During the first phase, the robot is trained for the ability to produce the required elementary behavior in variable robot/environmental conditions and to respond to the linguistic command (e.g. "reach-object" or "grasp") produced by the caretaker by eliciting the appropriate behavior. During the second phase, the robot is trained for the ability to produce complex integrated behaviors in a linguistic assisted condition (in which the caretaker produces the appropriate sequence of linguistic inputs: "move-object", "reach-object", open, "grasp", "reach-location", and "open") and for the ability to self-talk (i.e. for the ability to self-generate and anticipate the linguistic commands produced by the caretaker).

3   Results and Conclusions

The analysis of an initial set of experiments carried out by using a realistic simulation of the robot and of the physical robot/environmental interaction demonstrated how the robot can successfully acquire the chosen lower-level and higher-level behaviors. At the end of the training process the robot also shows an ability to produce higher level behaviors autonomously through the exploitation of self-talk (i.e. the ability to self-generate sequences of linguistic commands analogous to those provided by the caretaker during training).

More generally the comparison of different training conditions indicate that: (i) the realization of an incremental training process and the scaffolding role provided by the caretaker constitutes a crucial prerequisite for the ability to master the task, and (ii) the availability of linguistic input associated to goal oriented behavior promotes the emergence of behavioral units that can then be re-used and recombined to produce integrated higher-level behaviors.

References

Nolfi, S.: Behaviour as a complex adaptive system: on the role of self-organization in the development of individual and collective behaviour. ComplexUs 2(3-4), 195–203 (2006)

Mirolli, M., Parisi, D.: Towards a Vygotskyan Cognitive Robotics: The Role of Language as a Cognitive Tool. New Ideas in Psychology 9, 298–311 (2011)

Zhang, Y., Weng, J.: Task transfer by a developmental robot. IEEE Transactions on Evolutionary Computation 2(11), 226–248 (2007)

Insect-Like Odor Classification and Localization on an Autonomous Robot

Lucas L. López-Serrano[1], Vasiliki Vouloutsi[1], Alex Escudero Chimeno[1], Zenon Mathews[1], and Paul F.M.J. Verschure[1,2]

[1] Universitat Pompeu Fabra (UPF). Synthetic, Perceptive, Emotive and Cognitive Systems group (SPECS)
Roc Boronat 138, 08018 Barcelona, Spain
http://specs.upf.edu
[2] Institució Catalana de Recerca i Estudis Avançats (ICREA)
Passeig Llus Companys 23, 08010 Barcelona, Spain
http://www.icrea.cat

Abstract. The study of natural olfaction can assist in developing more robust and sensitive artificial chemical sensing systems. Here we present the implementation on an indoor fully autonomous wheeled robot of two insect models for odor classification and localization based on moth behavior and the insect's olfactory pathway. Using the biologically based signal encoding scheme of the Temporal Population Code (TPC) as a model of the antenna lobe, the robot is able to identify and locate the source of odors using real-time chemosensor signals. The results of the tests performed show a successful classification for ethanol and ammonia under controlled conditions. Moreover, a comparison between the results obtained with and without the localization algorithm, shows an effect of the behavior itself on the performance of the classifier, suggesting that the behavior of insects may be optimized for the specific sensor encoding scheme they deploy in odor discrimination.

Keywords: Artificial olfaction, Biomimmetics, Robotics, Temporal Population Code, Odor localization, Odor classification, Chemotaxis.

1 Introduction

Insect olfaction has been a topic in research over the last years. The neuronal pathways in insects has been deeply investigated and several theories about their way of functioning have been proposed [1][2][3]. Even though it is still not clear how information is processed and represented to perform odor classification, several models mimic biological of olfactory processing[4][5]. This is motivated by the fact that insects, with a relatively small brain, can perform complex odor processing in multi-dimensional spaces very accurately and in a short time [6].

In robotics, robust odor classification systems for real world applications are still in a preliminary developmental status. Here we address the question of how behavior and sensor encoding interact in the insect brain in the service of

T.J. Prescott et al. (Eds.): Living Machines 2012, LNAI 7375, pp. 371–372, 2012.

olfactory detection and mapping. We answer this question by investigating a robot based biologically constrained model of the moth.

The odor classification task has been traditionally approached through pattern recognition strategies such as K-nearest neighbors or Linear Discriminant Analysis (LDA) or a combination of both [7]. These usually share two common characteristics; first, they work with the absolute value received by the sensors paying no attention to temporal patterns and second, they depend on a significant input difference in the chemosensors to be able to perform classification. One of our objectives is to overcome these two barriers.

We show that the temporal structure of the odor induced signal contributes to classification. In addition, we show that discrimination can be further enhanced by adjusting the input sampling of the robot to the temporal dynamics of the classification system of the model antennal lobe. Based on these results we hypothesize that the specific cast and surge behavior observed in chemical search is not only tuned to the plume structure but also to to internal dynamics of the antennal lobe.

Acknowledgements. Supported by the European Community's Seventh Framework Programme (FP7/007-2013) under grant agreement no. 216916. Biologically inspired computation for chemical sensing (NEUROCHEM).

References

1. Knuessel, P., Carlsson, M.A., Hansson, B.S., Pearce, T.C., Verschure, P.F.M.J.: Time and space are complementary encoding dimensions in the moth antennal lobe. Computation in Neural Systems (2007)
2. Wyss, R., König, P., Verschure, P.F.M.J.: Invariant representations of visual patterns in a temporal population code. Proc. Natl. Acad. Sci. U S A, 324–329 (2003)
3. Martinez, D., Montejo, N.: A model of stimulus-specific neural assemblies in the insect antennal lobe. PLoS Computational Biology (2008)
4. Lechón, M., Martinez, D., Verschure, P.F.M.J., i Badia, S.B.: The role of neural synchrony and rate in high-dimensional input systems. the antennal lobe: a case study. In: International Joint Conference on Neural Networks (2010)
5. Huerta, R., Nowotny, T., García-Sánchez, M., Abarbanel, H.D.I., Rabinovich, M.I.: Learning classification in the olfactory system of insects. Neural Computation 16(8), 1601–1640 (2004)
6. Pearce, T.C., Chong, K., Verschure, P.F.M.J., i Badia, S.B., Carlsson, M.A., Chanie, E., Hansson, B.S.: Chemotactic Search in Complex Environments: From Insects to Real-World Applications. NATO Science Series II: Mathematics, Physics and Chemistry, vol. 159, pp. 181–207. Springer, Netherlands (2004)
7. Chen, H., Goubran, R.A., Mussivand, T.: Improving the classification accuracy in electronic noses using multi-dimensional combining (mdc). IEEE (2004)

Autonomous Viewpoint Control from Saliency

Shijian Lu and Joo Hwee Lim

Institute for Infocomm Research, A*STAR, Singapore

Abstract. Autonomous navigation has a high demand on controlling the viewpoint of the visual sensor so that it aims at the object of interest. Ideally, the viewpoint should be maneuvered in a similar way as our human gaze which is driven by either saliency or tasks. We study saliency modelling from image histograms. Our study shows that the visual saliency can be efficiently captured by the traditional 1D image histogram. Experiments over fixational eye tracking data also show that the histogram-based saliency achieves state-of-the-art performance.

Keywords: autonomous viewpoint control, saliency modeling.

1 Introduction

Viewpoint control is very important to the navigation of remote agents such as tele-robot. Due to various constraints in environments and communications, manual viewpoint control is costly and even impossible for applications such as extra-terrestrial exploration. In contrast, autonomous viewpoint control would be much more efficient and flexible as it requires little human involvement. It could be achieved through modeling of visual saliency that predicts where humans look when presented with a natural scene/image.

Computational saliency modelling has attracted increasing interests with different methods reported in recent years [1–3]. Basically, most methods are slow but saliency computation needs to be as fast as possible. In addition, most models are sensitive to variations in image scale that widely exists due to zooming-in/-out. We designed a histogram-based saliency model that is ultra-fast, tolerant to variations in image scale, and obtains state-of-the-art prediction accuracy.

2 Method

Based on the idea that saliency should be negatively correlated with frequency, we computes saliency from 1D image histograms. Given an image in YIQ space, an inverse histogram $\tilde{H}$ is first built for each channel as follows:

$$\tilde{H} = H_a - \overline{H} \tag{1}$$

where H is the histogram and H_a is the number of non-zero H elements:

$$H_a = \left(\sum_{x=x_l}^{x_u} \mathbb{NZ}\big(\overline{h}(x)\big) \right)^{-1} \tag{2}$$

T.J. Prescott et al. (Eds.): Living Machines 2012, LNAI 7375, pp. 373–374, 2012.

where x_l/x_u denote lower/upper image bound. $NZ(x)$ is a non-zero function that returns 1 if $x > 0$ otherwise 0. The image saliency can be computed by:

$$S(i,j) = \sum_{p=i-z}^{i+z} \sum_{q=j-z}^{j+z} \tilde{h}\big(x(p,q)\big) \tag{3}$$

where z is the size of a neighbourhood window. The overall image saliency can be determined by the average of the computed channel saliency.

Fig. 1. For six AIM [2] images in the first row, rows 2-3 show the corresponding histogram-based saliency maps and the human fixational attention maps, respectively

3 Results

The proposed technique is tested over the dataset reported in [2] through the analysis of Receiver Operating Characteristic. Experiments show that it obtains an Area Under Curve (AUC) of 70.16 that is comparable to 69.08, 69.90, and 68.13, respectively, as reported in [1–3]. Besides, it is scale-tolerant and AUCs of 70.32 and 70.58 are obtained when the saliency is computed at 0.6 and 0.3 of the original image scale. Finally, it is ultra-fast with an average speed of 0.04 seconds which is much faster than 0.18, 10.43 and 5.2 seconds, respectively, as reported in [1–3]. Figure 1 illustrates the proposed technique where the histogram-based saliency predicts the fixational attention map accurately.

References

1. Hou, X., Zhang, L.: Saliency Detection: A Spectral Residual Approach. In: CVPR, pp. 1–8 (2007)
2. Bruce, N., Tsotsos, J.: Saliency, attention, and visual search: An information theoretic approach. J. Vis. 9(4), 1–24 (2009)
3. Zhang, L., Tong, M.H., Marks, T.K., Cottrell, G.W.: SUN: A Bayesian framework for saliency using natural statistics. J. Vis. 8(7), 1–20 (2008)

Bio-inspired Design of an Artificial Muscular-Hydrostat Unit for Soft Robotic Systems[*]

Laura Margheri[1], Maurizio Follador[1,2], Matteo Cianchetti[1], Barbara Mazzolai[2], and Cecilia Laschi[1]

[1] The BioRobotics Institute, Scuola Superiore Sant'Anna, Viale Rinaldo Piaggio, 34 56025 - Pontedera (PI), Italy
[2] The Center for Micro-BioRobotics@SSSA, Istituto Italiano di Tecnologia (IIT), Polo Sant'Anna Valdera, Viale Rinaldo Piaggio, 34 56025 - Pontedera (PI), Italy
{laura.margheri,maurizio.follador,matteo.cianchetti, cecilia.laschi}@sssup.it, barbara.mazzolai@iit.it

Abstract. The octopus arms totally lack of rigid skeleton, and show unique motor and manipulation capabilities thanks to the skill of varying and controlling the stiffness. To take inspiration for the design of innovative technological actuators for soft robotic systems, we investigated the architecture of the muscle fibers in the octopus arm, and we measured their mechanical performance *in vivo*. The key features "extracted" from the octopus arm have been "translated" into engineering specifications, and the identified requirements have been used to design an artificial muscular hydrostat unit, obtaining an actuating component with controllable stiffness capabilities and various applications for a novel generation of soft-bodied robots.

Keywords: Bioinspired systems, soft robotics, octopus.

Octopuses show unique motor capabilities which are extraordinary to take inspiration for the design of innovative technological solutions to be used in soft robotics [1]. Octopuses have soft bodies with no rigid skeleton and are made of highly compliant tissues, with variable and controllable stiffness (*muscular hydrostat* [2]). The particular arrangement of the arms muscle fibres, their mechanical properties and the constant volume characteristic, allow these animals to move in the world and manipulate objects with unexpected dexterity. To take inspiration for the design of innovative technological actuators for soft robotic systems, we investigated the architecture of the muscle fibres in the arm, and we measured the performance of the elongation and shortening of the arms directly on moving octopuses.

The muscular fibres have been studied using histology and ultrasound, directly measuring the dimension of the bundles, their proportion in respect to the whole arm, as well as their orientation. Longitudinal muscles are grouped in four main bundles along the arm, and act to shorten the structure. As opposed, the transverse muscle fibres and trabeculae allow to obtain large elongation with small diameters reduction [2]. We found that the trabeculae play an important role to optimize the performance, thus their arrangement has been modelled and optimized to design the artificial

---

[*] This work was supported in part by the European Commission in the ICT-FET OCTOPUS Integrating Project, under contract no. 231608.

T.J. Prescott et al. (Eds.): Living Machines 2012, LNAI 7375, pp. 375–376, 2012.
© Springer-Verlag Berlin Heidelberg 2012

counterpart. Contemporaneously, using purposely designed tools we studied the elongation and shortening capability of the octopus arm, measuring that the arm can elongate up to 70% with a diameter reduction of 23%, and shorten up to 50% [3]. These measurements have been used as specifications to manufacture the artificial components. The actuating elements in the artificial muscular hydrostat unit (SMA springs shown in Table 1, first column) have been arranged in a transverse configuration (representing the transverse muscles and trabeculae) and in longitudinal direction (representing the longitudinal muscles).

Table 1. Scheme of the arrangement of the SMA actuators inside the unit (first column); pictures of the unit performing movements (second column); results of the shortening/ elongation tests (last column).

	Scheme	Real	Test results	
Shortening			length	61.2 mm
			diameter	21.2 mm
			percentage	25.1 %
			speed	9 mm/s
Elongation			length	43.3 mm
			diameter	28.3 mm
			percentage	41.3 %
			speed	6 mm/s

The springs have been mechanically anchored to an external braided supporting structure. This external braid, similarly to the connective tissue in the octopus arm, has the role to guarantee the control of shape changes and allows to optimize the performances during elongations and shortenings. The artificial *muscular hydrostat* unit is completely soft, able to elongate and shorten with the separate actuation mechanism of transverse and longitudinal elements, as well as to stiffen by using a simultaneous activation in both the directions. Pictures of the unit integrated in a soft robotic octopus-like arm are shown in Table 1 (second column). The performances of a single *muscular hydrostat* unit (Table 1, last column) do not completely cover the real octopus arm range yet, but the bio-inspired design that derives from the biological model allows obtaining an actuating component with controllable stiffness capabilities and various applications. Indeed, the unit can be integrated in different soft arms, with diverse dimensions, and it can be used in different environments (air or water). The proposed technological solution represents a good translation of biological key principles into robotics design, going beyond the traditional concept of classic robotic systems with rigid structure, and thus representing an innovative actuated component to be integrated in more complex soft robotic systems.

References

1. Laschi, C., Cianchetti, M., Mazzolai, B., Margheri, L., Follador, M., Dario, P.: A Soft Robot Arm Inspired by the Octopus. Adv. Robotics 26, 709–726 (2012)
2. Kier, W.M., Smith, K.K.: Tongues, tentacles and trunks: the biomechanics of move-ment in muscular-hydrostats. J. Linn. Soc. Lond. (Zool.) 83, 307–324 (1985)
3. Margheri, L., Laschi, C., Mazzolai, B.: Soft robotic arm inspired by the octopus. I. From biological functions to artificial requirements. Bioinspir. Biomim. 7, 025004 (2012)

Texture Classification through Tactile Sensing

Uriel Martinez-Hernandez[1], Hector Barron-Gonzalez[2], Mat Evans[2],
Nathan F. Lepora[2], Tony Dodd[1], and Tony J. Prescott[2]

[1] ACSE, University of Sheffield, U.K.
[2] Department of Psychology, University of Sheffield, U.K.
{uriel.martinez,h.barron,mat.evans,n.lepora,
t.j.dodd,t.j.prescott}@sheffield.ac.uk

To perform tasks in human-centric environments, humanoid robots should have the ability to interact with and learn from their environment through the sense of touch. In humans, the loss of this capability can be catastrophic – for instance, the absence of a proprioception sense of limb movement can result in a dramatic loss of the precision and speed of hand movements. Furthermore, not only humans but also animals use tactile sensing to explore their environment, with one notable example being the tapping exploration known as whisking performed by rats with their long facial vibrissae. The movement of tactile sensors against an object surface to generate tactile information is known as active tactile sensing because it relies on actively moving the sensor to generate the tactile sensations. Humans make use of different exploratory procedures (EPs) to extract key information of the objects (e.g. tapping, contour following). It is of great interest how humans and animals develop and select these EPs [6], which motivates the present study into integrating these biomimetic properties within a robotic system.

There have been some recent developments in enabling robotic hands with the sense of touch. One way of discriminating objects with tactile sensory data has used a self organising map [1]. Another study has presented a method for texture classification with a robotic finger using the Fourier coefficients [2]. Novel biologically-inspired methods for discriminating textures with artificial whisker sensors are proposed in [3,4], which employed a maximum likelihood method related to naive Bayes. The iCub humanoid has been equipped with a tactile sensory system to interact with its environment [5]. The sensors are capacitive, respond to pressure and are placed in the five fingertips and the palm (12 taxels per fingertip and 48 taxels per palm).

In this work we concentrate on texture classification through a tapping exploration procedure using the iCub fingertip, in contrast to previous methods that have used sliding exploration [2]. We mounted an iCub fingertip on an xy-robot that can move accurately in a horizontal plane to allow investigation of tactile sensing while moving over various textures. The communication and control of this experimental system was done through the YARP library, a standard protocol for the iCub platform. Figures 1a and 1b show the configuration for tapping exploration and the various test materials used for the classification. The iCub fingertip performed five taps per second over five different positions

T.J. Prescott et al. (Eds.): Living Machines 2012, LNAI 7375, pp. 377–379, 2012.

Fig. 1. Experimental setup for texture classification. (a) iCub fingertip tapping on the cardboard, carpet I, sandpaper and carpet II textures in (b).

for each of the textures with a sample rate of 50Hz. In our first experiment, twelve trials were collected, with the first trial used for training and the rest for testing. The maximum likelihood classifier [3,4] was used obtaining a 100% classification rate. In our second experiment the level of contact pressure was increased by changing the positional parameters of the xy-robot. The result of this increment was a change in classification accuracies: for the cardboard, sandpaper, and carpet II a 100% rate of classification was still obtained, whilst for carpet I the result decreased to 90%. These results indicate that texture classification based on tapping exploration depend on the contact pressure. The main objective of this work was to demonstrate on the iCub platform that an active tapping exploration inspired by animals, such as the whisking performed by rats and the tapping sometimes performed by humans, is appropriate for collecting and classifying sensory information. We validated the approach using the maximum likelihood method that has been previously found to be effective for artificial whisker data for texture classification [3,4]. Good results were found across different levels of contact pressure, although there was some dependence of accuracy on the pressure variable. This suggests that sensory information about pressure could be used to regulate the contacts so that better classification may be achieved, which is a topic currently under investigation in our laboratory.

Acknowledgments. This work was supported by FP7 grants BIOTACT (ICT-215910), EFAA (ICT-248986) and the Mexican National Council of Science and Technology (CONACYT).

References

1. Takamuku, S., Gomez, G., Hosoda, K., Pfeifer, R.: Haptic discrimination of material properties by a robotic hand. In: IEEE 6th International Conference on ICDL 2007, pp. 1–6 (July 2007)
2. Jamali, N., Sammut, C.: Majority voting: Material classification by tactile sensing using surface texture. IEEE Transactions on Robotics 27(3), 508–521 (2011)
3. Lepora, N.F., Evans, M., Fox, C.W., Diamond, M.E., Gurney, K., Prescott, T.J.: Naive Bayes texture classification applied to whisker data from a moving robot. In: IJCNN 2010, pp. 1-8 (July 2010)

4. Lepora, N.F., Fox, C.W., Evans, M., Diamond, M.E., Gurney, K., Prescott, T.J.: Optimal decision-making in mammals: Insights from a robot study of rodent texture discrimination. Journal of The Royal Society Interface (January 2012)
5. Schmitz, A., Maggiali, M., Natale, L., Bonino, B., Metta, G.: A tactile sensor for the fingertips of the humanoid robot icub. In: IEEE/RSJ International Conference on IROS 2010, pp. 2212–2217 (October 2010)
6. Lederman, S.J., Klatzky, R.L.: Extracting Object Properties Through Haptic Exploration. Acta Psychologica 84, 29–40 (1993)

Bio-inspiration for a Miniature Robot
Inside the Abdomen

Alfonso Montellano López[1], Robert Richardson[1], Abbas Dehghani[1], Rupesh Roshan[1],
David Jayne[2], and Anne Neville[1]

[1] Mechanical Engineering, University of Leeds, Leeds, United Kingdom
{mnaml,R.C.Richardson,A.A.Dehghani-
Sanij,R.Roshan,A.Neville}@leeds.ac.uk
[2] Academic Surgical Unit, St. James's Hospital, University of Leeds, Leeds, United Kingdom
D.G.Jayne@leeds.ac.uk

Abstract. Intra-body mobility is a promising and challenging development for future surgical robots. This poster presents the bio-inspired features of a novel intra-abdominal robot. This new development is adhesion-reliant and moves against gravity on the inside wall of the human abdomen. The adhesive pads on the robot use a micro-structured surface inspired by tree frogs to obtain wet adhesion. The robot integrates a detachment mechanism inspired by the way geckoes peel off their toes. Locomotion is inspired by amoebas, changing the shape of the robot's body and alternating adhesion between the moving part of the robot and the part that remains still.

Keywords: Bio-inspiration, Miniature robot, Surgical Robotics.

1   Motivation

Surgical procedures through small incisions reduce trauma for the patient and shorten recovery time in hospital. Robotic assistants are currently used for minimally invasive operations. However they are generally cumbersome, require a large capital cost and their manoeuvrability is limited by their anchorage external to the body. In order to overcome these limitations, over recent years robots to move fully inside the body have been developed. This poster explains the different sources of inspiration taken from nature for the design of a novel miniature intra-abdominal robot.

2   Prototype of the Intra-abdominal Robot

For a laparoscopic operation, the abdomen is inflated to create an operating cavity. The robot is designed to move inside the human abdomen during this type of procedure. The robot is inserted in the abdomen through a surgical port and carries a camera walking upside-down on the surface of the internal wall of the abdomen, the peritoneum. As Figure 2 shows, the prototype of the robot uses four adhesive pads actuated by eight very small and lightweight linear piezoelectric motors (Squiggle®). The robot moves upside-down detaching, moving and re-attaching one pad at a time.

T.J. Prescott et al. (Eds.): Living Machines 2012, LNAI 7375, pp. 380–381, 2012.
© Springer-Verlag Berlin Heidelberg 2012

3   Bio-inspiration for Adhesion Control and Locomotion

Geckoes attach to surfaces using tiny hairs on their toes which increase surface contact. Tree frogs use similar structures to attach to wet surfaces. To detach their feet, geckoes peel off their toes by curling them up. The robot uses a micro-structured surface inspired by tree frogs to obtain adhesion to the peritoneum. The mechanism between two adjacent pads of the robot (Figure 1 (a)) peels off the pad by twisting it. The twist is caused by actuating the vertical and horizontal motors involved in the motion of the pad (Figure 1 (b)).

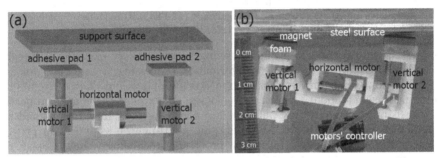

Fig. 1. (a) Components of the detachment mechanism between two pads of the robot, (b) mechanism of the prototype peeling off a magnetic pad with an attachment force similar to the bio-inpired surface

Amoebas deform part of their body (pseudopod) to reach a new location. The part of the body that does not move is adhered to the surface. The pseudopod attaches to the surface when it reaches the new position, then the rest of the body detaches so it can move as well. Motion through body deformation and alternating adhesion inspires the locomotion of the robot. To detach one pad, the previous peeling-off mechanism is used. After detachment, the horizontal motors move the pad across to a new position. The vertical motor is used to make contact with the surface and re-attach the pad (see Figure 2). The sequence is then repeated for the rest of the pads.

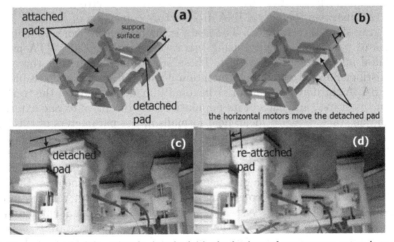

Fig. 2. Once one pad of the robot is detached (a), the horizontal motors can move it to a new position (b), (c) and (d) show the prototype of the robot moving the pad

Systematic Construction of Finite State Automata Using VLSI Spiking Neurons

Emre Neftci[1], Jonathan Binas[1], Elisabetta Chicca[2], Giacomo Indiveri[1], and Rodney Douglas[1]

[1] Insitute of Neuroinformatics, University of Zurich and ETH Zurich, Wintherthurerstrasse 190, 8057 Zurich
[2] Cognitive Interaction Center of Excellence (CITEC), University of Bielefeld, Universitätstrasse 21-23, Bielefeld, Germany

Spiking neural networks implemented using electronic Very Large Scale Integration (VLSI) circuits are promising information processing architectures for carrying out complex cognitive tasks in real-world applications. These circuits are developed using standard silicon technologies, and exploit the analog properties of transistors to emulate the phenomena underlying the computations and the communication in the brain. Neuromorphic multi-neuron systems can provide a low-power and scalable information processing technology, that is optimally suited for advanced and future VLSI processes [1].

However, because the relationship between function and neural architecture is still not understood, their configuration to perform target state-dependent computations is a challenging task. Moreover, due to the non-linear relationship between the chip parameters and those of theoretical neuron models, the configuration of the hardware neurons to match the target behavior is not straightforward.

We introduce a procedure to configure real-time VLSI spiking neural networks, which solves these two problems by mapping the computations of classical Finite State Automata (FSA) onto the neural circuit. The procedure systematically constructs a neural architecture based on a class of biologically inspired, stereotypical circuits expressing the Soft Winner–Take–All (sWTA) function [2]. Using a parameter translation technique [3], the procedure systematically maps the key characteristics of the sWTA network onto its electronic counterpart.

Fig. 1a illustrates the neural FSA architecture consisting of three sWTA networks (gray-shaded rectangles). The state is maintained in working memory in the form of persistent activity, using two recurrently coupled sWTA networks (horizontal and vertical rectangles). These are coupled such that there are as many distinct populations able to maintain persistent activity as possible states in the FSA. A third sWTA network (diagonally shaded) mediates the transitions between the states, and each of its populations are receptive to one external input and one state. Using a mean-field formalism, the parameters of this sWTA are computed such that only a single population activates when an input is provided [4]. The neural FSA is constructed by connecting the output of each transition population to the target states, according to the state diagram.

Fig. 1b shows the response of a neuromorphic chip implementing the neural FSA during an input string consisting of five X's, provided at random intervals (blue shadings). The neural system responds in a state-dependent manner, in that its activity is conditional on the input and the previous state.

In principle, this procedure can be applied to generate the circuitry for implementing automata of almost any complexity [2]. This achievement is an

T.J. Prescott et al. (Eds.): Living Machines 2012, LNAI 7375, pp. 382–383, 2012.
© Springer-Verlag Berlin Heidelberg 2012

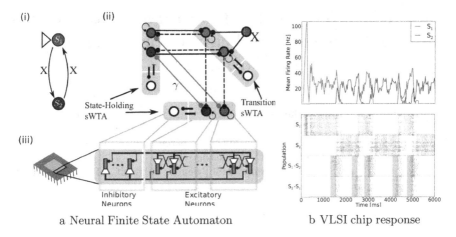

a Neural Finite State Automaton   b VLSI chip response

Fig. 1. (a) (i) State diagram where circles indicate states and arrows the transitions between them. The white triangle specifies the initial state. For this automaton, the active state flips between S_1 and S_2 in response to X. (ii) The neural circuit implementing the automaton in (i). (iii) The multi-neuron chips used in the neuromorphic setup feature low-power integrate-and-fire neurons with dynamic synapses. Each population of a sWTA network (circle) is represented in hardware by 16 recurrently coupled neurons, which compete against the other populations via global inhibition. (b)(top) Average firing rate one state-holding sWTA. The inputs were provided at arbitrary intervals to the transition neurons (blue shadings). (b)(bottom) Raster plot of the activity in the state-holding elements and the transitions networks (S_1 to S_2 and S_2 to S_1). Consistently with the state diagram, the neural system reliably flips its state at each presentation of X.

important milestone towards building generic-purpose neuromorphic systems that are able to interact with unconstrained real-world environments in space- and power-constrained applications, such as in robotic or mobile hand-held devices. In future work, we will couple these neuromorphic hardware with event-based vision sensors to achieve sophisticated sensorimotor tasks.

Acknowledgements. This work was supported by the EU ERC Grant "neuroP" (257219), the EU ICT Grant "SCANDLE" (231168) and the Excellence Cluster 227 (CITEC, Bielefeld University).

References

[1] Indiveri, G., Linares-Barranco, B., Hamilton, T., van Schaik, A., Etienne-Cummings, R., et al.: Neuromorphic silicon neuron circuits. Frontiers in Neuroscience 5, 1–23 (2011)

[2] Rutishauser, U., Douglas, R.: State-dependent computation using coupled recurrent networks. Neural Computation 21, 478–509 (2009)

[3] Neftci, E., Chicca, E., Indiveri, G., Douglas, R.: A systematic method for configuring VLSI networks of spiking neurons. Neural Computation 23(10), 2457–2497 (2011)

[4] Neftci, E.: Towards VLSI Spiking Neuron Assemblies as General-Purpose Processors. Ph.D. thesis, ETH Zürich (2010)

Self-burial Mechanism of *Erodium cicutarium* and Its Potential Application for Subsurface Exploration

Camilla Pandolfi[1], Diego Comparini[2], and Stefano Mancuso[2]

[1] Advanced Concepts Team, European Space Research and Technology Centre, Noordwijk, The Netherlands
[2] Department of Plant, Soil and Environmental Science, University of Florence, Italy
camilla.pandolfi@esa.int

Abstract. *Erodium cicutarium L.* plants disperse their seeds by a combination of two dispersal strategies: explosive dispersal, and self-burial dispersal. As the fruits dry, the stresses developed in the structure cause the sudden separation of the seeds that fly away from the plant. Then, once on the ground, the seeds respond to variations in the external humidity : their dispersal unit, is helical when dry and linear when wet. The day-night cycle of humidity results in a coiling and uncoiling motor action, that moves the seed across the surface and into the ground. The present study aims at getting a deeper insight into the self-burial strategy of *Erodium cicutarium* in order to implement this ability into mechanical structures that relying on changes in the external parameters (such as light, temperatures or humidity) can achieve passively the same goal.

Keywords: Hygroscopic movement, self-burial, subsurface exploration.

Seed dispersal represent an excellent source of inspiration for developing a new generation of biomimetic technologies. Plants evolved methods to improve their chance of success in many different environments, and their seeds can have wings to fly or buoyant structures to surf water. *Erodium cicutarium L.* plants are characterized by two coupled dispersal strategies: an explosive dispersal to fly the seeds away from the plant and a self-burial strategy of the seeds, that increase the chance of a safe germination [1]. Both these movements are possible thanks to hygroscopic tissues, that generate passive movement via changes in the hydration of the cell walls. As far as the self-burial strategy might concern, each seed has a special dispersal unit (the awn) that, once on the ground, respond to variations in the external humidity and change its configuration accordingly: the awn is helical when dry and linear when wet. The day-night cycle of humidity results in a coiling and uncoiling motor action, that, combined with other accessory structures, moves the seed across the surface and into the ground [2] [3].

The present study aims at getting a deeper insight into the self-burial strategy of *Erodium cicutarium* in order to implement this ability into mechanical

T.J. Prescott et al. (Eds.): Living Machines 2012, LNAI 7375, pp. 384–385, 2012.

structures that relying on changes in the external parameters (such as light, temperatures or humidity) can achieve passively the same goal. At the moment we are testing the performance of the seed in several experimental situations: different soil textures, modified seeds, and contribution of the launch. The self-burial behavior will be discussed in light of its potential application for biomimetic technologies such as penetrators for subsurface investigation. Our preliminary results reveal that given the small size and weight of the seed, it is reasonable to think that the system doesn't need high axial force applying a motion that requires no additional steady coupling with the surface. A system with this characteristics could be very useful in low gravity environments (such as asteroids, moons, and small planets) providing a different solution to the set of unconventional penetrators developed to address this scope [4]. Providing probes and sensors with this behavior would facilitate the exploration of the surface underneath regolith providing information related to the geology of the soil, the temperature or simply dig themselves to find shelter from the space environment for long term monitoring of specific targets of the environment.

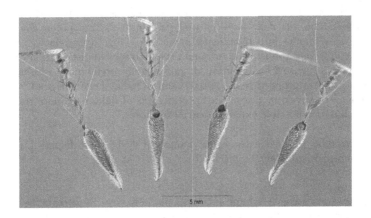

Fig. 1. The morphology of the seeds of Erodium cicutarium (Photo courtesy of Steve Hurts @ USDA-NRCS PLANTS database

References

1. Stamp, N.: Efficacy of explosive vs. hygroscopic seed dispersal by an annual grassland species. American Journal of Botany, 555–561 (1989)
2. Stamp, N.: Self-burial behaviour of erodium cicutarium seeds. The Journal of Ecology, 611–620 (1984)
3. Elbaum, R., Zaltzman, L., Burgert, I., Fratzl, P.: The role of wheat awns in the seed dispersal unit. Science 316(5826), 884 (2007)
4. Gao, Y., Ellery, A., Jaddou, M., Vincent, J., Eckersley, S.: A novel penetration system for in situ astrobiological studies. International Journal of Advanced Robotic Systems 2(4), 281–286 (2005)

Tragopogon dubius, Considerations on a Possible Biomimetic Transfer

Camilla Pandolfi, Vincent Casseau, Terence Pei Fu, Lionel Jacques, and Dario Izzo

Advanced Concepts Team,
European Space Research and Technology Centre, Keplerlaan 1, Noordwijk
The Netherlands
camilla.pandolfi@esa.int
http://www.esa.int/act

Abstract. *Tragopogon dubious* is a small herbaceous plant that uses the wind as dispersal vector for its seeds. The seeds are attached to stalked parachutes which increase the aerodynamic drag force on the seeds. This decreases their rate of descent, and hence increases the total distance traveled. The relatively large natural parachute of *Tragopogon dubious* is an ideal model in a biomimetic structure owing to its relative large size, sturdy and robust structure, and the hierarchical distribution of its fibers. The present contribution describes some preliminary results on the structural properties and aerodynamical behavior of this seed, with the goal of developing new stream of designs of lighter or more robust parachute for possible extra-terrestrial purposes.

Keywords: Plumed seeds, aerodynamics of seeds, wind dispersal, pappus morphology, plant biomechanics.

Dispersal of seeds away from the parental plant is very important to reach new habitats and to survive in changing environmental conditions. For these reasons, plants invest in a variety of strategies to exploit the most abundant form of energy available (i.e. wind, water, animals).

Tragopogon dubious, provides its seeds with probably the biggest parachute available in nature [1], and it achieves its final size by two hierarchical orders of branching: primary and secondary fibers arranged as shown in Fig.1. In a windy atmosphere, the horizontal distance travelled by plumed seeds is greatly influenced by the wind speed. For a lower wind speed, vertical descent velocity remains a good indicator of the horizontal flight capacity, but an increase of wind will result in an exponential increase of the horizontal dispersal distance [2]. This change in the flight behavior has not been fully explained, and a more thorough investigation could reveal some advantageous properties that have not been characterized yet.

The present contribution describes some preliminary results on the structural properties and aerodynamical behavior of the parachuted seed of *Tragopogon*, with the goal of developing new stream of designs of lighter or more robust

T.J. Prescott et al. (Eds.): Living Machines 2012, LNAI 7375, pp. 386–387, 2012.

parachute for possible extraterrestrial purposes. Space exploration requires simple solutions which maximize adaptability and robustness together with weight and size constrains. It also place additional challenges due to uncertain and unpredictable environmental conditions. Sending a swarm of sensors to explore atmosphere endowed celestial bodies has been proposed before [3] [4], and provide them with aerial-platforms derived from the cone-shaped parachute of Tragopogon could give some advantages. In the present contribution we are focusing on three main points in the model: the distribution of the fibers, the structural properties of the main ribs, and the overall aerodynamic behavior of the structure. The hierarchical distribution of its fibers could lead to a reduction of the mass required for the parachute. Furthermore, the cone-shaped parachute could have a better aerodynamic stability in a windy atmosphere and it will naturally tend to recover its orientation after being destabilized by strong gusts.

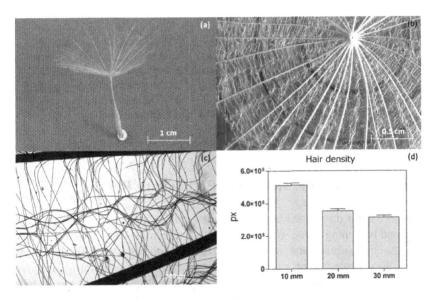

Fig. 1. (a) Plumed seed of *Tragopogon dubious*; (b) Optical scanned image of the natural parachute; (c) The fine mesh of the secondary hairs; (d) Secondary hairs density

References

1. McGinley, M., Brigham, E.: Fruit morphology and terminal velocity in tragopogon dubious (l.). Functional Ecology, 489–496 (1989)
2. Hensen, I., Müller, C.: Experimental and structural investigations of anemochorous dispersal. Plant Ecology 133(2), 169–180 (1997)
3. Quadrelli, M., Chang, J., Mettler, E., Zimmermann, W., Chau, S., Sengupta, A.: System architecture for guided herd of robots exploring titan. In: Proceedings of 2004 IEEE Aerospace Conference, vol. 1. IEEE (2004)
4. Southard, L., Hoeg, T., Palmer, D., Antol, J., Kolacinski, R., Quinn, R.: Exploring mars using a group of tumbleweed rovers. In: 2007 IEEE International Conference on Robotics and Automation, pp. 775–780. IEEE (2007)

Root-Soil Interaction Models for Designing Adaptive Exploring Robotic Systems

Liyana Popova[1,2], Alice Tonazzini[1,2], and Barbara Mazzolai[1]

[1] Center for Micro-Biorobotics@SSSA, Istituto Italiano di Tecnologia (IIT), Viale R. Piaggio 34, 56025 Pontedera, Italy
{liyana.popova,alice.tonazzini,barbara.mazzolai}@iit.it
[2] The BioRobotics Institute, Scuola Superiore Sant'Anna, Viale R. Piaggio 34, 56025 Pontedera, Italy
{l.popova,a.tonazzini}@sssup.it

Abstract. Plants represent an amazing source of inspiration for designing and developing adaptive robotic systems, representing plants the best example among living beings of efficient soil exploration. Influence of geometrical and mechanical properties of the living plant roots in the root-soil interaction was investigated in order to fully exploit these biological features in the robotic artefact. The study was performed by means of Finite Element method.

Keywords: root apex mechanical model, tip shape, root penetration strategy.

1 Introduction, FE Model and Preliminary Results

In order to exploit distributed environmental resources, plants develop a network of growing and branching roots, whose tips (apices) are highly sensorized and efficiently explore the soil volume and up-taking water and minerals to fulfil their primary functions. First prototype of a novel robotic system for soil exploration, inspired by capacity of plant root apices to manage and integrate multiple external signals without having brain-like structure, was recently developed [1]. Penetration strategies of living roots are still to be investigated and then implemented into robotic artefact.

Living roots elongate into the soil by expanding cells of the elongation region (ER) situated just behind the root tip and can generate axial penetration pressure up to 1 MPa. Roots with bigger diameter were found to penetrate better hard soils because of increased resistance to buckling. The diameter enlargement is inhomogeneous [2] and this may also be advantageous for soil penetration. In order to better understand the root penetration strategies, different models of soil deformation under the probe pressure are described in literature. Each type of model elucidates some of the aspects of root-soil interaction process. In this work Finite Element (FE) modelling was implemented focusing on the importance of cap shape and on the root enlargement strategy in the soil penetration.

Two FE static models with planar axial symmetrical geometry were built by using COMSOL Multiphysics® 4.1 program to address the problem [3]. In the first model a prescribed displacement was applied to the penetrating root tip to evaluate the soil

T.J. Prescott et al. (Eds.): Living Machines 2012, LNAI 7375, pp. 388–389, 2012.

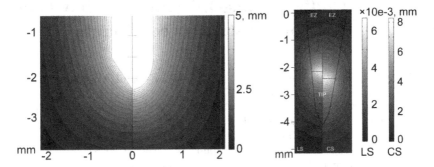

Fig. 1. Soil total displacement (a) caused by conical (on the left) and elliptic (on the right) tip profiles. Axial displacements (b) due to the expansion of ER in LS geometry (on the left) and CS geometry (on the right) [2]. All the constitutive materials were considered to be elastic isotropic. Young modulus of root and soil were of 80 MPa and 40 MPa respectively [3].

deformations. Conical and elliptic profiles were chosen to approximate the root cap shape in a robotic tip. In the second model 1MPa expanding pressure along the root axis was applied to the ER in order to simulate the root penetration. In this model two geometries of maize roots grown in the loose (LS) and compact (CS) sands were implemented [2].

Elliptic profile produced larger front of soil deformation around the tip with respect to the conical profile (Fig. 1a). This effect determines greater energy dissipation, correlated with reduced penetration performance of rounded tips. The second model showed that the tip axial displacement was approximately two times greater in the case of CS geometry (Fig. 1b), evaluated with the same soil elasticity and the same elongation pressure. Penetration performance of two geometries resulted to be approximately the same when the soil Young modulus was increased from 40 MPa to 70 MPa only for CS geometry. Thus the CS apex geometry resulted to be more efficient in soil penetration.

The results of this study are the first step towards the definition of appropriate geometry of the robotic root aimed at increasing the penetration performance. Fracture mechanics, geometrical parameters, and other plant penetration strategies (e.g., root anchorage and buckling) will be investigated as necessary steps to obtain an optimised root-soil interaction model.

References

1. Mazzolai, B., Mondini, A., Corradi, P., Laschi, C., Mattoli, V., Sinibaldi, E., Dario, P.: A miniaturized mechatronic system inspired by plant roots for soil exploration. IEEE/ASME Transactions on Mechatronics, 1–12 (2011)
2. Iijima, M., Barlow, P.W., Bengough, A.G.: Root cap structure and cell production rates of maize (Zea mays) roots in compacted sand. New Phytol. 160, 127–134 (2003)
3. Clark, L.J., Ferraris, S., Price, A.H., Whalley, W.R.: A gradual rather than abrupt increase in soil strength gives better root penetration of strong layers. Plant Soil 307, 235–242 (2008)

A Soft-Bodied Snake-Like Robot That Can Move on Unstructured Terrain

Takahide Sato[1], Takeshi Kano[1], Akihiro Hirai[1], and Akio Ishiguro[1,2]

[1] Research Institute of Electrical Communication, Tohoku University
2-1-1 Katahira, Aoba-ku, Sendai 980-8577, Japan
7, Goban-cho, Chiyoda-ku, Tokyo 102-0075, Japan
[2] Japan Science and Technology Agency, CREST
{sato,tkano,hirai,ishiguro}@riec.tohoku.ac.jp
http://www.cmplx.riec.tohoku.ac.jp/

Abstract. Snakes utilize terrain irregularities and attain propulsion force by pushing their bodies against scaffolds. We have previously proposed a local reflexive mechanism of snake locomotion that exploits its body softness. In this study, we develop a soft-bodied snake-like robot to investigate the validity of the proposed mechanism in the real world.

Keywords: Snake-like robot, Autonomous decentralized control.

Snakes actively utilize terrain irregularities and produce an effective propulsion force by pushing their body against scaffolds, in spite of its simple one-dimensional bodily structure (Fig. 1) [1]. Such dexterous locomotion can be achieved by coordinating the movement of many points along the body, with each having some degrees of freedom, adaptively to the environment in a decentralized manner. Clarifying this remarkable mechanism will lead to a better understanding of animal locomotion mechanisms as well as help to develop robots that work well in undefined environments. We have previously proposed a simple local reflexive mechanism of snake locomotion that exploits its body softness [2]. Our aim of this study is to develop a soft-bodied snake-like robot to investigate the validity of the proposed mechanism in the real world.

Overview of CAD image of the snake-like robot is shown in Fig. 2a. The robot consists of multiple homogeneous body segments that are concatenated one dimensionally. Figure 2b shows the photograph of two segments we developed. The size of each segment is 54 mm × 58 mm × 14 mm, and its weight is 0.03 kg. A motor is mounted on each segment, which is connected to the adjacent segment not directly but indirectly via an elastic element. The joint angle is measured by a potentiometer. The body surface is covered by silicone rubbers, in which pressure sensors are implemented. The motor angle is determined according to the local reflexive mechanism proposed previously [2]: it is determined by using information about the angle of the anterior joint and the local pressure on the body wall (see details in [2]).

We are now manufacturing multiple segments. Experimental results using the robot will be shown in the presentation.

T.J. Prescott et al. (Eds.): Living Machines 2012, LNAI 7375, pp. 390–391, 2012.

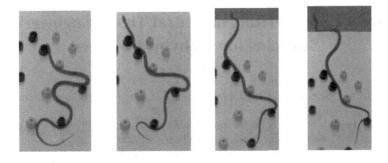

Fig. 1. Locomotion of a real snake (*Elaphe climacophora*) on unstructured terrain

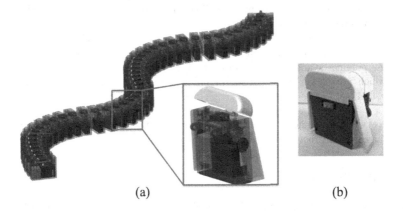

(a)   (b)

Fig. 2. (a) CAD image of a soft-bodied snake-like robot. (b) Two body segments we developed.

Acknowledgments. We acknowledge support from Grant-in-Aid for Young Scientists (B) No. 22760310 from the Ministry of Education, Culture, Sports, Science, and Technology (MEXT), Japan, and from the Global COE Program ("Basic & Translational Research Center for Global Brain Science" and "Center of Education and Research for Information Electronics Systems"). We thank the Japan Snake Institute for its cooperation.

References

1. Moon, B.R., Gans, C.: Kinematics, muscular activity, and propulsion in gopher snakes. J. Exp. Biol. 201, 2669–2684 (1998)
2. Kano, T., Sato, T., Kobayashi, R., Ishiguro, A.: Decentralized control of scaffold-assisted serpentine locomotion that exploits body softness. In: Proc. of IEEE/RSJ Int. Conf. Robot. Auto., pp. 5129–5134 (2011)

Direct Laser Writing of Neural Tissue Engineering Scaffolds for Biohybrid Devices

Colin R. Sherborne, Christopher J. Pateman, and Frederik Claeyssens

Dept. of Materials Science & Engineering, The Kroto Research Institute, North Campus,
University of Sheffield, Broad Lane, Sheffield S3 7HQ
{Mta06cs,dtp09cjp,f.claeyssens}@shef.ac.uk

Abstract. In this study we explore the manufacture of polymer based, 3D structures for neuronal guidance made by direct laser writing (DLW). This technique enables the optimisation of the support structure via utilising Computer Aided Design and Manufacturing (CAD/CAM). The mechanical and chemical properties can be tuned via custom made polymer constructs. We present results on the polymer synthesis, DLW structuring, neuronal cell culture and guidance which provides a proof-of-concept for these scaffolds for use in peripheral neural biohybrid devices.

Keywords: peripheral nerve repair, neuronal guidance, biohybrid devices, polymer micro-structuring, direct laser writing.

1 Introduction

Injuries to the peripheral nervous system (PNS) are extremely common and it is estimated that ~300.000 people per year incur a traumatic peripheral nerve injury in Europe alone. The implication of sustaining such an injury is considerable and is commonly associated with loss of sensory and/or motor function. An important part of bioengineering research in this field is directed towards reconnecting the severed nerve ends via an entubulation device called a nerve guidance conduit (NGC), this research has the potential to direct nerves to a biohybrid device. Previous work on three dimensional, patterned surface constructs [1] and cell alignment strategies [2] can be expanded upon to design a favorable topography for directed growth of neuronal cells. The topographical features can be finely adjusted to promote alignment and guide these cells by controlling the cell-material interaction relative to the size of the cell. In this study we highlight the use of direct laser writing (DLW) to produce and optimise the structural, mechanical and chemical properties of these nerve guidance conduits. We hope to encourage both contact guidance and cell attachment between the neuronal cells, and will highlight the diversity this emerging technology offers for directed neuronal growth.

2 Methods

DLW patterning is achieved by moving the photosensitised sample with a *xyz*-translational stage relative to the focal point of a high NA objective. A frequency

T.J. Prescott et al. (Eds.): Living Machines 2012, LNAI 7375, pp. 392–393, 2012.

doubled sub-nanosecond Nd:YAG laser (532nm, ~0.5ns) is used, and cross linking occurs only at the focused point of highest intensity. We explore the use of the commercially available prepolymer, polyethylene glycol p(EG), and the potential work with in-house biodegradable polymers, poly caprolactone p(CL) and poly lactic acid p(LA).

3   Results and Conclusion

Rapid microstructuring of photocurable polymers via DLW was achieved (Fig. 1). Additionally, preliminary peripheral guidance channels were structured with micron scale resolution. We observed directed outgrowth of rat derived explant dorsal root ganglion (DRG) cells. We will highlight the use of these materials and parameters for neuronal tissue engineering applications and orientated growth of primary Schwann cells and DRGs and the directed growth of NG108-15 neuronal cells, through topographical features.

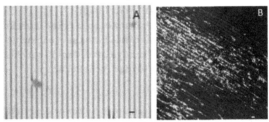

Fig. 1. A. 50 micron spaced lines fabricated by DLW. B. 10 day DRG explant outgrowth on 30 micrometre spaced lines, red indicates β tubulin III, green S100 and blue DAPI nuclei stain. Scale bars 50 and 30 microns

Acknowledgements. The authors thank J.W. Haycock and S. Rimmer (Sheffield). This work was supported by the EPSRC.

References

1. Melissinaki, V., Gill, A.A., Ortega, I., et al.: Direct laser writing of 3D scaffolds for neural tissue engineering applications (2011), doi:10.1088/1758-5082/3/4/045005
2. Jeon, H., Hidal, H., Hwang, D.J., et al.: The effect of microscale anistropic cross patterns on fibroblast migration. Biomaterials 21(2010), 4286–4929 (2010)

Biorobotic Actuator with a Muscle Tissue Driven by a Photostimulation

Masahiro Shimizu[1], Shintaro Yawata[2], Kota Miyasaka[3], Koichiro Miyamoto[4],
Toshifumi Asano[5], Tatsuo Yoshinobu[4], Hiromu Yawo[5], Toshihiko Ogura[3],
and Akio Ishiguro[2]

[1] Graduate School of Information Science and Technology, Osaka University, 2-1
Yamada-oka, Suita, 565-0871, Japan
[2] Research Institute of Electrical Communication, Tohoku University, 2-1-1 Katahira,
Aoba-ku, Sendai, 980-8577, Japan
[3] Institute of Development, Aging and Cancer (IDAC), Tohoku University, 4-1,
Seiryo, Aoba, Sendai, 980-8575, Japan
[4] Graduate School of Engineering, Tohoku University, 2-1-1 Katahira, Aoba-ku,
Sendai, 980-8577, Japan
[5] Graduate School of Life Sciences, Tohoku University, 2-1-1, Katahira, Aoba,
Sendai, 980-8577, Japan
m-shimizu@ist.osaka-u.ac.jp, yawata@cmplx.riec.tohoku.ac.jp,
asano@ige.tohoku.ac.jp, yawo@mail.tains.tohoku.ac.jp,
ishiguro@riec.tohoku.ac.jp, {k-miya,nov}@ecei.tohoku.ac.jp,
{ogura,k.miyasako}@idac.tohoku.ac.jp

Abstract. The authors aim to develop a biorobotic actuator with a
muscle tissue of developing chick embryos driven by a photostimulation.
The goal is to find an appropriate method for driving such bio-actuators.
To do so, we introduce channelrhodopsin-2 which works as both a photoreceptor and an ion channel. We here report on the deeply interesting
experimental results as follows: (1) We employed developing chick embryos, that was electroporated chopWR-venus to prospective limb region.
(2) Then, we observed that the muscle tissue actuator is contracted by
a blue light stimulation.

Keywords: Muscle tissue, Photostimulation, channelrhodopsin-2.

1 Introduction

Living organisms change their bones, muscles and neurons by self-organization in
order to adapt to their environment. This process occurs not only at the level of
the individual, but also on a tissue level, and cellular level, as significant property
such as self-reproduction, self-repair and self-assembly. However, most state-of-the-
art robots are made of metals and semiconductors which are unable to dynamically
change physical and chemical characteristics during operation so it is not possible
to show the adaptive functionality of structural changes in the body itself. Because
of this, to implement a robot with the adaptive functionality of a living organism derived from a mechanical system, robotics and molecular biology need to be combined
to achieve a flexible intelligent mechanical system like that of a living organism.

T.J. Prescott et al. (Eds.): Living Machines 2012, LNAI 7375, pp. 394–395, 2012.
© Springer-Verlag Berlin Heidelberg 2012

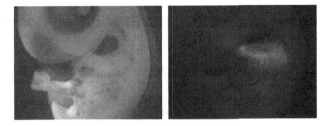

Fig. 1. Chick embryos with channelrhodopsin-2. Right one shows fluorescence imaging.

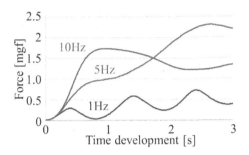

Fig. 2. Contraction force by a muscle tissue when photostimulation were applied

2   Experimental Results

This study intend to create a bio-robot which expresses the inherent superior characteristics of a living organism with less invasive control method. To do so, we develop a actuator for a robot with a muscle tissue of developing chick embryos. In this study, we drive the actuator by introducing "photostimulation" and "genetic engineering technology". Specifically, the activation of a muscle tissue is triggered with the protein channelrhodopsin-2, which functions as light-gated ion channel[1] and is electroporated to prospective limb region of the developing chick embryos(Fig. 1). So the developed muscle tissue is drived by stimulus pulse of blue light. Thus, unlike the existing driving technique with creating an electric field, the proposed method is expected less damage to cells. The muscle tissue utilized for a actuator not only contract with specific timing, but also can control the contraction force by frequency changes of photostimulation(Fig. 2).

Acknowledgments. This work has been partially supported by KAKENHI (24680023) and Global COE Program ("Basic & Translational Research Center for Global Brain Science", MEXT, Japan.

Reference

1. Ishizuka, T., Kakuda, M., Araki, R., Yawo, H.: Kinetic evaluation of photosensitivity in genetically engineered neurons expressing green algae light-gated channels. Neurosci., Res. 54, 85–94 (2006)

Shape Optimizing of Tail for Biomimetic Robot Fish

Majid Siami and Vahid Azimirad

The Center of Excellence for Mechatronics, School of Engineering Emerging Technologies,
University of Tabriz, Tabriz, Iran
azimirad@tabrizu.ac.ir

Abstract. This paper introduces a tail optimization method for robotic fish regarding to friction force decreasing. The method is based on Computational Fluid Dynamic (CFD). Hence the hydrodynamic model is derived, its motion directions (up, down, left and right) are analyzed and some simulation studies are done. Then results are presented and the shape and length of tail is optimized to have the minimum friction force in turning to left and right.

Keywords: robot fish, biomimetic robots, optimization.

1 Introduction

Hydrodynamic studying of robotic fish by (Computational Fluid Dynamic) CDF has application in evaluating and predicting drag and lift forces [1]. Some previous works in this area have been focused on the path tracking [2]. Some other studied hydrodynamic models optimization; Mohammadshahi et al. modeled robot fish calculated the forces in various conditions but they did not consider the optimality of tail shape [3]. In this work a robot fish is modeled and its motion in up, down, right and left directions are studied and verified by [3]. The best value for length of tail to have the minimum friction force in turning to left and right is calculated by simulation studies. Also the optimal shape of tail in order to decreasing friction is achieved.

2 Optimization of Tail Deviation Point and Shape in Moving to the Left and Right Directions

The modeling is done in 2D environment in which the flow is distributed and incompressible. The lift and drag coefficients are analyzed and shown in fig. 5. The equations of lift and drag are as Eqs 1 and 2:

$$L = (p_L - p_u)c\, Cos\alpha \quad \rightarrow \quad C_L = \frac{L}{\frac{1}{2}\rho_\infty V_\infty^2 c} \tag{1}$$

$$D = (p_L - p_u)c\, Sin\alpha \quad \rightarrow \quad C_D = \frac{D}{\frac{1}{2}\rho_\infty V_\infty^2 c} \tag{2}$$

In which P_L is the lower limit of pressure, P_u is the upper limit of pressure, ρ_∞ is density of flow, V_∞^2 is velocity of flow, C is chord of model, L is lift force and D is drag force. The velocity of robot is assumed to be 0.75 m/s. (Fig. 1)

T.J. Prescott et al. (Eds.): Living Machines 2012, LNAI 7375, pp. 396–397, 2012.

Fig. 1. x and *c* parameters in simulation study of robotic fish

In the Fig. 1, x is the length of fish mines the length of the tail and c is the whole length of the fish. Simulation are done for various values of x and c. finally the three models in which (x/c) = 0.2, 0.3 and 0.4 is considered (Fig. 2)

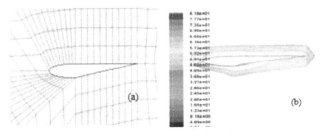

Fig. 2. (a) simulation model of robot fish (b) shear stress for (x/c) =0.3

Simulations showed that in the model with (x/c) = 0.3, the shear stress in the opposite side of turning, is lower than one in other models. In other simulation studies some curves with various positions and radii are considered on the tail and it is observed the existing a curve with radius of 0.2 in the position of (x/c) = 0.55 have an effectively decrease the shear stress and friction.

3   Conclusion

The optimized shape and length of tail in robotic fish is calculated to have the minimum friction force in turning to left and right. The best state is when the (x/c) =0.3 in which by the least drag force, the most efficiency is achieved. It is funded that the tail with a curve (radius of 0.2 in the position of (x/c) = 0.55) has better function in decreasing frictional forces than one with straight line form.

References

1. Zhen, Y., Chun, Y.: CFD Uncertainty Analysis of Unsteady Flows around submarine, China (2009)
2. Liu, J., Duks, I., Knight, R., Hu, H.: Development of fish-like swimming behaviors for an autonomous robotic fish. In: International Conference on Control, UK (September 2004)
3. Mohammadshahi, D., et al.: Design, Fabrication and Hydrodynamic Analysis of a Biomimetic Robot Fish. In: Conf. on Automatic Control, Modeling and Simulation, Istanbul (2008)

Intuitive Navigation of Snake-Like Robot with Autonomous Decentralized Control

Yasushi Sunada[1], Takahide Sato[2], Takeshi Kano[2],
Akio Ishiguro[2,3], and Ryo Kobayashi[1,3]

[1] Department of Mathematical and Life Sciences, Hiroshima University,
1-3-1 Kagamiyama, Higashi Hiroshima 739-8526, Japan
[2] Research Institute of Electrical Communication, Tohoku University,
2-1-1 Katahira, Aoba-ku, Sendai 980-8577, Japan
[3] Japan Science and Technology Agency, CREST,
Sanban-cho, Chiyoda-ku, Tokyo 102-0075, Japan
m116659@hiroshima-u.ac.jp,tsato@cmplx.riec.tohoku.ac.jp,
{tkano,ishiguro}@riec.tohoku.ac.jp,ryo@math.sci.hiroshima-u.ac.jp

Abstract. So far, we have developed snake-like robots whose locomotion is achieved primarily by ADC (autonomous decentralized control), with only the direction of travel and velocity given by radio control signals[3]. In this paper, we introduce an attempt to improve the navigation system of the robot by using a more intuitive eye wear controller. We will demonstrate its function using a simulation of the robotic system.

Keywords: autonomous decentralized control, snake-like robot, intuitive navigation.

Animals exhibit highly supple and agile motion, which is achieved by controlling a very large number of degrees of freedom (DOF) embedded in their bodies. Also, their motion is robust in uncertain surroundings. Such amazing abilities are hard to realize in robots today. One reason is that almost all existing robots are controlled in a completely centralized manner. It is known that the nervous systems of animals contain CPGs (central pattern generators), which work as local controllers[1]. This is our motivation for the concept of ADC. In this context, our group has developed an amoeboid robot (Slimy) controlled completely by ADC and inspired by actual slime mold[2]. We also developed a snake-like robot (HAUBOT) controlled by ADC with a remote control signal[3].

In this paper, we introduce our attempt to turn this snake-like robot into a more sophisticated human-robot system. Our design policies are as follows.

1. A human navigator gives control signals with a small number of DOF such as target direction and velocity.
2. The robot performs the instructed locomotion depending on the environment by controlling its large number of DOF in an ADC manner.
3. Interaction between the navigator and the robot are designed so that the navigator can become sensually synchronized with the robot.

T.J. Prescott et al. (Eds.): Living Machines 2012, LNAI 7375, pp. 398–399, 2012.

Our robot consists of a one dimensional array of segments connected by joints. Each joint generates torque proportional to the difference between the joint angle and the target angle, which is given by the anterior joint angle (or by the local oscillator). The target angle is also modified by the local pressure on the side of body in order to utilize bumps as scaffolds[4]. This is the ADC part of the locomotion of the snake-like robot. A human navigator uses the head-mount display, Vuzix iWear VR920, as a visual feedback and control device. The navigator sees the world around the robot in the wearable display, and simply turns their face in the desired direction of travel. Then the yaw angle of the display is sent to the robot, which modifies the target angle of the first anterior joint. The velocity signal is given by the keyboard, which modifies the frequency of oscillation of the target angle for the second anterior joint. Then the robot spontaneously generates motion in the desired direction at the signaled pace. We will demonstrate how the simple ADC rules and the navigation system act in harmony to attain adaptive locomotion in a complex environment.

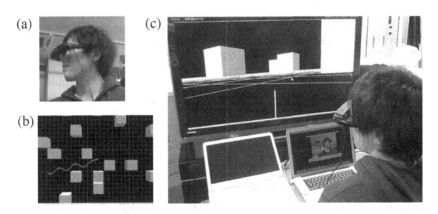

Fig. 1. (a) Eye wear controller Vuzix iWear VR920 (b) Top view of the simulated snake-like robot. (c) Navigators vision in the virtual world is shown in the front display.

References

1. Grillner, S.: Neural networks for vertebrate locomotion. Scientific American 274, 64–69 (1996)
2. Umedachi, T., Takeda, K., Nakagaki, T., Kobayashi, R., Ishiguro, A.: Fully decentralized control of a soft-bodied robot inspired by true slime mold. Biol. Cybern. 102, 261–269 (2010)
3. Sato, T., Kano, T., Ishiguro, A.: A decentralized control scheme for an effective coordination of phasic and tonic control in a snake-like robot. Bioinspr. Biomim. 7, 016005 (2012)
4. Kano, T., Sato, T., Kano, T., Kobayashi, R., Ishiguro, A.: Decentralized control of scaffold-assisted serpentine locomotion that exploits body softness. In: Proc. IEEE/RSJ Int. Conf. Robot Auto., pp. 5129–5134. IEEE, New York (2011)

Design of Adhesion Device Inspired by Octopus Sucker

Francesca Tramacere[1,2,*], Lucia Beccai[2], and Barbara Mazzolai[2]

[1] BioRobotics Institute, Scuola Superiore Sant'Anna, Pisa, Italy
[2] Center for Micro-BioRobotics@SSSA, Istituto Italiano di Tecnologia, Pontedera, Italy
{francesca.tramacere,lucia.beccai,barbara.mazzolai}@iit.it

Abstract. In this work we present the design of a new actuated adhesion device inspired by octopus suckers. The octopus suckers are very interesting because they are able to attach in wet conditions on different surfaces, and (as explained in the Kier and Smith hypothesis) the connective tissue fibers of the sucker may store elastic energy, allowing to maintain attachment over extended periods. These features represent a great source of inspiration to conceive innovative adhesion systems working in the same environmental conditions of the biological counterpart. Starting from these premises, we have designed a novel bioinspired adhesion device which exploits the incompressibility of water and a low energy consuming strategy.

Keywords: Adhesion, suction cup, sucker, octopus, wet.

1   Introduction

The octopus suckers are muscular-hydrostats that consist of two parts connected by a constricted orifice: the infundibulum, namely the exposed disk-like portion of the sucker, and the acetabulum, namely the upper hollow portion. The infundibular surface bears a series of radial grooves and it is encircled by a rim. This particular structure plays an important role during adhesion process. In fact, in accordance with Kier and Smith's adhesion model [1], octopus suckers make adhesion in two steps: 1) suckers form a seal at the rim around infundibulum and 2) they reduce the pressure in the acetabular cavity by contracting the acetabular muscles. The dense network of radial grooves present on the infundibular portion helps water movement inside the acetabular cavity, which is filled with water. The ability of water (being incompressible) to withstand this decrease in pressure is essential to sucker function. Also, it was hypothesized that octopus sucker may store elastic energy in its connective tissue; thus, the sucker obtains adhesion without any muscular effort, this resulting in efficient adhesion (i.e. characterized by low energy consumption) for an extended period. While recently we have developed octopus sucker-like adhesion cups, with passive properties and based on soft and compliant materials [2], the present work describes the design of the first active prototype in accordance with the morphological structure of the octopus sucker.

---

* Corresponding author.

T.J. Prescott et al. (Eds.): Living Machines 2012, LNAI 7375, pp. 400–401, 2012.
© Springer-Verlag Berlin Heidelberg 2012

2   Design of the Artificial Suction Cup

The design of the active suction cup prototype consists of two main parts: acetabular chamber (ac) and infundibular portion (in) (Fig. 1). The acetabular chamber, made of hard material (e.g. resin), has embedded a sliding piston (p) that creates water tension (mimicking the acetabular muscles contraction). The infundibular portion, instead, needs to be soft (i.e. silicone), thus compliant to the external surface, and capable of mimicking the radial grooves and rim of the biological suckers. An orifice connects acetabulum and infundibulum and includes a valve (v) in resin material, which slides vertically in its support. In the radial peripheral region of the valve there are holes, which play an important role in the adhesion process. When the drilled disk (valve) is in contact with the upper surface of its support, the holes allow water flowing (up-phase, Fig.1 (c)); while, this is not possible when the drilled disk is in contact with lower surface of its support which occludes the holes (down-phase, Fig.1 (d)). Thus, when suction is induced by pulling the piston, the valve is passively brought in up-phase, and the acetabulum communicates with infundibulum presenting the same low pressure. At this time, if the valve is brought in the down-phase and the suction gene-rator is removed, the pressure in the acetabular chamber increases but it is still low in the infundibular portion (since the two compartments have been previously isolated). Due to such configuration, the valve is maintained in the down-phase passively by means of the pressure difference created between acetabulum and infundibulum. Here, in contrast with solutions in which air is used, we exploit the incompressibility of water and its capability of withstanding negative pressures. Considering that in the natural sucker the lowest reported negative pressure is of ~0.168 MPa, an artificial sucker mimicking the natural behavior, when having a diameter of 1 cm, would reach a 21 N adhesion force. Hence, we can envisage that the conceived design is the first step towards a new generation of efficient artificial suction cups.

Fig. 1. Qualitative images. (a) Suction cup in rest state; (b) Suction cup in adhesion state; (c) valve during suction, up-phase; (d) valve in passive adhesion state, down-phase.

References

1. Kier, W.M., Smith, A.M.: The Structure and Adhesive Mechanism of Octopus Suckers. In-tegr. Comp. Biol. 42, 1146–1153 (2002)
2. Tramacere, F., Beccai, L., Mattioli, F., Sinibaldi, E., Mazzolai, B.: Artificial Adhesion Me-chanisms inspired by Octopus Suckers. In: IEEE International Conference on Robotics and Automation. IEEE Press, St. Paul (2012)

Author Index

Akkad, Rami 365
Akrami, Athena 341
Antontelli, Marco 83
Arienti, Andrea 347, 349
Asano, Toshifumi 394
Ayers, Joseph 1
Azimirad, Vahid 339, 396

Baddeley, Bart 216
Balampanis, Fotios 333
Barron-Gonzalez, Hector 335, 377
Basil, Jennifer 286, 352
Beccai, Lucia 400
Betella, Alberto 238
Binas, Jonathan 382
Blustein, Daniel 1
Bobo, Luis 13
Bonser, Richard H.C. 355, 357, 359
Bönzli, Eva 343
Bortoletto, Roberto 26
Boxerbaum, Alexander S. 38
Boyer, Frédéric 50

Calisti, Marcello 337, 347
Caluwaerts, Ken 62
Carpi, Federico 74
Casseau, Vincent 386
Chicca, Elisabetta 382
Chiel, Hillel J. 38
Chinellato, Eris 83, 192
Cianchetti, Matteo 347, 375
Claeyssens, Frederik 392
Clément, Romain O. 365
Comparini, Diego 384
Cruse, Holk 120
Cutkosky, Mark 95

Daltorio, Kathryn A. 38
Dastoor, Sanjay 95
Dehghani, Abbas 380
del Pobil, Angel P. 83
Demiris, Yiannis 192
De Rossi, Danilo 74

Diamond, Mathew E. 341
Djordjevic, Goran S. 345
Dodd, Tony 377
Douglas, Rodney 382
Duff, Armin 156
Dumitrache, Ioan 204

Ejaz, Naveed 107
Escudero Chimeno, Alex 371
Eshgi, Sadjad 339
Esmaeili, Vahid 341
Evans, Mat 335, 352, 377

Fassihi, Arash 341
Favre-Félix, Antoine 62
Felderer, Florian 274
Fellermann, Harold 343
Ferri, Gabriele 347
Follador, Maurizio 375
Frediani, Gabriele 74
Fremerey, Max 345
Fu, Terence Pei 386

Ghirlanda, Stefano 298
Giorelli, Michele 337, 347
Giorgio Serchi, Francesco 349
Girard, Benoît 62
Girardi, Stefano 274
Graham, Paul 216
Grand, Christophe 62
Grasso, Frank W. 286, 298, 352

Hadorn, Maik 343
Hajimohammadi, Hamid 339
He, Fuben 26
Herreros, Ivan 13
Hirai, Akihiro 390
Hoinville, Thierry 120
Hou, Jinping 355, 357, 359
Husbands, Philip 216

Idei, Ryo 262
Indiveri, Giacomo 382

Ishiguro, Akio   262, 361, 363, 390, 394, 398
Izzo, Dario   386

Jacques, Lionel   386
Jayne, David   380
Jeronimidis, George   355, 357, 359

Kano, Takeshi   361, 363, 390, 398
Khamassi, Mehdi   62
Kirbach, Andreas   132
Kobayashi, Ryo   361, 398
Krapp, Holger G.   107
Krause, Jens   365

Landgraf, Tim   132, 365
Laschi, Cecilia   337, 347, 349, 375
Lebastard, Vincent   50
Lepora, Nathan F.   335, 367, 377
Leugger, Tobias   369
Lim, Joo Hwee   373
López-Serrano, Lucas L.   371
Lu, Shijian   373
Luvizotto, André   144, 238

Mahmud, Mufti   274
Mancuso, Stefano   384
Manzino, Fabrizio   341
Marcos, Encarni   156
Margheri, Laura   375
Martinez-Hernandez, Uriel   335, 377
Maschietto, Marta   274
Mathews, Zenon   371
Mazzolai, Barbara   375, 388, 400
Menzel, Randolf   132
Mitchinson, Ben   168
Miyamoto, Koichiro   394
Miyasaka, Kota   394
Mohan, Vishwanathan   180
Montellano López, Alfonso   380
Morasso, Pietro   180

Neftci, Emre   382
Neville, Anne   380
Nguyen, Hai   365
N'Guyen, Steve   62
Nolfi, Stefano   250, 369

Oertel, Michael   132
Ognibene, Dimitri   192
Ogura, Toshihiko   394

Pagello, Enrico   26
Pandolfi, Camilla   384, 386
Pateman, Christopher J.   392
Pavel, Ana Brânduşa   204
Pearson, Martin J.   168
Philippides, Andrew   216
Pipe, Anthony G.   168
Popova, Liyana   388
Prescott, Tony J.   168, 335, 352, 367, 377

Quinn, Roger D.   38

Rabinovich, Mikhail I.   228
Rasmussen, Steen   343
Renda, Federico   347
Rennó-Costa, César   144, 238
Richardson, Robert   380
Rojas, Raúl   132, 365
Roshan, Rupesh   380

Sánchez-Fibla, Martí   156, 238
Sarabia, Miguel   192
Sartori, Massimo   26
Sato, Takahide   390, 398
Savastano, Piero   250
Sherborne, Colin R.   392
Shimizu, Masahiro   394
Siami, Majid   396
Staffa, Mariacarla   62
Stuart, Hannah   95
Sunada, Yasushi   398
Suzuki, Shota   363

Tanaka, Reiko J.   107
Tonazzini, Alice   388
Tramacere, Francesca   400

Umedachi, Takuya   262

Varona, Pablo   228
Vasile, Cristian Ioan   204
Vassanelli, Stefano   274
Verschure, Paul F.M.J.   13, 144, 156, 238, 321, 333, 367, 371
Volz, Stephen G.   286
Vouloutsi, Vasiliki   371

Wehner, Rüdiger   120
Weiss, Sam   95

Westphal, Anthony   1
Witte, Hartmut   345

Yaegashi, Kazuyuki   361
Yawata, Shintaro   394
Yawo, Hiromu   394

Yoshinobu, Tatsuo   394
Young, Briana   298

Zipser, David   309
Zucca, Riccardo   321